Principles
of
Environmental Sampling

Principles
of
Environmental Sampling

Lawrence H. Keith, Editor

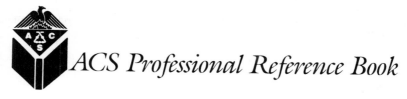

ACS Professional Reference Book

AMERICAN CHEMICAL SOCIETY   1988

Library of Congress Cataloging-in-Publication Data

Principles of environmental sampling/Lawrence H. Keith, editor.

p.   cm.

Includes bibliographies and indexes.

ISBN 0-8412-1173-6.   ISBN 0-8412-1437-9 (pbk.)

1. Pollution—Measurement. 2. Environmental chemistry. 3. Environmental monitoring. 4. Quality assurance.

I. Keith, Lawrence H., 1938–   . II. American Chemical Society.

TD193.P75   1987
628.5′028′7—dc19   87-22975
                                                                            CIP

About the Editor

Lawrence H. Keith has been involved with the principles and problems of environmental chemistry for over 20 years. Beginning with the U.S. Environmental Protection Agency and continuing with Radian Corporation, he has contributed to environmental sampling and analysis involving method developments, the priority pollutant list, national drinking water surveys, chlorinated dioxin and dibenzofuran analyses, and many other similar programs. He is an author of over 50 papers and a dozen books encompassing the areas of environmental chemistry, expert systems, and chemical health and safety.

A member of the American Chemical Society (ACS) since college days, Dr. Keith has served as past chairman of the Environmental Chemistry Division and the Central Texas Section. He is an ACS delegate to the U.S. National Committee as part of the International Association for Water Pollution Research and Control; chairman of the ACS Committee on Environmental Improvement, Subcommittee on Environmental Monitoring and Analysis; and board member of the *Journal of Chemical Health and Safety*. At Radian Corporation in Austin, TX, Dr. Keith serves as a Senior Program Manager and Principal Scientist.

Contents

Sampling Waters

Sampling Air and Stacks

Sampling Biota

Sampling Solids, Sludges, and Liquid Wastes

Glossary and Indexes

Contributors

Albert, Richard   page 337
Center for Food Safety and Applied Nutrition
Food and Drug Administration
Washington, DC 20204

Baldocchi, D. D.   page 297
Atmospheric Turbulence and Diffusion Division
National Oceanic and Atmospheric Administration
Oak Ridge, TN 37831

Barcelona, Michael J.   page 3
Aquatic Chemistry—Water Survey Division
Illinois Department of Energy and Natural Resources
Champaign, IL 61820-7495

Black, Stuart C.   page 109
Environmental Monitoring Systems Laboratory
U.S. Environmental Protection Agency
Las Vegas, NV 89114

Bollinger, Mark   page 221
Rocky Mountain Analytical Laboratory
A Division of Enseco, Inc.
4955 Yarrow Street
Arvada, CO 80002

Bone, Larry I.   page 409
Dow Chemical Company
Baton Rouge, LA 70816

Borgman, Leon E.   page 25
Statistics Department
University of Wyoming
Laramie, WY 82071

Bourke, John B.   pages 355, 375
Analytical Laboratories
New York State Agricultural Experiment Station
Cornell University
Geneva, NY 14456

Bryant, Mark A.   page 255
Harding Lawson Associates
5580 Havana Street, Suite 5A
Denver, CO 80239

Callaway, Owen   page 221
Rocky Mountain Analytical Laboratory
A Division of Enseco, Inc.
4955 Yarrow Street
Arvada, CO 80002

Carlberg, Kathy   page 221
Rocky Mountain Analytical Laboratory
A Division of Enseco, Inc.
4955 Yarrow Street
Arvada, CO 80002

Clements, John B.   page 287
Environmental Monitoring Systems Laboratory
U.S. Environmental Protection Agency
Research Triangle Park, NC 27711

Cowgill, U. M.   page 171
Mammalian and Environmental Toxicology
Dow Chemical Company
Midland, MI 48674

Englund, Evan J.   page 73
Exposure Assessment Research Division
U.S. Environmental Protection Agency
Las Vegas, NV 89114-5027

Flatman, George T.   page 73
    Exposure Assessment Research Division
    U.S. Environmental Protection Agency
    Las Vegas, NV 89114-5027

Garner, Forest C.   page 363
    Lockheed Engineering and Management Services Company
    P.O. Box 15027
    Las Vegas, NV 89109

Hicks, B. B.   page 297
    Atmospheric Turbulence and Diffusion Division
    National Oceanic and Atmospheric Administration
    Oak Ridge, TN 37831

Holcombe, Larry J.   page 395
    Radian Corporation
    Austin, TX 78766

Horwitz, William   page 337
    Center for Food Safety and Applied Nutrition
    Food and Drug Administration
    Washington, DC 20204

Jackson, Larry P.   page 415
    Western Research Institute
    Laramie, WY 82071
    Current address: GT Environmental Laboratories
    4 Mill Street
    Greenville, NH 03048

Johnston, M. Timothy   page 85
    Radian Corporation
    Austin, TX 78720-1088

Journel, Andre G.   page 45
    Department of Applied Earth Sciences
    Stanford University
    Stanford, CA 94305-2225

Keith, Lawrence H.   page 85
Radian Corporation
Austin, TX 78720-1088

Kent, Robert T.   page 231
IT Corporation
2499-B Capital of Texas Highway
Austin, TX 78746

Kulkarni, Shrikant   page 157
Research Triangle Institute
Research Triangle Park, NC 27709

Lewis, David L.   pages 85, 119
Radian Corporation
Austin, TX 78766

Lewis, Robert G.   page 287
Environmental Monitoring Systems Laboratory
U.S. Environmental Protection Agency
Research Triangle Park, NC 27711

Liggett, Walter S.   page 191
National Bureau of Standards
Gaithersburg, MD 20899

Malley, Michael J.   page 255
Harding Lawson Associates
5580 Havana Street, Suite 5A
Denver, CO 80239

Maskarinec, M. P.   page 145
Oak Ridge National Laboratory
Oak Ridge, TN 37831-6120

Messner, Michael J.   page 157
Research Triangle Institute
Research Triangle Park, NC 27709

Meyers, T. P.   page 297
Atmospheric Turbulence and Diffusion Division
National Oceanic and Atmospheric Administration
Oak Ridge, TN 37831

Moody, R. L.   page 145
Oak Ridge National Laboratory
Oak Ridge, TN 37831-6120

Myers, Lawrence E.   page 157
Research Triangle Institute
Research Triangle Park, NC 27709

Newburn, Lorance H.   page 209
Environmental Division
Isco, Inc.
531 Westgate Boulevard
Lincoln, NE 68528-1586

Norris, James E.   page 247
BCM Converse, Inc.
108 St. Anthony Street
Mobile, AL 36602

Parr, Jerry   page 221
Rocky Mountain Analytical Laboratory
A Division of Enseco, Inc.
4955 Yarrow Street
Arvada, CO 80002

Payne, Katherine E.   page 231
IT Corporation
2499-B Capital of Texas Highway
Austin, TX 78746

Peters, James A.   page 317
Monsanto Company
River Road
Addyston, OH 45001

Quimby, William F.   page 25
Department of Mathematical Sciences
Montana State University
Bozeman, MT 59717

Smith, Franklin   page 157
Research Triangle Institute
Research Triangle Park, NC 27709

Smith, James S.   page 255
Walter B. Satterthwaite Associates, Inc.
720 North Five Points Road
West Chester, PA 19380
Current address: Trillium, Inc.
Coatesville, PA 19320

Spittler, Terry D.   pages 355, 375
Analytical Laboratories
New York State Agricultural Experiment Station
Cornell University
Geneva, NY 14456

Stapanian, Martin A.   page 363
Lockheed Engineering and Management Services Company
P.O. Box 15027
Las Vegas, NV 89109

Steele, David P.   page 255
Walter B. Satterthwaite Associates, Inc.
720 North Five Points Road
West Chester, PA 19380

Tanner, Roger L.   page 275
Environmental Chemistry Division
Department of Applied Science
Brookhaven National Laboratory
Upton, NY 11973

Taylor, John K.   page 101
National Bureau of Standards
Gaithersburg, MD 20899

Triegel, Elly K.   page 385
Triegel & Associates, Inc.
190 South Warner Road
Wayne, PA 19087

Watson, John G.   page 263
Desert Research Institute
University of Nevada System
P.O. Box 60220
Reno, NV 89506

Williams, Llewellyn R.   page 363
Quality Assurance and Method Development Division
Environmental Monitoring Systems Laboratory
U.S. Environmental Protection Agency
944 East Harmon Avenue
Las Vegas, NV 89114

Yfantis, Angelo A.   page 73
Exposure Assessment Research Division
U.S. Environmental Protection Agency
Las Vegas, NV 89114-5027

Young, Susan J.   page 355
Food and Drug Administration
Washington, DC 20204

Preface

The goal of this book is to ensure consideration of the many variables and special techniques that are needed to plan and execute reliable sampling activities. Specific needs will dictate which techniques are actually incorporated in sampling plans and which are rejected; the key point is to be certain that those variables and techniques not selected were rejected because of the sampling goals rather than because of being overlooked.

Obtaining reliable environmental samples is a difficult process. Generally, the objective is to take representative samples of a heterogeneous and changing piece of our world in order to analyze for components that constitute a very tiny fraction of the samples (often at or below the parts-per-billion level). Other complicating factors are that the matrix is usually very complex, thereby facilitating analytical interferences such as masking and false positives. And once the sample is taken, other interferences can be introduced during transport or preservation. Furthermore, the analytes of interest are sometimes unstable. No wonder sampling is often considered to be the weakest link in the chain of planning–sampling–analysis–reporting activities.

This discussion shows that the reliability of the overall data cannot be greater than that of the reliability of the weakest part of the chain of events constituting an environmental sampling and analysis effort. What good is a precise analytical report if the samples are not representative of their source?

The American Chemical Society (ACS) Committee on Environmental Improvement recognized this problem and sponsored a symposium upon which this book is based, in conjunction with the Divisions of Agrochemistry, Analytical Chemistry, and Environmental Chemistry, to bring together a distinguished group of experienced scientists to present the benefits of their knowledge in this field. The group was charged with presenting their advice and recommendations on *what* should be considered when planning an environmental sampling task. This information is different from that on *how* to conduct environmental sampling, which is very detailed and specific depending on the objectives. However, some procedures almost always should be followed, and these do represent "how to" recommendations for conducting specific tasks.

Certain principles of planning, sample design, and quality control prevail over all the special considerations that matrix variations impose. These principles are discussed in the first section. Special matrix requirements (e.g., sampling equipment and techniques, and preservation) are then discussed along with the principles of sampling that involve them. These topics are discussed in sections involving water; air and stacks; biota; and solids, sludges, and liquid wastes.

This book also serves as the basis for an ACS short review on the subject. This review will not be as detailed as this work and will be a companion to the ACS short review "Principles of Environmental Analysis" (*Anal. Chem.* **1983**, *55*, 2210–2218).

LAWRENCE H. KEITH

PLANNING AND SAMPLE DESIGN

Chapter 1

Overview
of the Sampling Process

Michael J. Barcelona

The involvement of chemists, and analytical chemists in particular, can substantially improve the result of sampling and analytical programs. The development of meaningful sampling protocols demands careful planning of the actual procedures used in sample collection, handling, and transfer. Critical aspects of sampling protocol development cannot be covered by traditional field and laboratory quality control measures. Criteria for sample representativeness must be developed with careful attention to the physical, chemical, and biological dynamics of the environment under investigation. Preliminary sampling and a well-conceived sampling experiment can provide the validation and experience necessary to design efficient sampling protocols that will meet program needs.

SAMPLING IN THE ENVIRONMENT for chemical analysis is a complex subject, and can be as varied and complicated as the objects that must be sampled to investigate the environmental effects and fates of chemical species on the planet. The subject does not lend itself to textbook treatments for students and has long been given diminished importance relative to the improvement and verification of analytical methods for environmental applications. Indeed, in reviewing environmental opportunities in chemistry, a recent National Research Council report (1) repeatedly cited the need for improved sensitivity and selectivity of analytical techniques and only in passing mentioned improved sampling for chemical analysis. However, that which cannot be reliably sampled is seldom worth the care and expense of analysis. Also, numerous analytical problems exist with chemicals in the environment at the parts-per-billion (μg L^{-1}) and parts-per-trillion (ng L^{-1}) concentration levels without delving further into the frontiers of ultratrace analyses of air, water, or land environments.

1173–6/88/0003$06.25/0 © 1988 American Chemical Society

Many scientists, however, recognize the need for accurate and precise environmental sampling as well as analysis.

This chapter has two objectives for improved experimental design for field sampling problems. The main objective is to encourage the development of accurate and precise sampling protocols that go beyond the confusion of general methods, techniques, and procedures available at the present time. *Protocols* are thorough, written descriptions of the detailed steps and procedures involved in the collection of samples. This sampling validation objective is analogous to that described by Taylor (2) for analytical protocols to avoid the serious consequences of systematic error (i.e., bias or inaccuracy) on the results and conclusions of environmental studies. These studies may be undertaken to investigate contaminant distributions or the potential for exposure to contaminants for a variety of purposes.

Field handling and procedural blank determinations that permit meaningful evaluations of contaminant exposures have been the focus of several recent discussions. These determinations are particularly useful in interpreting the effects of systematic errors on interlaboratory analytical comparisons (3–5). This chapter supports the identification and control of sources of sampling error when such errors exceed those inherent in analytical determinations.

The more subtle objective of this chapter is to encourage a realistic appraisal of the practical limits that systematic sampling errors and resultant bias place on the purpose, results, and conclusions of studies in environmental chemistry. This objective has been included because although the actions of sampling may be deceptively simple, they must be carefully planned, refined, and documented if truly representative samples are to be provided for the purposes of the investigation. Furthermore, the results of environmental studies are frequently extended to purposes other than those of the original investigation. These objectives apply equally to specialized research as well as routine regulatory or compliance monitoring efforts because the results of research investigations are often generalized to a variety of environmental conditions.

Review of the Literature

Sampling for chemical analysis has been reviewed critically in the chemical literature. The excellent works of Kratochvil and co-workers (6, 7) provide a sound basis for chemists interested in recognizing and exploring chemical sampling problems. Their main emphasis is quite practical and the review article (6) includes valuable citations of past work for a variety of environmental matrices. These authors carefully point out the roles that statistics (8) and chemometrics (9, 10) can play in resolving sampling problems. They further underscore the need for the involvement of chemists, and more specifically analysts, in the planning, execution, and interpretation of the results of sampling and analytical efforts.

An environmental scientist's view of the sampling process is often quite different from that of a statistician. The scientist may be interested in representative samples of water from the hypolimnion of a particular type of lake. The statistician, on the other hand, may envision samples as a subset of the universe of all reducing surface water samples. In the environment, collecting samples (objects) from a largely uncharacterized universe of objects is often required. Only through some prior sampling experience can a sample or sample population be related to the universe or parent population that is the territory of statistical theory. This prior sampling experience can also permit some generalization of the results if the experiment is properly designed. The distinction between samples and parts of a parent population can be better illustrated by a reexamination of the types of objects and samples that are ultimately collected for analysis.

Kateman and Pijpers (*11*) categorized objects (e.g., a well-mixed fluid) from which samples are derived on the basis of the degree of homogeneity and the nature of the spatial change in a particular quality (e.g., dissolved lead content). The aim of their categorization was to lead into the corresponding types of samples and sampling strategies that are needed for analytical quality control. Their scheme has been expanded in Figure 1 to include temporal as well as the spatial change inherent in environmental sampling. Spatially heterogeneous objects or sample origins present a much greater challenge to accurate sampling than homogeneous objects that may only exist in the laboratory. This greater challenge comes from the need for both more specific criteria for representativeness in heterogeneous populations and for more detailed characterization of the conditions under which sampling takes place. For example, representative sampling of dissolved lead in a well-mixed solution in a beaker is rather simple compared to sampling dissolved lead in a variable mixture of reactive aqueous effluents entering a treatment plant operation. In the treatment plant, the effluent mixture is a much more complex object both in composition and extent as well as in variability in space or time. Useful criteria for representative effluent samples would necessarily include qualifiers of flow rate, process status, time, and perhaps other physical and biological variables. Kateman and Pijpers (*11*) provided a general review of the relationships between object types and theoretically optimal sample sizes, numbers, and frequencies for a variety of applications.

For each type of object or sample origin, corresponding types of samples or subsamples can be found that result from sampling, pooling, or compositing, as well as the reduction or preparation steps prior to analysis. Samples suitable for analysis must be representative parts of the object. Figure 2 contains an expanded overview of the nomenclature of sample types suggested by Kateman and Pijpers (*11*) for field sampling applications. Increments (or grabs) issue from the object or parent population. They are parts of the object, but because they are not representative parts of the object, they are not called samples. The *gross sample* (bulk sample) may be seen as a pool of two increments that is reduced or prepared as subsamples for analysis. Field control samples and laboratory

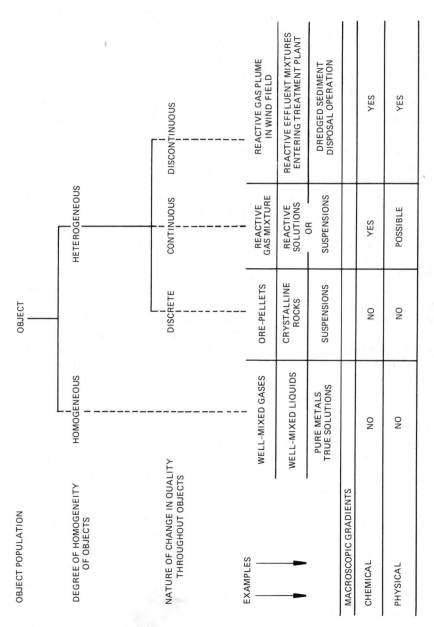

Figure 1. Types of macroscopic objects or sample origins.

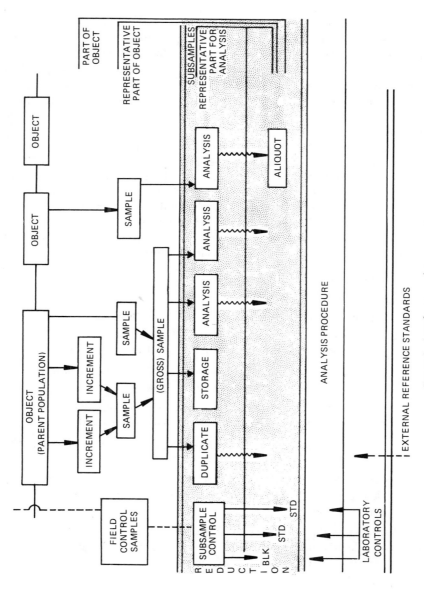

Figure 2. Sample nomenclature overview.

controls enter into the analysis stream with aliquots of the samples. Once these controls and the subsamples have been prepared, negligible sampling error is assumed to occur when aliquots are taken. For the purposes of the present work, the division between sampling and analytical errors is placed after the reduction step that is shaded in Figure 2.

Representativeness in sampling, therefore, presumes that the analysis of the sample or duplicate samples shows the same results as would the object itself. In more practical terms, the characteristic analyte or quality of the sample must be identical to, or minimally disturbed from, the object's quality. Sampling any object will be subject to random variations. In representative sampling efforts, however, we strive to identify and control systematic deviations, or determinate error, caused by sampling.

Planning for representative sampling should be made an integral part of the design of environmental studies. A number of references treat statistical sampling designs within study designs that are worth careful attention. These references include works that deal with biological (12-14), geochemical (15), and water quality monitoring studies. Among these, the methodology presented by Green (12) is particularly well-suited to environmental sampling for chemical analysis.

Green's methodology rests on a clear conception of the problem or question that needs to be answered. His suggestions included the following:

1. replicate samples within each combination of time, location, or other controlled variable;

2. an equal number of randomly allocated replicate samples;

3. samples collected in the presence and absence of a condition in order to test whether the condition has an effect;

4. preliminary sampling to provide a basis for evaluation of sampling design and statistical analysis options;

5. verification of the efficiency and adequacy of the sampling device or method over the range of conditions to be encountered;

6. proportional focusing on homogeneous subareas if the sampling areas have large-scale environmental patterns;

7. verification of sample unit size as appropriate for the size, densities, and spatial distributions of the objects that are being sampled; and

8. testing of the data to establish the nature of error variation to decide whether to transform the data, utilize distribution-free statistical analysis procedures, or test against simulated null-hypothesis data.

Green concluded his suggestions with seasoned advice: Once the best statistical method has been chosen and has provided the test of the hypothesis, accepting the result is preferable to rejecting it and searching for a "better" method.

More theoretical sampling design references may also be useful, and several have been reviewed in detail (7). The theoretical work of Gy (19) (dealing with solid sampling) is interesting, particularly one of his more recent publications (20). In the more recent work, he makes the distinction between sample handling and sampling as an error-generating operation. This statement is an acknowledgment that elements of a sampling operation may cause serious errors and cannot be treated strictly by statistics. Examples of such error-prone elements of the sampling operation are sampling locations and sampling mechanisms or materials. Unlike sample size, number, or sampling frequency, which are elements that can often be handled statistically, identifying and controlling the previously mentioned sources of systematic error are very difficult without prior sampling experience (12, 20, 21). Furthermore, in regard to location; sampling devices; or mechanisms, materials, and handling operations, the expertise of the chemist can be most fruitfully employed.

This area has been recognized by the American Chemical Society Committee on Environmental Improvement (22), which suggested the following minimum requirements for an acceptable sampling program:

1. a proper statistical design that takes into account the goals of the study and its certainties and uncertainties;

2. instructions for sample collection, labeling, preservation, and transport to the analytical facility; and

3. training of personnel in the sampling techniques and procedures specified.

To bolster these suggestions, the committee emphasized that all sampling procedures should be written into detailed protocols similar to the need for analytical protocols in a quality assurance program. Furthermore, the committee suggested documentation of decisions as to what methods, techniques, procedures, or materials are to be included in the protocols for sampling particular matrices for particular constituents.

These suggestions for sampling protocols are really expressions of professional accountability analogous to those expected from the peer-reviewed literature. Few chemical professionals would expect to publish a procedure for an organic synthesis or for the analysis of an exotic chemical species without carefully documenting the exact steps, preparations, and performance measures for such a procedure. Yet, the environmental literature contains numerous instances of striking phenomena "observed" in samples from one locale that were collected by largely undocumented procedures. Many research and monitoring efforts would materially benefit from improvement in the documentation of sampling and analytical work.

If an overall study program is viewed as a hypothesis to be tested by the scientific method of observation (i.e., sampling and analysis), which is followed by interpretation and reevaluation of the hypothesis, the value of detailed protocols can be readily appreciated. This parallel development of program purpose or hypothesis testing is depicted in Figure 3. The figure shows the progression from mere methods, techniques, or procedures to detailed protocols that in turn will often need to be refined to adequately test the hypothesis or achieve the purpose of a program. The value of experimentation will be to strengthen the results and conclusions of environmental sampling and analysis programs. This value is especially true for identification and control of systematic error. The literature of environmental chemistry also provides a number of examples where recognition of systematic sampling problems has led to significant advances in our understanding of environmental processes and chemical fates. The framework of a sampling protocol provides a basis for the application of growing sampling experience to a variety of present and future problems.

Elements of Environmental Sampling Protocols

There are far too many potential types and purposes of investigations in environmental chemistry to present a generally applicable strategy or formula for preparing sampling protocols. General guidance for combined sampling and analysis quality assurance (QA) program planning is available for a number of monitoring applications (23-28). However, guides that deal with protocols and QA procedures for air, water, wastewater, seawater, and hazardous wastes analyses are far more numerous than references that include the practical aspects of sampling. Thus, the need is for successively refined sampling experiments in specific sampling applications.

The historical strength of science as opposed to other fields of human endeavor (e.g., politics and religion) is the scientific or experimental method bolstered by peer review. Experimentation and experimental design skills are not widely taught in science or engineering curricula. For this reason, among others, environmental chemistry is often viewed by our basic chemical colleagues as a business of comparative, rather than absolute, measurements or observations. Much of the recent literature in environmental chemistry shows that this view is truly not the case, and examples of nearly absolute measurements of chemical contaminants in admittedly nonequilibrium systems are not hard to find. The fact that experimental design skills are necessary for reliable studies suggests that the subject should be taught as are other details of experimental or analytical work. A good source book for basic experimental design that is particularly useful for students is that of Wilson (29), among others (30, 31). A generalized sampling protocol for environmental applications presented here highlights the results of successful sampling experiments.

Table I contains an outline for a sampling protocol that would have general application in environmental research or monitoring. The protocol begins with the purpose of the overall program and the specific purpose or purposes of the

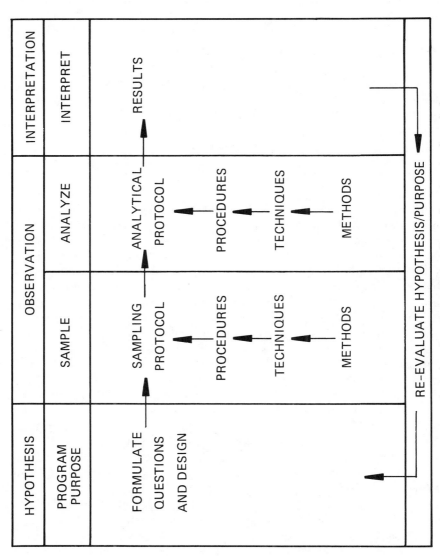

Figure 3. Relationship of program purpose and protocols to the scientific method.

Table I. Outline of a Generalized Sampling Protocol

Main Point (Program Purpose)	Subelements
Analytes of interest	Primary and secondary chemical constituents and criteria for representativeness
Locations	Site, depth, and frequency
Sampling points	Design, construction, and performance evaluation
Sample collection	Mechanism, materials, and methodology
Sample handling	Preservation, filtration, and field control samples
Field determinations	Unstable species and additional sampling variables
Sample storage and transport	Preservation of sample integrity

sampling effort. From this point, the specifics of what the samples are to be analyzed for and the questions "how many", "where", "when", and "how" are addressed in order.

A seemingly simple series of tasks in an initial sampling effort would be to first estimate the variability and mean value(s) of the analytes in the samples and apply statistics to estimate the number of samples and frequency of sampling necessary to achieve the acceptable confidence levels and to fulfill the program purposes. Then, the volumes or types of sample necessary for the specific determinations would be identified. Finally, graduate students or laboratory technicians would be sent out to the field armed with homemade or commercial sampling gear and having a firm resolve to "drown a few worms" and return with the needed samples.

This final step is often the weakest link in the sampling operations. Although a sufficiently large number of samples can normally account for random errors (32), serious systematic or determinate sources of error may be involved in the use of certain sampling devices. These problems mainly affect sampling accuracy or the relation between the analytical result (presuming that the analysis is perfectly accurate) and the actual composition of the environmental medium being sampled. As Lodge (33) pointed out, a clear distinction exists between sampling accuracy and representativeness. Sampling inaccuracy can often be minimized by sampling experiments and technological refinements. *Representativeness* is the correspondence between the analytical result and the actual environmental quality or the condition experienced by a contaminant receptor. Regardless of the purpose of the study or investigation, laboratory-oriented QA measures can only account for errors that occur after sample collection. The planning decisions, identification of sites, and procedures for sample collection must also be subject to QA and peer review.

Certainly, scientists must exercise care in the sample collection step of a sampling protocol and ensure that the sampling point and mechanism are not subject to serious systematic error. Accepting manufacturers' claims of the performance of "representative" sampling devices may be particularly imprudent without careful experimentation. Similar precautions may apply to the other elements in a sampling protocol.

Program Purpose

The goals or purposes of an environmental program or study are implicit in the task of sampling protocol preparation. The cost, time frame, and overall goals of a particular study may override efforts to carefully plan and conduct an adequate sampling operation. However, the long-term consequences of the quick answer should be considered. Environmental scientists can also argue that the cost of planning and conducting a limited sampling experiment can save considerable expense as well as "face" in studies that deal with trace chemical constituents of health concern. All individuals involved in the effort should understand the overall and immediate purposes of the study and recognize that the data must be well-documented as to quality.

Analytes of Interest

The selection of chemical constituent(s) of interest to a particular study may be categorized as primary and secondary. The *primary chemical constituents* may represent known species that have been identified in the sampling matrix or are required to be determined by regulation. Examples of these species include those required by drinking water regulations, maximum contaminant level rules, air quality regulations, or a variety of monitoring programs. *Secondary chemical constituents* may include transformation products of primary chemical species, environmental variables needed to characterize conditions or meet criteria for representativeness, and other chemical species that may be indicators of sample integrity. The secondary category is no less important than the primary grouping.

Consider a study that required a survey of priority pollutant compounds in groundwater. The concentrations of these trace compounds and species provide little insight into the bulk chemical composition and geochemical conditions in the environmental matrix. Many of the 129 priority pollutants undergo substantial chemical or biochemical transformations in air, water, and soils and result in the formation of other compounds and products (34). A more extreme example may result from restricting analytical determinations to the parent compound (e.g., aldicarb) in a pesticide formulation for an aquatic fate or transport study only to learn later that its transformation products are far more mobile, persistent, and toxic (e.g., the aldicarb sulfoxide and sulfone) (35). Allocating large amounts of funds and human resources for sampling and analysis of natural waters for selected or suspected toxic compounds is unrealistic without also considering pH, major cations and anions, total organic carbon, and alkalinity determinations. A more complete analysis in many cases could provide the means to check the consistency of selected samples by mass or charge balance methods and may aid in determining controlling reactions or conditions important for treatment or remediation efforts.

Careful consideration should also be given to the choice of analytical methods for specific determinations. This consideration is not only important for optimization of sample collection and handling procedures, but also to avoid matrix interferences for certain types of samples. Analytical methods developed and validated for drinking water or industrial wastewaters may not be applicable to samples that represent mixtures of these matrices. Also, the aims of research projects frequently require the use of more specific and sensitive types of analytical procedures than those in "standard" references.

The minimum sample volumes and types of sample preservation and handling procedures also depend on the detail and specificity of the proposed analytical program. Sample volumes necessary for specific analyses must be identified with careful consideration of representativeness. For example, both organic (36) and inorganic (37) chemical species in drinking water may vary significantly with the volume of the sample depending on whether sampling is done immediately on opening the tap or after allowing the water to flow for a time.

In addition to the specifics of analyte and analytical method selection, recognizing that physical, meteorological, or hydrologic variables may also need to be measured or determined is important. In many cases, the usefulness of chemical results is fully realized only when the evidence of environmental chemical processes can be linked to other controlling variables. This data is therefore essential to the interpretation of the chemical results and should be included in sampling protocol planning.

Sampling Location and Frequency

The sampling location in space and time can have a very real effect on the quality and usefulness of data in environmental chemistry. Site selection, of course, should be made primarily on the basis of the study goals and the nature of the environmental phenomenon or process under consideration. The optimum number, spacing, and frequency of samples at a coarse scale can best be estimated after a preliminary sampling experiment. Geostatistical and kriging approaches can be most helpful in working with this data, particularly where large numbers of heterogeneous samples may be needed to meet minimum confidence levels in sample results (38-41). This type of problem has been encountered frequently in geochemical studies of soils and subsurface solids. This problem is essentially the difficulty in obtaining a representative sample of a powdered solid material (2, 6, 19, 20, 21, 42) complicated by the problems of obtaining representative solid samples over a large geographic region. Geochemists have applied methods of hierarchical analysis of variance for preliminary geochemical studies to optimize the sampling designs of large-scale efforts (43–45). For example, optimum numbers, discrete depths, and locations have been chosen in lakes and within geographic zones of state or provincial regions in selected studies. The power of these approaches is that they allow the scientist and the program consumer (i.e., government agency or official) to qualitatively discuss the trade-offs involved in changes of such designs.

From a practical point of view, formal design plans have real potential in hazardous waste site assessments and in cleanup evaluations where the challenge of inhomogeneous sample matrices may be further complicated by spatial variability and the need to know suspect constituent concentrations at the parts-per-million (mg L^{-1}) level or below (46, 47).

Besides the numbers of sampling locations and samples to be collected, the frequency of sampling is the most significant cost multiplier in a sampling operation. Certain frequencies of sampling are set by regulation. More often in research and other investigations, relatively short-term studies are undertaken with little concern for the dynamics of the environment. If the purposes of the program include optimization of detection or sensitivity of the results of sampling operations to trends (i.e., long-term changes in environmental quality) or periods (i.e., short-term changes in environmental quality), this aspect of the sampling protocol needs very careful attention.

The distinction between long- and short-term changes is usually relative to the time period over which measurements are made. Periodic changes may be of the order of seconds at a sampling frequency of minutes. Trends are normally considered changes that may occur over several periods or sampling runs. If the environmental value or quality of interest varies with a certain frequency, the sampling frequency must be at least twice the frequency of that variation.

The problem of optimizing sampling frequency within time, program purpose, and cost constraints has been addressed for many environmental media (18, 48–53). The results of a "one-shot" or single-period sampling operation cannot be expected to represent average conditions in most instances.

Sampling Points.   Samples may be collected from bridges, banks, stationary towers, or platforms as well as ships, airplanes, or passive personnel monitors. The important considerations here should be minimal disturbance of the samples. The design and construction of sample access points should disturb the local environment as little as possible or the collection of biased results may be inevitable. The literature for specific applications should be considered in the design of new operations.

Scale problems can also arise in the definition of the sampling point. Collection of a surface water sample from a lake or the ocean seemed to be the simplest operation one could perform with a bucket. In time, limnologists and oceanographers recognized that surface water samples collected with buckets were quite different in organic or trace metal content than samples collected by screens (54). The surface microlayer, enriched from fivefold to several hundred-fold in lipids, Pb, Ni, Cu, and bacterial content, provides very different results from water samples taken just below the surface of water bodies (55-57). The selection of the sampling point must be done with an appreciation that certain microenvironments may either be created or traversed by sampling gear that may yield results quite different from the desired point in air, water, or soil environments. In the case of the sea-surface microlayer, the discovery of this microenvironment proved embarrassing to previous work; however, an exciting new area of research was opened with implications for air–sea gas exchange, bubble and aerosol formation, and chemical fractionation investigations (58). Too often, the slow recognition of systematic sources of error, like picking up the "slick" from the "surface" in water sample collection, may delay significant advances in environmental science. The value of a critically evaluated sampling experiment cannot be underestimated in such instances. Even though the chemical or biochemical constituents of such microenvironments may not be the object of specific studies or investigations, they may cause matrix effects or other analytical interferences for the chemical constituents of interest. Field control samples (i.e., blanks, spiked samples, and colocated samples) can be extremely helpful in evaluating these analytical effects on chemical results.

Sample Collection.   *Sample collection* involves contact of the sample with the sampling device and its materials of construction. Often, the sample may contact the sampling technician or adjacent materials as well. The entire path of sample retrieval should be scrutinized to minimize systematic sources of error that cannot be accounted for by conventional laboratory-oriented QA and quality control (QC). The same care required for sample storage vessels should be extended to the selection of materials, methods, and devices for sample collection.

The history of trace metal sampling and analysis in the environment provides classic examples of the value of recognition and control of systematic sampling error. Basically, the modern environment has levels of Pb in the air, dust, and soil that may bias improperly collected samples by orders of magnitude. The pioneering work of Patterson and co-workers (59, 60) on environmental Pb sampling and analysis spurred an avalanche of research that resulted in the well-known "decrease" in levels of trace metals in the oceans over the past 20 years (61-64). These references show that previous attempts to unravel the geochemistry and environmental fates of trace metals were stymied because of biased sample collection and handling techniques. The contamina-tion of samples of air, water, and biota by artifact trace metals had obscured the true environmental distributions and controlling processes. Far worse, however, was the delayed realization that not all environmental samples in geologic time had equivalent industrial metal levels.

Even though sample collection and handling techniques at the "clean-room" level may not be necessary for routine monitoring or investigative efforts, very careful consideration must be given to the choice of sampling mechanism and methodology for chemical contaminants in the parts-per-billion (μg L^{-1}) range. References 59–64 represent carefully documented attempts to control systematic error.

Here again, as in the case of recognizing the impact of the sea-surface microlayer on surface samples, the control of industrial Pb contamination has had a great effect on the results and interpretation of environmental chemistry research. It is sobering to contrast the modern practice for sampling seawater for Pb in its diligence and detail with the thousands of somewhat haphazardly collected samples of groundwater delivered under strict chain of custody to a variety of contract laboratories under various hazardous waste monitoring programs. We will undoubtedly discover subtle aspects of waste-constituent geochemistry in the future once sample collection details are as carefully documented as are laboratory QA and QC procedures (65).

Poor sample collection procedures can seriously bias chemical results. Documentation of sampling procedures in protocols can help to identify and control such errors. The core of the sampling operation is the mechanism or device used to collect the sample. Because we rarely know the true value for the concentration of a particular chemical constituent in the environment at any single time, we must either test sampling mechanisms under controlled laboratory conditions or intercompare them simultaneously in the field. In the laboratory, the concentration of the chemical constituent can be controlled, but the exact environmental conditions are difficult to simulate. In the field, depending upon whether loss (e.g., volatile compounds) or contamination (e.g., airborne Pb) is a likely source of systematic mechanism-related error, the highest or lowest result, respectively, is normally accepted as the most reliable for certain applications. Device-related errors cannot be accounted for by blanks, standards, or replicate control samples. The actual performance of sampling mechanisms must be verified in critical sampling applications. Numerous examples in the literature serve as a guide for future efforts. Many of these are provided in Table II. These examples demonstrate that sampling mechanisms and materials can contribute relatively large errors in comparison to analytical errors in environmental chemistry results, particularly for trace level concentrations (i.e., < 1 mg L^{-1}) of chemical constituents.

Poor recoveries and positive bias caused by contamination are not the only problems that may arise from the use of certain types of samplers. The efficiency with which a person can control the operation of the sampler; maintain stable, reproducible operating conditions; and recognize a malfunction all play a significant part in the actual performance of environmental sampling equipment. Unfortunately, few manufacturers provide detailed operation or calibration instructions. More frequently, the principle of operation is referenced to a literature citation and specific precautions for safe use may be provided. Individual investigators, with the aid of the analytical staff, should plan to evaluate sampling performance in order to at least estimate the degree of systematic error as well as the routine precision that may be expected under

Table II. Selected Sampling Mechanism and Materials Evaluations

Environmental Matrix	Sample	Reference
Air	Aerosol metals	66
	Aerosol SO_4^{2-} and NO_3^-	95
	Precipitation	67–69
Water	Volatile organic compounds	70
Groundwater	Dissolved gases and volatile organic compounds	71
	Ferrous iron	72
	Miscellaneous	73
Seawater	Trace metals	74–76
	Surface microlayer	77
Soil	Pore water	78, 80
	Pore water organic compounds	79

field conditions. A general validation scheme for sampling devices is shown in the box on page 19. Although this scheme may be too involved for individual projects, large field programs would benefit from the use of such validation activities. If sampling errors are less than those that may be contributed by the analytical operation, reducing them further is probably not worth the effort in most cases.

Materials selection can be a potential pitfall for both sampling operations and achieving the purpose of an investigation. In situations where the sample remains in contact with a potentially reactive, sorptive, or leaching material, the opportunities for gross systematic errors exist. Sampling tubing (70, 81, 82), gaskets (36), plated surfaces (37, 83), and a variety of other such exposures have resulted in serious errors in sampling. The selection of a general sampling mechanism or a specific device made of appropriate materials for the application should be based on the most sensitive (i.e., labile, volatile, and reactive) chemical constituents under investigation. Once the sampling evaluation and experiment have been conducted, documenting the procedures and format field books to record deviations from the sample collection part of the sampling protocol is a relatively simple task (84).

Sample Handling, Field Determinations, Storage, and Transport

Most of the recent QA and QC guidance manuals available for the national environmental monitoring programs provide sound guidance for planning the procedures for sample preservation and handling. Because many sampling sites are exposed to rain, wind, sunlight, and temperature extremes, field transfers and

---

Sampling Device Validation Scheme Protocol

Laboratory: Operating range (ruggedness), recovery (accuracy), and precision

1. Generate known atmosphere, solution, etc.
2. Pump or collect multiple samples (and backup if possible) with device and collect controls by reference (direct) method.
3. Repeat step 2 holding concentration fixed and varying humidity, depth, lift, submergence, volume, or flow rate.
4. If possible, repeat step 3 over a range of concentrations for the expected range of variables.
5. Repeat steps 1–4 with interferences present, if possible.

Field: Estimation of recovery (accuracy) and precision

1. Use reference method, if possible, to establish the background, stable concentration.
2. Pump or collect multiple samples with device; if necessary, use backup samples to check breakthrough.
3. Repeat steps 1 and 2 over the range of field variables.

---

manipulations should be kept to an absolute minimum. Filtration procedures should be streamlined and protected so as to minimize sample transfers in the open air. Filter media selection and pretreatment steps should be planned and documented with the reduction of bias in mind (85, 86). A number of researchers performed filtration experiments in order to identify and control sample contamination or loss for trace metal determinations due to filtration apparatus or procedures (87-90). The detailed publications of Hunt and Gardner (90-92) are particularly useful toward minimizing systematic errors in filtration steps.

The sampling staff should be aware of routine laboratory procedures for both personnel and sample integrity protection. Gloves and eye protection provide valuable safeguards and can minimize artifact trace metals (93) or organic contamination of samples (94). Determinations of unstable chemical species or those that are difficult to preserve and store should be completed as soon as possible. A rapid field analytical method may be preferable to a laboratory method that requires extraordinary or complex sampling and handling procedures.

An effort should be made to handle and preserve field control samples (i.e., blanks, spikes, and colocated samples) in the same manner as the environmental samples. This precaution will allow more effective identification and control of postsample collection errors. Although not always possible, the experience of running the sampling device evaluation and sampling experiment should be useful in planning the sample handling and preservation procedures.

Once the samples have been delivered to the laboratory, the sampling operation extends to the point of sample reduction prior to analysis. The analyst who is consulted on sampling protocol development can materially improve the overall reliability of the results and conclusions of the program. A willingness to experiment (i.e., make and acknowledge errors) will provide the balance between statistical and experimental concerns needed to design and execute a cost-effective sampling effort.

Acknowledgments

I am thankful to a number of my past and present colleagues for their insight, help, and advice in the preparation of this chapter. The comments of the reviewers are also very much appreciated. Thanks also go to Pam Beavers who typed the manuscript and to Lynn Weiss who drafted the figures.

References

1. *National Research Council Opportunities in Chemistry*; National Academy Press: Washington, DC, 1985; p 18, pp 195–208.
2. Taylor, J. K. *Anal. Chem.* **1983**, *55*(6), 600A–608A.
3. King, D. E. In National Bureau of Standards Special Publication 422; La Fleur, Philip D., Ed.; National Bureau of Standards: Gaithersburg, MD, 1976; pp 141–150.
4. Youden, W. J. *Statistical Techniques for Collaborative Tests*; Association of Official Analytical Chemists: Washington, DC, 1973.
5. Murphy, T. J. In National Bureau of Standards Special Publication 422; La Fleur, Philip D., Ed.; National Bureau of Standards: Gaithersburg, MD, 1976; pp 509–539.
6. Kratochvil, B.; Taylor, J. K. *Anal. Chem.* **1981**, *53*(8), 924A–938A.
7. Kratochvil, B.; Wallace, D.; Taylor, J. K. *Anal. Chem.* **1984**, *56*(5), 113R–129R.
8. Youden, W. J. *J. Assoc. Off. Anal. Chem.* **1967**, *50*(5), 1007–1013.
9. Kowalski, B. R. *Anal. Chem.* **1980**, *52*(5), 112R–122R.
10. Frank, T. E.; Kowalski, B. R. *Anal. Chem.* **1982**, *54*(5), 232R–243R.
11. Kateman, G.; Pijpers, F. W. In *Chemical Analysis*; Elving, P. J.; Winefordner, J. D., Eds.; Wiley–Interscience: New York, 1981; Vol. 60, pp 15–69.
12. Green, R. H. *Sampling Design and Statistical Methods for Environmental Biologists*; Wiley: New York, 1979.
13. Bernstein, B. B. *J. Environ. Manage.* **1983**, *16*, 35–43.
14. Electric Power Research Institute. *Sampling Design for Aquatic Ecologic Monitoring*; Electric Power Research Institute: Palo Alto, CA, 1985; Vols. 1–5; EPRI EA–4302.
15. Garrett, R. G.; Goss, T. I. In *Geochemical Exploration*; Watterson, J. R.; Theobald, P. K., Eds.; Association of Exploration Geochemistry: Toronto, Ontario, Canada, 1978.
16. Liebetrau, A. M. *Water Resour. Res.* **1979**, *15*(6), 1717–1725.
17. Ellis, J. C.; Lacey, R. F. In *River Pollution Control*; Stiff, M. J., Ed.; Ellis Horward: Chichester, England, Chapter 17, pp 247–274.
18. Sanders, T. G.; Ward, R. C.; Loftis, J. C.; Steele, T. D.; Adrian, D. D.; Yevjevich, V. *Design of Networks for Monitoring Water Quality*; Water Resources Publications: Littleton, CO, 1983; p 328.
19. Gy, P. M. *Sampling of Particulate Materials: Theory and Practice*; Elsevier: New York, 1979.

20. Gy, P. M. *Analusis* **1983**, *11(9)*, 413–440.
21. Brands, G. *Fresenius' Z. Anal. Chem.* **1983**, *314*, 646–651.
22. American Chemical Society Committee on Environmental Improvement. *Anal. Chem.* **1980**, *52(14)*, 2242–2249.
23. *Fed. Regist.* **1978**, *43(132)*, 29696–29741.
24. Riggan, R. M. *Technical Assistance Document for Sampling and Analysis of Toxic Organic Compounds in Ambient Air*; Battelle Laboratories, Columbus, OH; prepared for U.S. Environmental Protection Agency–Environmental Monitoring and Support Laboratory: Research Triangle Park, NC, 1983; EPA 600/4–83–027.
25. *Sampling and Analysis of Rain*; Campbell, S. A., Ed.; ASTM Special Technical Publication 823; American Society for Testing and Materials: Philadelphia, PA, 1983.
26. Peden, M. E. *Development of Standard Methods for the Collection and Analysis of Precipitation*; Illinois State Water Survey CR 381; U.S. Environmental Protection Agency–Environmental Monitoring and Support Laboratory: Cincinnati, OH, 1986.
27. Peden, M. E. In *Sampling and Analysis of Rain*; Campbell, S. A., Ed.; ASTM Special Technical Publication 823; American Society for Testing and Materials: Philadelphia, PA, 1983; pp 72–83.
28. Topol, L. E.; Levon, M.; Flanagan, J.; Schwall, R. J.; Jackson, A. E. *Quality Assurance Manual for Precipitation Measurement Systems*; U.S. Environmental Protection Agency–Environmental Monitoring and Support Laboratory: Research Triangle Park, NC, 1985.
29. Wilson, E. B. *An Introduction to Scientific Research*; McGraw–Hill: New York, 1952.
30. Box, G. E. P.; Hunter, W. G.; Hunter, J. S. *Statistics for Experimenters*; Wiley: New York, 1978.
31. Mace, A. E. *Sample-Size Determination*; Reinhold: New York, 1964.
32. Janardan, K. D.; Shaeffer, D. J. *Anal. Chem.* **1978**, *51(7)*, 1024–1026.
33. Lodge, J. P. In National Bureau of Standards Special Publication 422; La Fleur, Philip D., Ed.; National Bureau of Standards: Gaithersburg, MD, 1976; pp 311–320.
34. Kobayashi, H.; Rittmann, B. E. *Environ. Sci. Technol.* **1982**, *16*, 107A–183A.
35. Cohen, S. Z.; Creeger, S. M.; Carsel, R. F.; Enfield, C. G. In *Treatment and Disposal of Pesticide Wastes*; Krueger, R. F.; Seiber, J. N., Eds.; ACS Symposium Series 259; American Chemical Society: Washington, DC, 1984; pp 297–325.
36. LeBel, G. L.; Williams, D. T. *J. Assoc. Off. Anal. Chem.* **1983**, *66(1)*, 202–203.
37. Neff, C. H.; Schock, M. R.; Marden, J. I. *Relationship Between Water Quality and Corrosion of Plumbing Materials in Buildings*; State Water Survey; prepared for U.S. Environmental Protection Agency–Water Engineering Research Laboratory: Cincinnati, OH, 1987; EPA 600/S2–87–036.
38. Hughes, J. P.; Lettenmaier, D. P. *Water Resour. Res.* **1981**, *17(6)*, 1641–1650.
39. Flatman, G. T.; Yfantis, A. A. *Environ. Monit. Assess.* **1984**, *4*, 335–349.
40. Gilbert, R. A.; Simpson, J. C. *Environ. Monit. Assess.* **1985**, *5*, 113–135.
41. Brown, K. W.; Mullins, J. W.; Richitt, E. P.; Flatman, G. T.; Black, S. C.; Simon, S. J. *Environ. Monit. Assess.* **1985**, *5*, 137–154.
42. Grant, C. L.; Pelton, P. A. *Role of Homogeneity in Powder Sampling*; Remedy, W. R.; Woodruff, J. F., Eds.; ASTM Special Technical Publication 540; American Society for Testing and Materials: Philadelphia, PA, 1972; pp 16–29.
43. Tourtelot, H. A.; Miesch, A. T. In *Trace Element Geochemistry in Health and Diseases*; Freeman, J., Ed.; Geological Society of America Special Paper 155; Geological Society of America: Denver, CO, 1975; pp 107–118.
44. Connor, J. J.; Myers, A. T. *Role of Homogeneity in Powder Sampling*; Remedy, W. R.; Woodruff, J. F., Eds.; ASTM Special Technical Publication 540; American Society for Testing and Materials: Philadelphia, PA, 1972; pp 32–36.
45. Garrett, R. G.; Goss, G. E. In *Proceedings of the 7th International Geochemical Exploration Symposium*; Watterson, J. R.; Theobald, P. K., Eds.; Association of Exploration Geochemistry: Toronto, Ontario, Canada, 1978; pp 371–383.

46. Perket, C. L.; Barsotti, L. R. In *Multilaboratory Analysis of Soil for Lead*; Petros, J. K.; Lacy, W. J.; Conway, R. A., Eds.; ASTM Special Technical Publication 886; American Society for Testing and Materials: Philadelphia, PA, 1986; pp 121–138.
47. Loehr, R. C.; Martin, J. H., Jr.; Newhauser, E. F. Ibid., pp 285–297.
48. Madsen, B. C. *Atmos. Environ.* 1982, *16(10)*, 2515–2519.
49. Shaw, R. W.; Smith, M. V.; Pour, R. J. *J. Air Pollut. Control Assoc.* 1984, *34(8)*, 839–841.
50. Nelson, J. D.; Ward, R. C. *Ground Water* 1981, *6*, 617–625.
51. Loftis, J. C.; Ward, R. C. *Water Resour. Bull.* 1980, *16*, 501–507.
52. Sanders, T. G.; Adrian, D. D. *Water Resour. Res.* 1978, *14*, 569–576.
53. Lettenmaier, D. P. *Water Resour. Res.* 1976, *12(5)*, 1037–1046.
54. Garrett, W. D. In *Proceedings of the Symposium on Organic Matter in Natural Waters*; Hood, D. W., Ed.; Institute of Marine Science Publication No. 1; University of Alaska, AK, 1968.
55. Duce, R. A.; Quinn, J. G.; Olney, C. E.; Piotrowicz, S. J.; Ray, B. J.; Wake, T. L. *Science (Washington, DC)* 1972, *176*, 161–163.
56. Zsolnay, A. *Mar. Chem.* 1977, *5*, 465–475.
57. Morris, R. J.; Culkin F. *Nature (London)* 1974, *250(5468)*, 640–642.
58. Macintyre, F. In *The Sea*; Goldberg, E. D., Ed.; Wiley–Interscience: New York, 1974; Vol. 5, *Marine Chemistry*; pp 245–299.
59. Patterson, C. C.; Settle, D. M. In National Bureau of Standards Special Publication 422; La Fleur, Philip D., Ed.; National Bureau of Standards: Gaithersburg, MD, 1976; pp 321–351.
60. Patterson, C. C.; Settle, D. M.; Schaule, B. K.; Burnett, M. In *Marine Pollutant Transfer*; Windom, H. L.; Duce, R. A., Eds.; Lexington Books: Lexington, KY, 1976.
61. Schaule, B. K.; Patterson, C. C. *Earth Planet. Sci. Lett.* 1980, *47*, 176–198.
62. Burnett, M.; Patterson, C. C. In *Proceedings of an International Experts Discussion on Lead: Occurrence, Fate and Pollution in the Environment*; Branica, M., Ed.; Pergamon: Oxford, England, 1978.
63. Bruland, K. W.; Franks, R. P.; Knauer, G. A.; Martin, J. H. *Anal. Chim. Acta* 1979, *105*, 233–245.
64. Windom, H. L.; Smith, R. G. *Mar. Chem.* 1979, *7*, 157–163.
65. *Quality Control in Remedial Site Investigations*; Perket, C., Ed.; ASTM Special Technical Publication 925; American Society for Testing and Materials: Philadelphia, PA, 1986.
66. Milford, J. B.; Davidson, C. I. *J. Air Pollut. Control Assoc.* 1985, *35(12)*, 1249–1260.
67. Slanina, J.; Van Raaphorst, J. G.; Zijp, W. L.; Vermuelen, A. J.; Rolt, C. A. *Int. J. Environ. Anal. Chem.* 1979, *6*, 67–81.
68. Chan, W. H.; Lusis, M. A.; Stevens, R. D. S.; Vet, R. J. *Water, Air, Soil Pollut.* 1983, *23*, 1–13.
69. Schroder, L. J.; Linthurst, R. A.; Ellson, J. E.; Vozzo, S. F. *Water, Air, Soil Pollut.* 1985, *24*, 177–187.
70. Ho, J. S-Y. *J. Am. Water Works Assoc.* 1983, *12*, 583–586.
71. Barcelona, M. J.; Helfrich, J. A.; Garske, E. E.; Gibb, J. P. *Ground Water Monit. Rev.* 1984, *4(2)*, 32–41.
72. Stoltzenburg, T. R.; Nichols, D. G. *Preliminary Results on Chemical Changes in Ground Water Samples Due to Sampling Devices*; prepared for Electric Power Research Institute: Palo Alto, CA, 1985; EA–4118.
73. Gibb, J. P.; Schuller, R. M.; Griffin, R. A. *Procedures for the Collection of Representative Water Quality Data from Monitoring Wells*; Coop. Ground Water Report 7; Illinois State Water and Geological Surveys: Champaign, IL, 1981.
74. Segar, D. A.; Berberian, G. A. In *Analytical Methods in Oceanography*; Gibb, T. R. P., Jr., Ed.; Advances in Chemistry 147; American Chemical Society: Washington, DC, 1975; Chapter 2, pp 9–15.
75. Brewers, J. M. ; Windom, H. L. *Mar. Chem.* 1982, *11*, 71–86.

76. Spencer, M. J.; Betzer, P. R.; Piotrowicz, S. *Mar. Chem.* **1982**, *11*, 403–410.
77. Van Vleet, E. S.; Williams, P. M. *Limnol. Oceanogr.* **1980**, *25*, 764–770.
78. Brown, K. W. In *Land Treatment, A Hazardous Waste Management Alternative*; Loehr, R. C.; Malina, J. F., Jr., Eds.; Center for Research in Water Resources, University of Texas: Austin, TX, pp 171–185.
79. Barbee, G. L.; Brown, K. W. *Water, Air, Soil Pollut.* **1986**, *29*, 321–331.
80. Evberett, L. G.; McMillion, L. G. *Ground Water Monit. Rev.* **1985**, *5(3)*, 51–60.
81. Barcelona, M. J.; Helfrich, J. A.; Garske, E. E. *Anal. Chem.* **1985**, *57(2)*, 460–464. (Errata, *Anal Chem.* **1985**, *5713*, 2752.)
82. Barcelona, M. J.; Gibb, J. P.; Miller, R. A. *A Guide to the Selection of Materials for Monitoring Well Construction and Ground Water Sampling*; State Water Survey Publication 327; U.S. Environmental Protection Agency: Cincinnati, OH, 1983; EPA 600/S2-84-024.
83. Andersen, K. E.; Nielsen, G. D.; Flyvholm, M.; Fregert, S.; Gruvberge, B. *Contact Dermatitis* **1983**, *9*, 140–143.
84. Barcelona, M. J.; Gibb, J. P.; Helfrich, J. A.; Garske, E. E. *Practical Guide for Ground-Water Sampling*; State Water Survey Publication 374; U.S. Environmental Protection Agency: Cincinnati, OH, 1985; EPA 600/S2-85-104.
85. Quinn, J. G.; Meyers, P. A. *Limnol. Oceanogr.* **1971**, *16(1)*, 129–131.
86. Maienthal, E. J. Presented at the 157th National Meeting of the American Chemical Society, Minneapolis, MN, April 1969.
87. Wagemann, R.; Brunskill, G. J. *Int. J. Environ. Anal. Chem.* **1975**, *4*, 75–84.
88. Truitt, R. E.; Weber, J. H. *Anal. Chem.* **1979**, *51(12)*, 2057–2059.
89. Laxen, D. P. H.; Chandler, I. M. *Anal. Chem.* **1982**, *54(8)*, 1350–1355.
90. Gardner, M. J.; Hunt, D. T. E. *Analyst (London)* **1981**, *106*, 471–474.
91. Hunt, D. T. E. *Filtration of Water Samples for Trace Metal Determinations*; Technical Report TR 104; Water Research Centre, Medmenham Laboratory: Marlow, Bucks, England, 1979.
92. Gardner, M. J. *Adsorption of Trace Metals from Solution During Filtration of Water Samples*; Technical Report TR 172; Water Research Centre, Medmenham Laboratory: Marlow, Bucks, England, 1982.
93. Berman, E. In *National Bureau of Standards Special Publication 422*; La Fleur, Philip D., Ed.; National Bureau of Standards: Gaithersburg, MD, 1976; pp 715–719.
94. Hamilton, P. B. *Anal. Chem.* **1975**, *47(9)*, 1718–1720.
95. Milford, J. B.; Davidson, C. I. *J. Air Pollut. Control Assoc.* *37(2)*, 125–134, in press.

RECEIVED for review February 11, 1987. ACCEPTED April 24, 1987.

Chapter 2

Sampling for Tests of Hypothesis When Data Are Correlated in Space and Time

Leon E. Borgman and William F. Quimby

Sampling design and data collection are strongly influenced by (1) the purposes for which the data will be used; (2) the time, money, and effort allocated to the sampling task; and (3) the prior information available for planning. These requirements are examined relative to (1) geostatistical sampling and hypothesis testing, which are defined as procedures predicated by prior knowledge or initial estimation of the covariance structure in space and time, and (2) random or stratified sampling. Nonparametric tests for both types of procedures are outlined.

TWO ALTERNATE COMPETING PROCEDURES for sampling and performing tests of hypothesis on space–time data are available. The older procedure is based on the classical random sampling or stratified random sampling familiar in experimental design. The newer procedure is based on random field theory, which has developed primarily over the last 20 years. This newer procedure is still the subject of ongoing research and debate.

The situation is very similar to the controversy that began 20 years ago between classical hypothesis testing and Bayesian methods. For the classical procedures, prior data and judgment were used only to develop the model framework and to plan the collection of new data. The estimation and tests of hypothesis were performed only with the new data within the framework of the model.

1173–6/88/0025$06.00/0 © 1988 American Chemical Society

For the Bayesian procedures, the prior data and judgments played a much more active role throughout estimation and testing. The prior information was used not only in the model design but also in the establishment of a priori probability laws for parameters, which then played an intrinsic role in later computations.

For most statisticians, the controversy between classical and Bayesian procedures is no longer a matter of deciding which is right and which is wrong. Each is useful under some circumstances. The questions and arguments have shifted to the specifics of a given problem. Which method is best for the data at hand?

Classical random sampling (*1*) is similar to the classical hypothesis testing in that prior information and judgment are used to select the regions for sampling, the need for stratification or subregions, the variance as it relates to initial sample size, and similar questions of sampling framework. However, once the model is established, the newly collected data are used for estimation and testing of conjectures.

Random field sampling (geostatistical sampling) is similar to the Bayesian procedures in that prior judgment as to the stationarity of the random field, the covariance structure model, applicability of prior data from other similar areas, and so forth are intrinsically combined with the new data to produce the statistical conclusions. A much greater willingness exists to incorporate engineering judgment into the statistical process.

Of course, hybrid or sequential versions of each method have been used from time to time when the initial information was particularly inadequate. Random sampling may proceed in stages; improved estimates of the variance are made at each stage, and all data from the beginning are used in the computations at each stage. Geostatistical sampling is particularly powerful when applied sequentially. Indeed, geostatistical sampling is really the natural mode of use from which many of the methods developed in mining engineering. New results from mine sampling were folded in with the previous data to continually update the estimates of covariance structure, the panel averages for impending mining operations, and the errors of estimation. The analogy to Bayesian updating is difficult to avoid.

In this chapter, the model assumptions for each approach will be examined. The advantages and disadvantages of each technique will be examined candidly. The general framework for hypothesis testing in both cases will be outlined, and the problems of estimation of covariance functions in multidimensional space for the geostatistical techniques will be examined. The structure of nonparametric (i.e., distribution-free) methods will be reviewed, and applications to each approach will be considered. Because random sampling is a fairly well-known and established procedure, and because some difficulties are inherent with traditional nonparametric methods for the correlated data in the geostatistical model, special attention will be given at the end of the chapter to nonparametric geostatistical testing.

Model Assumptions

Random sampling and geostatistical (random field) sampling are based on very different concepts or models of the underlying space–time reality. To a large extent, these differences in fundamental perspective are the cause of most disagreements between advocates of the two methods. Once each group understands the rationale of the other, most arguments reduce to a discussion of the relevant advantages and disadvantages for specific applications.

Random Sampling

The random sample technique is well-characterized as the "basket of apples" method. Suppose the apples on a tree are to be sampled. The quantity to be estimated, and perhaps to be the parameter in subsequent tests of hypothesis, is the average value of the attribute of interest for the total assemblage of apples. Because there are a finite number of apples, this parameter could be determined exactly if all the apples were picked and the attribute measured on each and every apple. The "true" or theoretical parameter value is just the arithmetic average over the finite population.

From this viewpoint, the finite population is actually deterministic. Nothing is random about the total finite population. The total population just consists of N numbers, if there are N apples on the tree. The randomness lies in the sample. If a sample of n apples is selected for measurement, the arithmetic average for the n apples in the sample is random in accordance with the randomness involved in choosing n apples out of the N apples in the total finite population. The goal of the random sampling technique is to arrange the sampling so as to make every apple of the total population equally likely to be included in the sample. If this goal is achieved, every possible sample of size n taken from the N things is equally likely and has a probability equal to the reciprocal of $C(n, N)$, where $C(n, N)$ denotes the number of combinations of N things taken n at a time. Conceptually, the random sampling could be achieved by picking all the apples, mixing them thoroughly in a basket, and then picking out the sample of n apples. In practice, random numbers are used to determine which n of the N apples is to be included in the sample.

A variety of probability relations, including the variance of the sample average and its properties as an estimate of the average of the finite population, can be derived from the random sampling premise by combinatorial analysis (1). Numerous extensions are also possible. The probability of inclusion in the sample need not be equal as long as it is known. This possibility leads to so-called probability sampling. Another direction of extension is to subdivide the apples into subpopulations of various, more homogeneous groups (e.g., with factors such as sun exposure, size of branch, and nearness to nutrients arising from the roots up the main trunk). Because the subgroups, or *strata*, are more homogeneous, the within-strata variances are smaller, and each strata mean can be estimated with smaller variance from samples within that strata. The estimates of strata means can be combined to yield estimates of the population mean. This general body of techniques is called *stratified random sampling*.

The extension of random sampling to space–time data consists conceptually in imagining all possible samples of material for analysis being placed in "bags" of material. (Here, N would be quite large, but with some mental effort, N could be visualized as finite.) The procedure would then need to be organized so that every "bag" is equally likely to be included as one of the n samples selected for analysis. This randomness usually is achieved by stratification with respect to time and the selection of samples over the space within each time strata through the use of pairs of random numbers. Each pair of numbers gives the north–south and east–west coordinate for a sample to be collected. The spatial dimensions may also be divided into subregions, or additional strata, to obtain greater homogeneity and reduced standard deviations for the estimates to be made from the data.

Geostatistical Sampling

In contrast to the deterministic, finite sample perspective of random sampling, geostatistical sampling is based on the assumption that the reality being sampled is one realization of a random field. The significance and meaning of this premise is philosophically identical to the general concept of a random population. What is meant by the statement that the outcome of a game of chance is random? Generally, this statement is interpreted as meaning that the outcome is unknown, but that some probability law exists that describes the likelihood of various possible outcomes. The focus of interest is shifted to an attempt to develop a model for the probability law through analytical reasoning, collection of data, or some combination of the two.

In modern physics, space and time are visualized as a four-dimensional continuum. The outcome of the game of chance is just an event that occurs at some point in the continuum other than our present location. If we move to the space–time location where the outcome is observable, the event is no longer random. The event is a deterministic outcome that has happened. In this framework of thought, no fundamental difference exists between lack of knowledge of what occurs at another location in space or what occurs at some future time. Each case is just a location in the space–time continuum other than the present position. Consequently, modeling the probability of events at some other location is just as sensible as developing probabilities for future events.

This point has been discussed in so much detail because a first, very typical objection to the geostatistical approach for geological measurements is to claim that the geological phenomenon is not really random. Whatever is at the site is already there. One just has to go and see what it is. The point is that this statement is equally true for the outcome of the game of chance. One just has to go in time to that location and see what the outcome is. If decisions have to be made from the present location about outcomes at other locations, the probability model represents a consistent, rational framework for proceeding.

Most probability models are fundamentally centered around the joint probability law for the various measurable aspects of the outcome of concern. With luck, many of the aspects will be independent of each other, and the model can be subdivided into independent parts, each of which can be modeled separately. If the various aspects are intercorrelated, this subdivision is not possible. Then, attention is focused on characterizing the structure of the joint probability law as simply as possible.

A random field, $V(\mathbf{x}, t)$, is a collection of random variables, one for each vector of space coordinates $(\mathbf{x})$ and time (t) over some region of space and time locations. $V(\mathbf{x}, t)$ may be interpreted as the random variable at the location given within the parentheses. The observation at location $(\mathbf{x}, t)$ will be indicated by a lower-case letter [i.e., $v(\mathbf{x}, t)$]. If, for any collection of n locations, $(\mathbf{x}_1, t_1)$, $(\mathbf{x}_2, t_2)$, ... $(\mathbf{x}_n, t_n)$, the joint probability for V at these locations is the same as that of V at any other set of n locations obtained by making a fixed vector shift of $(\mathbf{X}, T)$ on each of the original n locations, then the random field is said to be *stationary* in the strong sense. If the expected value of $V(\mathbf{x}, t)$ is constant over the region, and the covariance of $V(\mathbf{x}, t)$ with $V(\mathbf{x} + \mathbf{X}, t + T)$ depends only on the shift vector $(\mathbf{X}, T)$, then the random field is said to be *covariance-stationary*, or stationary in the weak sense. For a covariance-stationary random field, the covariance is said to be isotropic if the covariance function depends only on the length of the vector, $(\mathbf{X}, T)$, and not on the direction of the vector.

In modeling a space–time random field for geostatistical sampling, a first task is to make judgments as to the stationarity or nonstationarity of the random field and then to characterize the assumed structure as thoroughly as possible from prior information and data. Much can be achieved without actually specifying the joint probability laws for the random field. Indeed, the many varieties of kriging (2) proceed only from estimates of the covariance function (or a related function called the *variogram*, which is slightly more general), together with appropriate assumptions concerning stationarity.

From the perspective of these concepts, *geostatistical sampling* can be defined as those sampling procedures that are guided by the assumed properties of the random field and prior, or early, estimates of the covariance or variogram functions. *Geostatistical estimation* is the process of using space–time data, together with the assumptions and structural characterization of the random field, to estimate parameters or characteristics (properties) related to the random field. *Geostatistical test of hypothesis methods* are procedures for making decisions concerning conjectures about the parameters of the random field from statistics computed from field data, together with the assumed random field characteristics and estimated covariance or variogram. Because the initial estimation of the covariance structure plays such a central role in all these procedures, substantial time is spent in the initial phases of a geostatistical study making estimates of the covariance or variogram functions and validating these estimates by various methods. The effort continues as data accumulate through periodic updates of the estimates and the associated validation and reliability studies.

Important Considerations

A sampling plan and the associated choices of estimation and hypothesis-testing techniques can be evaluated by several criteria. First, any sampling plan is aimed at the acquisition of data describing or characterizing some target population. For example, characterizing the lead content of the soil along some interstate highway might be desirable. Usually, not all of the target population is accessible. For one reason or another, parts of the target population may be unavailable for sampling. For the highway example, perhaps sections pass through private property where access is denied, or traverse very mountainous terrain where physical problems are encountered with reaching the desired sample locations. Restrictions of cost and time are very real in most sampling operations, so one more reduction usually occurs in the size of the population from which samples will be gathered to match the monetary restraints. In summary, three populations must be defined and considered in planning most sampling plans. These populations are (1) the target population, (2) the accessible population, and (3) the actually sampled population. The sampling plan should be organized so that valid inferences can be drawn from data taken from the actually sampled population that apply correctly to the target population. At the very least, the likely differences between the data and the target population should be studied and pointed out to potential users of the results.

Various options for sampling can be evaluated from other perspectives. Some of these options are (1) the primary purpose for which the data will be used, (2) the diversity of applications in which the data will be required to serve, (3) the type and availability of initial information that can be obtained to plan and guide the sample design, (4) the cost-effectiveness of the sampling options as it relates to the budgeted time and effort, and (5) the quantity or relations to be evaluated. Decisions concerning these aspects will affect which sampling procedures have the most desirable characteristics in a given application. It would be delightful if the sample plan could excel from all perspectives. The time and effort limitations involved in option 4 almost always make such a perfect plan impossible. In the real world, the sampling plan is usually the result of many compromises.

Some of the primary, or first priority, purposes for which the data are being collected are (a) for use in legal or regulatory activities, (b) as process information in engineering and scientific studies, (c) for routine monitoring and control of industrial or other operations, and (d) as an initial survey to plan future work, such as site cleanup. The collection of data for legal action often requires extreme conservatism and the selection of the sampling plan that is the least subject to objections under cross-examination. Such sampling is typically quite expensive and not cost-effective relative to obtaining the most information for the dollar spent. Data collected for engineering and scientific studies generally allows and even requires a maximal use of prior information to obtain as much useful information as possible with the allocated budget. Engineering judgment is an expected major component of sample plans for such a purpose. Sampling for routine monitoring applications is usually very constrained by cost and time considerations. Information of very specialized types is sought at minimal cost on a continuing basis. The collection of data for initial or baseline studies makes special demands on the planner. Ordinarily, very little initial information is available, and the sample plan is really a reconnaissance operation. Bayesian decision theory and other decision-theoretical methods may be appropriate to use in guiding the planning.

Data to be collected in the sampling may be required only for a single high-priority purpose. In that case, the design can be directed to be most efficient and cost-effective for that purpose. More commonly, the data are scheduled for a number of uses. The sampling plan is then a compromise between the various demands. Careful planning and control is needed to guarantee that each use is adequately satisfied. Often, the various applications of the data conflict with each other, and arbitrary decisions must be made concerning the priorities that are to be assigned to each application in the subdivision of the budgeted time and money.

The initial information available to plan and guide the organization of a sampling effort may vary through the following categories: (a) little or no information, as in the early reconnaissance phase of an investigation; (b) sparse but at least some data, as in the exploration phases of a study; (c) reasonable but not voluminous data, as in the development phases of an investigation; and (d) substantial data, as in the late development and production phases of mineral industry operations.

The time and cost limitations may be very extreme for short, "quick and dirty" sampling efforts made on short notice for some special immediate need. More commonly, a reasonable budget is available and time is allocated for adequate planning. Occasionally, sampling plans are encountered where the purpose is very important, such as in multimillion dollar litigation, and any reasonable amount of money is available provided that the data are expected to help advance the purpose.

Some of the quantities or relations that the sampling plan may be required to estimate are as follows:

- the regional mean and variance;
- the regional trend in mean;
- the differences in means and variances between areas;
- an estimation of probability laws for various sampled variabilities;
- the detailed contour maps on selected variables;

- the spatial correlation functions, such as the variogram and covariance functions;

- the probabilities for extremes and near extremes;

- the fraction of a given area that has values higher than a selected critical value;

- the confidence intervals and bounds on various sampled variables;

- the test statistics for hypothesis tests related to particular parameters of the random field;

- the regional location for some spatial pattern such as a plume of contamination in groundwater; and

- the likely future behavior of the processes present within the region being sampled.

Generally, the demands on the sampling plan and the associated costs increase down the list. For many of the items near the end of the list, substantial engineering and scientific judgment are involved.

Advantages and Disadvantages

Critique of Random Sampling

The major advantage of random sampling procedures in space–time data is the simplicity of the assumptions that must be made concerning the population to be sampled. Because the real world is considered deterministic and the randomness is purely related to the choices in sampling, all of the random field assumptions, such as stationarity, probability law, and covariance function, are irrelevant. No random field exists within the random sampling model.

This simplicity of assumptions protects the user from the "sin" of wishful thinking or poor judgment. Most personal opinions do not enter into the computations in any way as long as the sampling is correctly randomized. Some personal choices may be involved in the selection of the best stratification and the initial estimation of population standard deviations for sample size selection, but these choices, if wrong, are not crucial to the procedure in the sense of invalidating the method. Rather, these choices, if wrong, may make the sampling estimates slightly less reliable than if better choices were made.

A consequence of this insensitivity of the random sampling procedures to personal biases is that the estimates are easy to defend against most criticisms. In particular, data collected by random sampling are recommended for legal actions where the opposing counsel will seek any possible objections that can be found to discredit the survey results. This recommendation does not mean that other types of sample plans cannot be used in court actions, but only that quite a bit more effort may be required in preparing and justifying the assumptions and choices made.

If the random sampling methods are so defensible, why are other procedures used at all? The answer is that random sampling also has a variety of disadvantages for some applications. To some extent, good judgment is a substitute for money. Because random sampling avoids the use of most engineering judgment, more money must be spent usually to obtain this freedom from personal bias. The necessity of taking samples only at sites specified by random numbers precludes the possibility of sampling where good judgment suggests some significant spatial event is likely. The only random sample alternative is to take such a large number of samples that measurements near the site in question are present in the observations. Even then, the character of random numbers is such that usually sample sites cluster into bunches at some spatial regions and leave other areas with few or no observations.

Several ways around the difficulty of even coverage of the region under study are available. One way is to divide the region into rectangular cells with a selected Cartesian grid, and then to use each cell as a stratum for sampling. A specified number of samples are randomly selected from each stratum. The cells do not have to be of equal size if other choices are more desirable. However, enough random samples should be collected from each cell to enable a reliable estimate of the cell mean for that cell.

Another way to obtain even regional coverage is to use a random sampling technique called *systematic sampling*. Here, random numbers are used to lay a Cartesian grid over the region so that the coordinate origin and axis orientation are random. Then, samples are taken at each grid intersection. However, in this approach, the grid must not be placed by using personal judgment or for geographic convenience to minimize cost. The grid must be placed randomly. One advantage of systematic sampling is that the data obtained are very convenient for analysis by both the random sampling and the geostatistical sampling methods. The sampling scheme "lives" in both worlds.

In practice, random sampling procedures experience numerous logistic difficulties related to the necessity of letting sample sites be determined by random numbers. In addition to the difficulty of locating a specified map site accurately in the field, the sites picked by the random numbers are often in areas hard to reach. This situation, of course, serves the good purpose of making sure that no portions of the region are avoided in the sampling simply because of the access difficulties. On the other hand, traveling substantial distances just to take one sample is annoying. Having the sample sites possess some structure allowing a reasonable sequence in sampling is less difficult logistically and cheaper.

Also, what does one do if the site selected by the random number is covered by something other than what is to be sampled? After traveling some distance on foot to collect a soil sample at the designated site, finding that site is all rock outcrop with no exposed soil is annoying. Some users have suggested an on-the-site randomization for this case consisting of shutting your eyes; spinning around until you're dizzy; and then, with your eyes still closed, throwing a rock hammer or other object as far as you can. The substitute sample site is the location where the hammer lands. This on-the-site randomization is close enough in spirit to the random sampling method so that the data are probably satisfactory for use in the computations.

Most participants in a field sampling will have strong personal opinions as to where significant changes are taking place. Avoiding sampling at these sites is difficult. As a result, the actual samples usually consist of the randomly placed measurements, together with a fair number of discretionary, or personally placed, measurements. The discretionary samples cannot be used validly in the random sample estimation and testing. If the sample processing is expensive, the presence of a substantial number of discretionary samples can interfere with the planned estimate reliability because analyzing the additional samples diverts money from the intended use. Therefore, planning for a reasonable number of personally placed measurements from the beginning is desirable. These measurements can be used in contour maps and other characterizations of spatial distribution, but not in the formal mathematics of the random sample computations. As a matter of experience, these discretionary samples often prove to have crucial importance in following the process behavior within the region.

Critique of Geostatistical Sampling

A primary advantage of geostatistical sampling methods is that sample sites may be selected by personal judgment to best cover the area in a cost-effective and expeditious way. The primary restriction is that the space–time arrangement of sample sites should allow estimation of the covariance structure, and should cover the region in an even way and avoid the deletion of any significant portions of the region.

Responsibility is associated with this relatively free use of personal judgment. The reasons for each decision should be documented so that later charges of personal bias or capricious judgments can be refuted. Litigation is becoming so prevalent in today's society that every researcher should try to be prepared to demonstrate that the actions taken were in accordance with what a prudent engineer or scientist would have done under similar circumstances. Introducing some randomization of sample sites in the geostatistical sampling procedures is often possible without incurring later analytical difficulties. Where this is possible without causing unacceptable additional cost, incorporation of randomization is a desirable action that serves as further evidence that a serious attempt was made to avoid possible personal bias.

The random field structure taken as being present, after a suitable careful investigation of available information, provides a very rich model for various applications. Estimates can be made by a procedure, called *kriging*, of the average value for subplots within the region without even having any actual samples within the subplot. These estimates are possible from the assumed spatial persistence of the random field implicit in the assumed covariance function. Confidence intervals can be derived for these estimates from the covariance structure.

Statistical simulations conditioned on a relatively sparse data set can be generated to produce alternate, equally likely scenarios of what the spatial and temporal distribution of values may be. This method, called *conditional simulation*, allows the consequences of various random field assumptions to be studied in the computer before actual field data are collected. Conditional simulation is particularly useful in operations research studies of difficulties that can arise related to the sequencing of field operations.

Geostatistical sampling almost always provides more information for dollar spent than do the random sampling procedures. Furthermore, the geostatistical model accommodates itself well to many diverse applications with great flexibility. What then are the difficulties and drawbacks of the method?

Probably the greatest shortcoming is the difficulty in knowing when various assumptions are acceptable or not acceptable in a particular application. The experience and judgment of the user may be quite significant. Fortunately, many of the geostatistical procedures are fairly forgiving. Reasonable answers often are obtained even with the wrong assumptions. A desirable early phase of most geostatistical computations consists of sensitivity studies to determine which assumptions are most critical and which are relatively insignificant.

An aspect of the random field, which is often of some importance in the geostatistical computations, is the stationarity, or lack of stationarity, that can be assumed. The stationarity critically affects the averaging to be used in the estimation of the covariance or variogram function. If intrinsically nonstationary, the random field will require substantially more samples to characterize the region and make the planned estimates and tests of hypothesis. Careful determination of the types of nonstationarity present and planning the sampling around this structure will be necessary.

A final disadvantage of geostatistical sampling procedures is that more effort is required to validate the choices made and to prevent future criticism of the survey results. Court cases can be based effectively on geostatistical methods, but it is really necessary to do your homework. The arbitrary choices and personal judgments in the sample plan must be documented and justified. Of course, this procedure is no different from other engineering activities where personal judgments play a substantial role.

Hypothesis Testing

The emphasis in this chapter is placed on nonparametric tests and their application to random sample and geostatistical sample data. Nonparametric is taken to mean distribution-free. That is, *nonparametric procedures* are those statistical techniques that can be applied without concern for the actual distribution of the underlying population from which the data were collected. The techniques are organized so as to be very insensitive to the probability distributions of the actual population. The parametric case for geostatistical data has also been studied recently for some of the standard tests (3) but will not be considered in this chapter.

If reasonable judgments can be made concerning the true probability law for the population, then the parameters of the population can be estimated, and tests based on that probability model usually will be better and more powerful than the nonparametric tests. However, if the probability law and associated population assumptions are uncertain, the nonparametric tests represent procedures that are fairly dependable despite the lack of certainty concerning the population characteristics. Routine performance of both parametric (i.e., procedures based on the assumed probability law with its parameters) and nonparametric tests is often good practice. If too different from each other, the conclusions may indicate unexpected data behavior that should be investigated.

The tests that will be examined in this section for applicability to random and geostatistical samples are (1) the runs test, (2) the sign test, (3) the two-sample Wilcoxon or Mann–Whitney test, (4) the one-sample signed Wilcoxon test, and (5) the two-sample matched-pair Wilcoxon test. Full details on these methods for the independent case can be found in reference 4 and other standard nonparametric textbooks. Tables for probabilities and significance levels of the nonparametric tests are given in numerous handbooks (5, 6). The emphasis in this section will be on discussing how the tests are applied to the sample data, and in particular, how problems of interdependence may be treated in the random field, or geostatistical, case. The general framework and assumptions for each test will be discussed briefly without going into the full mathematics. The mathematics for the proposed geostatistical version of the tests is given in substantially more detail in reference 7.

The *runs test* is applicable to a sequence of independent random variables that may be having a change in mean value as the index of the sequence increases. The test proceeds by first determining the median of the set of numbers in the sequence, and then subtracting the median from each value in the sequence. Values that exceed the median are replaced by a plus symbol, and values that are less than the median are replaced by a negative symbol. Any values that equal the median are ignored. A run is a sequence of like symbols. For example, in the sequence +++−−−++−−++, three runs of pluses and two runs of minuses occur to yield a total of five runs. If the mean is steadily increasing through the sequence, the pattern might look like −−−++−+++++++, and there would tend to be fewer runs than would be expected if the values randomly wandered above and below the median throughout the sequence. Tables are published for the probability law for the number of runs when the mean is constant and the deviations above and below the median are purely random.

The *sign test* is applicable to any set of independent, identically distributed random variables for which testing is desired regarding whether the true theoretical median equals a specified value. The identically distributed assumption can be replaced by the requirement that all the random variables have the same theoretical median without affecting the test. The sign-test statistic is computed by counting the number of sample values, K, that are greater than the assumed median and the total number of values, N, that are not equal to the median. If the hypothesis that the true median equals the specified value is true, then K behaves as a random variable from a binomial population having parameters N and 0.5. If K is too different from the expected value of $N/2$, the null hypothesis is rejected.

The *two-sample Wilcoxon test* is designed to test whether the mean of one population differs from the mean of another population. A sample of size M is taken from the first population, and a sample of size N is taken from the second. Both samples are assumed to consist of independent, identically distributed observations, and the samples are assumed independent from each other. The test statistic is computed by combining the two samples into a single sample of size $M + N$, and then ranking this combined sample in order of increased size. The sum of ranks for the positions occupied by the first sample is computed. A similar sum of ranks is computed for the second sample. The test statistic is the smaller of the two rank sums. This rank sum value is compared with published critical values to evaluate whether the two samples appear to have the same mean (i.e., ranks are randomly intermingled) or are shifted relative to each other (i.e., low ranks belong predominantly to one sample, and high ranks belong predominantly to the other sample).

The *signed Wilcoxon test* is another test for the median. The sample is taken to contain independent, identically distributed observations from a reference population having a symmetric probability density. Under the null hypothesis, the reference population has a median equal to some specified value. Two new samples are artificially created from the original sample. The first of these consists of all the observations in the original sample that are larger than the specified value. The second artificial sample consists of the absolute values of all the observations in the original sample that are smaller than the specified value. The two-sample Wilcoxon statistic is computed for these two artificial samples, and the value is compared with published critical values for the signed Wilcoxon statistic.

The *matched-pair Wilcoxon test* is designed to test whether a treatment effect is significant as compared with another observation that is similar in all ways with the first except that the treatment was not received. The two observations, one with treatment and the other serving as control, are called a *matched pair*. A number of matched pairs are collected. Each matched pair is assumed to be independent from the other matched pairs. A population of differences is formed by subtracting the control observation from the treatment observation within each matched pair. Under the null hypothesis, the treatment effect is assumed to be zero, and the control and treatment observations are really taken from a common population. Under this assumption, the probability law for the difference is symmetric about zero. The signed Wilcoxon statistic is computed on the populations, and the null hypothesis rejected if the test statistic value is too small compared to the published tables for the signed Wilcoxon critical values.

Hypothesis Testing within the Random Sampling Framework

This case will primarily be considered for the situation where the size, N, of the finite population is very large with respect to the sample size, n. In this case, the sampling of the finite population, which is actually sampling without replacement, is asymptotically equivalent to sampling with replacement. Consequently, the observations obtained as a random sample from the very large finite population may be treated as independent, identically distributed random variables. The observations from the random sampling can be treated as a classical random sample in subsequent test and estimation procedures.

All of the nonparametric tests discussed apply with no modifications. The important concept to grasp is that within the random sampling scheme, the collection of possible samples is disassociated from their site of occurrence within the strata being sampled, and the samples collected are just so many "apples in a big basket" as was discussed previously. The spatial intercorrelation that may have been present for the in situ sample values is destroyed conceptually by the randomization of sample selection. Relative to the estimation of the population average value for that strata, this disassociation of samples from the site is unimportant. However, information has been discarded, and this loss is the ultimate reason for the generally higher cost of estimation and testing using the random sampling procedures as compared with the geostatistical procedures.

Covariance Structure for Geostatistical Procedures

The interdependence taken as present in the random field model causes problems in traditional hypothesis testing where some form of independence is the basis for most test statistics. If classical test procedures are to be applied to random field data, some initial data processing is required to produce an equivalent data set that possesses the required independence. Various tricks will be outlined for achieving this transformation.

The characterization of the multidimensional covariance function is central to the independence transformations. Without getting too deeply into the mathematics, a brief introduction to a possible scheme of characterization will be presented in this section. In geostatistics, the most common choice for a model of the covariance function in one dimension is probably the spherical formula having the following forms:

$$C(h) = B, \text{ if } h = 0 \tag{1}$$

$$C(h) = (B - A)\left\{\frac{1 - 1.5h}{H} + 0.5(h/H)^3\right\}, \text{ if } 0 < h < H \tag{2}$$

$$C(h) = 0, \text{ if } h > H \tag{3}$$

$$C(h) = C(-h) \tag{4}$$

where $C(h)$ is the covariance function of lag, h, in one dimension; B is the variance of the random field, or sill height; A is the nugget effect; and H is the zone or range of influence. These parameters (A, B, and H) are assumed to be positive.

One way to extend the one-dimensional covariance to two or more dimensions is to take the zone of influence in any given direction as defined by the quadratic form

$$\mathbf{H}^T \mathbf{M} \mathbf{H} = Q \tag{5}$$

where Q is some positive number, $\mathbf{H}$ is the vector in the given direction having length equal to the zone of influence in that direction, M is a square matrix, and the superscript T denotes the transpose of the vector $\mathbf{H}$. The quadratic form specifies a multidimensional ellipsoid having a size scaled by the magnitude of Q and an orientation determined by the matrix M.

What is the zone of influence in the direction specified by the unit vector $\mathbf{u}$? This problem may be easily solved by expressing $\mathbf{H}$ as $\mathbf{u}H$, where H denotes the zone of influence in the designated direction. If $\mathbf{u}H$ is substituted into the formula for Q, the value of H may be solved as

$$H = \left[\frac{Q}{\mathbf{u}^T \mathbf{M} \mathbf{u}}\right]^{1/2} \tag{6}$$

By using this specification of H, a formula for the covariance function $C_m(\mathbf{u}h)$ defined in multidimensional space is

$$C_m(\mathbf{u}h) = C(h) \tag{7}$$

where $C(h)$ is the one-dimensional spherical formula defined previously, except that the directional value of H has been introduced. This covariance function is one example of what has been called a *radial covariance function* with an elliptical base (8).

Any collection of observations taken from a stationary random field can be considered as a random vector, $\mathbf{V}$, whose covariance matrix is fully defined by the $C_m(\mathbf{u}h)$ formula. Let D represent this covariance matrix. If U is the matrix of eigenvectors for D, and L is the diagonal matrix of eigenvalues for D, then (9)

$$DU = UL \qquad (8)$$

and the transformed data vector, $\mathbf{V}_*$, is defined as

$$\mathbf{V}_* = L^{-\frac{1}{2}}U^T\mathbf{V} \qquad (9)$$

The value of $\mathbf{V}_*$ will have uncorrelated components having unit variance (10) and possess a vector expected value as follows:

$$\text{expected value of } \mathbf{V}_* = L^{-\frac{1}{2}}U^T\mathbf{1}m \qquad (10)$$

where m is the expected value for the original random field, and $\mathbf{1}$ is a vector, all of whose components are 1.0.

The eigen transformation produces a new data vector with uncorrelated components, each having unit variances and possibly unequal expected values. This statement is true even if the random field is not Gaussian. If, in addition, the original random field is Gaussian, the transformed data vector is multivariate-normal with independent components. This structure has various implications for both parametric and nonparametric testing. The consequences will be explored in substantial mathematical detail in papers planned for later presentation (7, 11). Meanwhile, some of the general details as they relate to nonparametric tests will be discussed in the next section.

Some Approximate Nonparametric Tests for Geostatistical Data

Data Sparse in Space and Time

The word "approximate" is included in the section title because the components of $\mathbf{V}_*$ may not really be identically distributed if the original random field is not Gaussian. However, the components all have unit variance and should be similar enough in shape to allow most nonparametric tests. If the univariate density for the random field is highly skewed, an initial symmetrizing transformation will be assumed to be made on the data, such as the logarithmic transformation for the log-normal case. The mean value will then become the median also for the transformed data. Suppose it is desired to test the hypothesis H_0: mean $= m_0$, where H_0 is the null hypothesis. This hypothesized mean is subtracted from the random field data before the eigen transformation. Because the original data have mean zero, if the null hypothesis is true, then the transformed data will also have mean zero. The original data were made roughly symmetric, and so the transformed data will be roughly symmetric about zero if H_0 is true. Consequently, the sign test or the signed Wilcoxon test can be used to test for the centering about zero.

What about a paired comparison between two areas, one of which is a control and the other has suffered possible contamination? Suppose n samples are collected from the control area; n samples are taken from the treatment area; and the spatial pattern of sample sites are identical in geometry, spacing, and orientation in the two areas. Corresponding samples from the same geometric locations are paired together. Let V denote the vector of differences for the n pairs of samples. The covariance matrix for V [cov(V)] can be expressed as

$$\text{cov}(V) = C_{11} + C_{22} - C_{12} - C_{21} \tag{11}$$

where $C_{11} = C_{22} =$ the covariance of the data vector in each area, and $C_{12} = C_{21}{}^T =$ cross-covariance of the control data vector with the treatment data vector.

The matrix cov(V) is then analyzed for eigenvalues and eigenvectors, and the vector of differences is eigen-transformed as before to yield the vector with uncorrelated components having variance zero. If no difference exist between the control and treatment areas, the mean of the eigen-transformed vector of differences will be zero. This situation can be tested with a signed Wilcoxon test.

If the sample sizes and geometries are not the same from the control area to the treatment area, a type of two-sample Wilcoxon test may be used as a test of "no treatment" effect. This test is particularly effective if the control and treatment areas are far enough separated from one another so that they are uncorrelated with each other. The full details of this test are given in reference 7. As a general summary of the approach, two data vectors stacked into one vector are rotated by the principal component transformation derived from their covariance matrix. The vectors are then adjusted with a special orthogonal matrix generalized from the Helmert matrix. If the null hypothesis of no treatment effect is true, the components of both transformed data vectors will center around the original (common) mean and have unit standard deviations. If the null hypothesis is false, the mean of the components for the transformed treatment data vector will have various shifts from the common value. The two-sample Wilcoxon test statistic is sensitive to these differences.

Long-Time Series at Selected Spatial Locations

If the sample sites are close enough together to be correlated, one approach to decouple the time dimension of the data is through the discrete Fourier transform of each time series. The Fourier coefficients for circular periodic time series are known to be uncorrelated and related to periodic influences such as seasonal and tidal variations. The Fourier coefficients at a given frequency for the combination of all the spatial sites form a correlated data vector that can be studied by the procedures mentioned previously. Tests of hypothesis for equality of response at that frequency can be performed.

If the correlation scale is small relative to the length of the time series, the stationarity of the time series relative to the mean, variance, or other values can be investigated with a subdivision of the time axis into subintervals that are large with the temporal zone of influence but small with respect to the length of the time series. Then the quantity of interest is computed for each subinterval. These subinterval averages will be very nearly uncorrelated with each other. Consequently, the runs test can be used to evaluate if there is reason to suspect that the time series is nonstationary relative to that quantity. The same technique can be used on voluminous spatial data that have a small range or zone of influence. A sequence of cell averages along a given directional line is submitted to the runs test to determine if trends are indicated as likely.

Abbreviations and Symbols

$\mathbf{1}$	vector having all components equal to 1.0
$+$	place symbol for a positive quantity
$-$	place symbol for a negative quantity
A	nugget effect in the formula for the covariance function
B	sill height or variance in the formula for the covariance function
$C_{11}, C_{12}, C_{21}, C_{22}$	partitions of cov($\mathbf{V}$)
$C(h)$	covariance function of lag in one dimension
$C(n, N)$	combinations of N things taken n at a time
$C_m(\mathbf{u}h)$	covariance function at lag h in direction $\mathbf{u}$ in multidimensional space
cov($\mathbf{V}$)	covariance matrix for the random vector $\mathbf{V}$
D	particular covariance matrix
h	lag or spatial distance in the covariance function
H	range or zone of influence in the formula for the covariance function
$\mathbf{H}$	multidimensional range or zone of influence
H_0	null hypothesis
L	diagonal matrix of eigenvalues
m	mean or expected value of the random field
Q	quadratic form
t	time of sampling
T	shift amount in time
T (superscript)	matrix transpose
$\mathbf{u}$	unit vector in multidimensional space
U	matrix whose columns are the eigenvectors of a specified covariance matrix
$V(\mathbf{x}, t)$	value (random) at site $\mathbf{x}$ and time t

V   random vector of data
V.   data vector after the principal components transformation
x   space coordinates of a sample site
X   vector shift in space

Acknowledgments

The research reported in this chapter was motivated in part from research projects sponsored at the University of Wyoming by the Environmental Monitoring Systems Laboratory of the U.S. Environmental Protection Agency (EPA) in Las Vegas, NV, and in part by our work for the Western Research Institute, a nonprofit corporation affiliated with the University of Wyoming in Laramie, WY (12). Some of the concepts for nonparametric testing of time series grew out of concurrent studies of hypothesis testing for ocean wave data in research sponsored by the Continental Shelf Institute (Sintef Group) of the Norwegian government in Trondheim, Norway. The support of the studies by these various groups is gratefully acknowledged.

Although the research described in this chapter was supported in part by the EPA through assistance agreement CR811243 to the University of Wyoming, this chapter has not been subjected to agency review and therefore does not necessarily reflect the views of the agency; thus, no official endorsement should be inferred.

References

1. Cochran, W. G. *Sampling Techniques*, 3rd ed.; Wiley: New York, 1977.
2. Journel, A. G.; Huijbregts, C. J. *Mining Geostatistics*; Academic: New York, 1978.
3. Quimby, W. Ph.D. Thesis, University of Wyoming, Laramie, 1986.
4. Lehmann, E. L.; D'Abrera, H. J. M. *Nonparametrics*; Holden–Day: San Francisco, CA, 1975.
5. Beyer, W. H. *Handbook of Probability and Statistics*; The Chemical Rubber Company: Cleveland, OH, 1966.
6. Owen, D. B. *Handbook of Statistical Tables*; Addison–Wesley: Palo Alto, CA, 1962.
7. Borgman, Leon. Presented at the Conference of Mathematical Geologists of the United States, Redwood City, California, April 1987, and scheduled for publication by the International Association of Mathematical Geologists.
8. Borgman, L. E.; Taheri, M.; Hagan, R. *Geostatistics for Natural Resource Characterization*; Verly, G.; David, M.; Journel, A. G.; Marechal, A. NATO Advanced Science Institute Series 122; D. Reidel: Dordrecht, Holland, 1984; pp 517–541.
9. Searle, S. R. *Matrix Algebra Useful for Statistics*; Wiley: New York, 1982.
10. Rao, C. R. *Linear Statistical Inference and Its Applications*, 2nd ed.; Wiley: New York, 1973.
11. Quimby, W. F.; Borgman, L. E. Montana State University, in preparation.
12. Borgman, L. E.; Quimby, W. F. *A Sampling Manual*; Western Research Institute: Laramie, WY, in press.

RECEIVED for review January 28, 1987. ACCEPTED July 14, 1987.

Chapter 3

Nonparametric Geostatistics for Risk and Additional Sampling Assessment

Andre G. Journel

Determination of probability distributions to characterize the uncertainty about any unknown is done prior and independently of the estimate(s) retained. These distributions are given through a series of quantile estimates and are not related to any particular prior model or shape. Determination of these distributions accounts for the data configuration, data values, and data quality. In environmental applications, availability of such distributions at each location over the site allows mapping of isopleth maps of the probability of exceedance, estimation of risks of false positives (α) and false negatives (β), and the assessment of need for additional sampling.

R IGOROUS AXIOMATIC NOTATIONS AND SPECIFIC JARGON of probabilistic inference may sometimes appear uninviting to practitioners. Yet the principle and sequence of that inference process is often remarkably simple: collect relevant data, observe the phenomenon of interest through these data, build a model, and then use the model beyond the data by assuming stationarity.

This chapter assumes the availability of a data set distributed in space where not all the data are of the same quality. The problem addressed is that of spatial interpolation and its reliability. For the case of a toxic chemical site, the problems of targeting and ranking areas for cleaning, assessing the risks of misclassification (i.e., false positives and negatives), and the need for additional sampling are addressed.

1173–6/88/0045$08.00/0 © 1988 American Chemical Society

Traditional interpolation techniques, including triangulation and inverse distance weighting, do not provide any measure of the reliability of the estimates; thus, an assessment of the risks imparted by interpolation error cannot be made. The main advantage of geostatistical interpolation techniques, which are essentially ordinary *kriging,* is that an estimation variance is attached to each estimate (1). Unfortunately, unless a Gaussian distribution of spatial errors is assumed, an estimation variance falls short of providing probability intervals and the error probability distribution required for risk assessment.

Regarding the characterization of uncertainty, most interpolation algorithms, including kriging, are *parametric*; a model for the distribution of errors is assumed, and parameters of that model, such as the variance, are provided by the algorithm. Most often that model is assumed normal or at least symmetric. Such congenial models are perfectly reasonable to characterize the distribution of measurement errors in the highly controlled environment of a laboratory. However, these models are questionable when used for spatial interpolation errors, which are the kind under study in this chapter.

The newly developed nonparametric geostatistical techniques set the modeling of the uncertainty as a priority, not the derivation of an "optimal" estimator. Indeed, the uncertainty model is independent of the particular estimate retained and depends only on the information (data) available. The uncertainty model takes the form of a probability distribution of the unknown rather than that of an estimation error.

In a nutshell, the probability that the unknown at a certain unsampled location is greater than a given threshold value is related to the proportion of data above that same threshold. That relationship accounts for the proximity of each datum to the unsampled location, the particular threshold considered, and the specificity of the data used.

In this chapter, I will present the concepts of nonparametric geostatistics by using a minimum of probabilistic formalism. For example, *stationarity* will be introduced informally as the decision to proceed with averaging and inference over a predetermined population or area in space. This definition is sufficient for the purpose of implementation. Inquisitive readers will notice that this chapter could be rewritten entirely without a probabilistic notation.

Probabilistic Assessment of Uncertainty

Consider an attribute value u to be measured. Any particular measurement would provide an estimated value u_j, likely to be in error. Thus, whenever possible, several measurements are performed yielding several estimates u_j; for example, $j = 1, \dots n$. Provided that the measurement device and its accuracy have remained constant over the sequence $j = 1, \dots n$, which implies no potential for trend in the n outcomes u_j, the distribution of these n outcomes u_j can be used to model the uncertainty about the attribute u.

For example, someone may say that "the probability that the unknown u is less than or equal to a threshold u_0" is modeled by the corresponding proportion of outcomes $u_j \leq u_0$. By this statement, the unknown value u has been elevated to the status of a random variable U. The cumulative distribution function (CDF) of U is modeled[†] by

$$\text{prob } \{U \leq u_0 | (n)\} = \text{proportion of } u_j \leq u_0 \tag{1}$$

The indicator function of the threshold value u_0 can be written as follows:

$$i(u_0; u_j) = \begin{cases} 0, \text{ if } u_0 < u_j \\ 1, \text{ if } u_0 \geq u_j \end{cases} \tag{2}$$

The previous model for uncertainty can now be written as equation 3:

$$F(u_0|(n)) = \text{prob } \{U \leq u_0|(n)\} = \frac{1}{n} \sum_{j=1}^{n} i(u_0; u_j) \in [0,1] \tag{3}$$

Again, the notation $F(u_0|(n))$ serves to recall that this probability model is a function of both the threshold value u_0 and the information set (n) constituted by the n outcome values u_j, where $j = 1, \ldots n$.

The model described by equation 3 corresponds to an equal-weighted average (averaged by $\frac{1}{n}$) of the indicator data $i(u_0; u_j)$. Under the previous hypothesis that the measurement accuracy has remained stable over the sequence $j = 1, \ldots n$, no reason exists to over- or underweight any particular outcome. Otherwise, an alternative model corresponding to an unequal weighting scheme may be considered. This alternative model is

$$F(u_0|(n)) = \text{prob } \{U \leq u_0|(n)\} = \sum_{j=1}^{n} a_j \, i(u_0; u_j) \in [0,1] \tag{4}$$

where $a_j \geq 0$ for all j, and $\Sigma_{j=1}^{n} a_j = 1$.

Equations 3 and 4 are different models of the uncertainty about the unknown value u. These expressions are not different estimates of some elusive "true" probability distribution. In particular, one cannot say that the model described by equation 4 is better than the model described by equation 3, before defining what a "good" model should feature. Also, such a definition would be needed to determine the set of weights a_j, where $j = 1, \ldots n$.

[†] The true value u, although unknown, is unique and not a variable distributed according to some probability distribution. Equation 1 represents a model for the uncertainty about u, not an estimate of an elusive CDF of a nonexistent random variable U. However, this model clearly is not unique and depends on the set (n) of measurements u_j; thus, the notation $|(n)$, which can be read loosely as "conditional to the information set (n)", can be written.

The conditions $a_j \geq 0$ and $\Sigma_j\, a_j = 1$ ensure that the function $F(u_0|(n))$ is an acceptable CDF (i.e., a nondecreasing function valued between [0,1]).

Probability Intervals

Availability of CDFs for models of uncertainty allows the derivation of probability intervals:

$$\text{prob } \{U \in [a, b]|(n)\} = F(b|(n)) - F(a|(n)) \tag{5}$$

Such a probability interval is a model figuring the uncertainty around the unknown u. Questions such as "How reliable is this probability interval?" amounts to asking "How reliable is the model $F(u_0|(n))$?". These questions cannot be answered unless a model for the distribution of CDF models is built. Statisticians have such second-level models, but most often they do not bother qualifying the uncertainty attached to models of uncertainty.

These probability intervals can be established prior to the choice of any particular estimate for the unknown value u.

Estimates for u

Beyond the assessment of uncertainty, a unique estimated value for u may be required for decision-making or engineering-design purposes. If no reason exists to over- or underweight any of the n outcomes, a "reasonable" estimate is the equal-weighted arithmetic average:

$$\hat{u}^{(1)} = \frac{1}{n} \sum_{j=1}^{n} u_j \tag{6}$$

An equally reasonable estimate is the median value of the n outcomes, (i.e., the value $\hat{u}^{(2)}$ that would leave approximately one-half of the outcome values below it and one-half above it):

$$\hat{u}^{(2)} = F^{-1}(0.5|(n)) \qquad \text{such that} \qquad F(\hat{u}^{(2)}|(n)) \approx 0.5 \tag{7}$$

If the n outcomes are not equally accurate, they should be weighted differently. The two following reasonable estimates for u were derived by using the weights a_j retained for the CDF model described by equation 4:

$$u^{*(1)} = \sum_{j=1}^{n} a_j u_j \tag{8}$$

$$u^{*(2)} = F^{-1}(0.5|(n)) \tag{9}$$

where $F(u_0|(n))$ is defined by equation 4.

However, other estimates can be derived independently of the CDF model. Examples include the following:

$$u^{*(3)} = \max \{u_j, \text{where } j = 1, \ldots n\} \tag{10}$$

$$u^{*(4)} = \frac{\max (u_j) + \min (u_j)}{2} \tag{11}$$

$$u^{*(5)} = \hat{u}^{(1)} \tag{12}$$

which is the equal-weighted arithmetic average, and

$$u^{*(6)} = \frac{1}{n-2} \sum_{j=1}^{n-2} u_j \tag{13}$$

which is the arithmetic average eliminating the lowest and highest observed outcomes.

All previous estimates can be considered reasonable, although they can be quite different from one another. Thus, a need arises to go beyond the adjective "reasonable" and define precisely worded criteria for retaining a single value for estimation of the unknown value u.

Because there is no unique "best in all cases" estimate for the unknown u, assessments of uncertainty such as probability intervals of the type given by equation 5 should depend not on the particular estimate chosen but solely on the available information (n). In other words, the uncertainty linked to an unknown value u is a function of the information available, not the estimate retained.

Consider n analyses u_k for a particularly toxic (lethal) substance; for cleaning decisions, the maximum value estimate $u^{*(3)}$ may be retained. Such an estimate, however reasonable, will be outside most probability intervals of the type described by equation 5 based on the same information set (n).

A corollary of the previous remark is that traditional 95% probability intervals, leaving 2.5% probability below and above the value, need not be centered or even contain the estimated value retained.

A major exception corresponds to the case of a mean estimate of the type described by equation 6 associated with a normal (Gaussian) distribution $F(u_0|(n))$. However, outside the environment of controlled experiments in a sterile laboratory, the reality of spatial distributions of contaminants rarely provides independent normally distributed data. This topic will be discussed further in the next section.

We have established the need for the following:

- defining a model for the uncertainty about the unknown value given the available information (n) and

- defining criteria for retaining a unique value for estimate of the unknown.

These two tasks don't have to be related, nor should they call for any Gaussian assumption. But before these needs can be discussed, the discussion on probabilistic assessment of uncertainty needs to be broadened to the case of nonrepetitive measurements at different locations of a given space.

Spatial Distributions

In the previous section, the case of repeated measurements, u_j, where $j = 1, \ldots$ n, of a unique attribute value u was considered. The CDF model $F(u_0|(n))$, described by equations 3 or 4, provided an assessment of the measurement uncertainty.

In this section, the uncertainty linked to spatial interpolation is considered when an attribute value $u(\mathbf{x})$ at location $\mathbf{x}$ is to be estimated from measurements $u(\mathbf{x}_j)$, where $j = 1, \ldots n$, made at different locations $(\mathbf{x}_j \neq \mathbf{x})$.

For the sake of simplicity, the data $u(\mathbf{x}_j)$ are considered error-free. This limitation is removed later in the section on soft kriging.

The previous space of measurement variability is now replaced by the physical space of locations $\mathbf{x}$, but otherwise the approach used in the previous section is the same.

Spatial independence is the state whereby the attribute value $u(\mathbf{x})$, at any location $\mathbf{x}$, is not influenced in any way by the attribute values $u(\mathbf{x}_j)$ at other locations $\mathbf{x}_j \neq \mathbf{x}$, no matter how close the attribute values are. If the $(n + 1)$ values $u(\mathbf{x}_j)$ and $u(\mathbf{x})$ relate to the same attribute U, but otherwise could be considered independent from each other, a possible model for the uncertainty about the unknown value $u(\mathbf{x})$, given the n data $u(\mathbf{x}_j)$, is the distribution of these data. More precisely, the probability that the unknown $u(\mathbf{x})$ is less than or equal to a given threshold u is the corresponding proportion of data values $u(\mathbf{x}_j)$. Therefore, the unknown value $u(\mathbf{x})$ has been elevated to the status of a random variable $U(\mathbf{x})$, the cumulative distribution of which is modeled by

$$\text{prob } \{U(\mathbf{x}) \le u|(n)\} = \text{proportion of data } u(\mathbf{x}_j) \le u, \text{ where } j = 1, \ldots n \quad (14)$$

If the indicator variables, one for each datum location $\mathbf{x}_j$ and each threshold value u are used, then equation 14 can be written as equation 16, which is similar to equation 3.

$$i(u; \mathbf{x}_j) = \begin{cases} 0, \text{ if } u < u(\mathbf{x}_j) \\ 1, \text{ if } u \ge u(\mathbf{x}_j) \end{cases} \quad (15)$$

$$F(u; \mathbf{x}|(n)) = \text{prob } \{U(\mathbf{x}) \le u|(n)\} = \frac{1}{n} \sum_{j=1}^{n} i(u; \mathbf{x}_j) \in [0,1] \quad (16)$$

By pooling all n data $u(\mathbf{x}_j)$ into the same CDF (equation 16), the n data $u(\mathbf{x}_j)$ are assumed to have something in common: they relate to the same attribute U. This relationship is the source of the *stationarity hypothesis*, which states that the $(n+1)$ random variables, $U(\mathbf{x})$ and $U(\mathbf{x}_j)$, where $j = 1, \ldots n$, have the same distribution that can be estimated by an average of the type presented in equation 16. The hypothesis of stationarity could be formally introduced, and the statistical models of jointly related random variables could proceed. However, the deterministic stance presented so far can still be informative.

Dependence

In the general case, some pattern of dependence exists between the attribute values $u(\mathbf{x})$, $u(\mathbf{x}')$, and $u(\mathbf{x}'')$ at the different locations $\mathbf{x}$, $\mathbf{x}'$, and $\mathbf{x}''$. Thus, the datum value that is the closest, or more generally, the most related to the unknown value, should be weighted more. This weighting leads to an unequal weighted average:

$$F(u; \mathbf{x}|(n)) = \text{prob}\,\{U(\mathbf{x}) \le u|(n)\}$$

$$= \sum_{j=1}^{n} a_j(\mathbf{x})\, i(u; \mathbf{x}_j) \in [0, 1] \qquad (17)$$

where $a_j \ge 0$ for all j, and $\sum_{j=1}^{n} a_j(\mathbf{x}) = 1$.

Because the probability value $F(u; \mathbf{x}|(n))$ needs to be calculated for each different threshold value u, the weights can be made dependent on u. The new CDF model can then be defined as follows:

$$F(u; \mathbf{x}|(n)) = \sum_{j=1}^{n} a_j(u; \mathbf{x})\, i(u; \mathbf{x}_j) \qquad (18)$$

The weights $a_j(u; \mathbf{x})$ need not be any more nonnegative nor sum to 1, as long as equation 18 remains that of a CDF (i.e., provided that the following order relations are verified):

$$F(u; \mathbf{x}|(n)) \in [0, 1] \qquad (19a)$$

$$F(u; \mathbf{x}|(n)) \ge F(u'; \mathbf{x}|(n)), \text{ for all } u \ge u' \qquad (19b)$$

Probability Intervals

The uncertainty about the value $u(\mathbf{x})$ can be assessed through probability intervals derived directly from any of the CDF models described by equations 16, 17, and 18.

$$\text{prob } \{U(x) \in [a, b] | (n)\} = F(b; \mathbf{x} | (n)) - F(a; \mathbf{x} | (n)) \tag{20}$$

Of particular importance for environmental applications is the probability of exceedance of a threshold b:

$$\text{prob } \{U(x) > b | (n)\} = 1 - F(b; \mathbf{x} | (n)) \tag{21}$$

Again, such probability intervals and probabilities of exceedance can be established prior to the choice of any particular estimate of the unknown value $u(\mathbf{x})$.

Isopleth curves for the probability of the exceedance described in equation 21 can be contoured for $\mathbf{x}$ varying in space and b fixed. For many decision-making processes, these isoprobability maps suffice; thus, no need arises for the additional hypotheses required to retrieve unique-valued estimates of $u(\mathbf{x})$.

Exactitude Requirement

Whatever CDF model $F(u; \mathbf{x} | (n))$ is retained to assess the uncertainty about the unknown $u(\mathbf{x})$, the model should be consistent with the fact that at data locations, no uncertainty exists.

$$
\begin{aligned}
F(u; \mathbf{x}_j | (n)) = & \quad \text{prob } \{U(\mathbf{x}_j) \leq u | (n)\} \\
\equiv i(u; \mathbf{x}_j) = & \begin{cases} 0, \text{ if } u < \text{datum value } u(\mathbf{x}_j) \\ 1, \text{ otherwise} \end{cases}
\end{aligned} \tag{22}
$$

for all data locations $\mathbf{x}_j$, where $j = 1, \ldots n$.

Equation 22 assumes that the data $u(\mathbf{x}_j)$ are exact. If the data are not exact, then the CDF model $F(u; \mathbf{x} | (n))$ need only be consistent with what is actually known at location $\mathbf{x}_j$, which is possibly only a prior distribution for the value $u(\mathbf{x}_j)$. This situation will be discussed in the section on soft kriging.

Equations 18 and 22 show that the CDF model $F(u; \mathbf{x} | (n))$ may be seen as an estimate of the unknown indicator function $i(u: \mathbf{x})$ by using weighted averages of the known indicator values $i(u; \mathbf{x}_j)$:

$$F(u; \mathbf{x} | (n)) = [i(u; \mathbf{x})]^* = \sum_{j=1}^{n} a_j(u; \mathbf{x}) \, i(u; \mathbf{x}_j) \tag{23}$$

Under a rigorous probability setting, the uncertainty model presented by equation 18 or 23 is indeed shown to be a linear estimate of the conditional distribution of $U(\mathbf{x})$ given the n data $U(\mathbf{x}_j) = u(\mathbf{x}_j)$ and is a linear estimator of the random variable (RV) $I(u; \mathbf{x})$ (2, 3).

The problem of determining the weights $\{a_j(u; \mathbf{x}), \text{ where } j = 1, \ldots n\}$ is addressed in the next section.

Determination of the Cumulative Distribution Function Uncertainty Model

The reason for considering an unequal weighting of the data, as in equations 4, 17, 18, or 23, is the recognition of the existence of patterns of spatial dependence between the data $u(x_j)$ and the unknown $u(x)$. Thus, some dependence is entailed between the indicator data $i(u; x_j)$ and the unknown indicator function $i(u; x)$. Hence, these patterns must be identified. But first, the notion of spatial dependence must be defined and measured.

Euclidian Measure

A datum value $u(x_j)$ can be seen as two pieces of information: one related to the datum location x_j, the second related to the attribute value. Common measures of proximity between two values $u(x)$ and $u(x + h)$ are linked to the Euclidian distance $|h|$, or modulus of the interdistance vector h; examples include the following:

- a constant $- |h|$, or
- the reciprocal of $|h|$ to some power $\frac{1}{|h|^w}$, where usually $w = 1$ or 2.

Such measures account for only the location information x_j of a datum $u(x_j)$, not for the attribute information u. Indeed, the Euclidian distance $|h|$ is attribute U independent; this distance is the same whether very continuous British thermal unit (BTU) grades in a coal seam or very erratic pH values over an urban area are considered. Also, the Euclidian distance is the same for the two indicator pairs $[i(u; x_j), i(u; x_j + h)]$ and $[i(u'; x), i(u'; x_j + h)]$, where $u \neq u'$.

Variogram Distance Measure

The traditional, although not unique, distance measure used in geostatistics is the variogram function $2\gamma(h)$ modeled from the experimental average squared discrepancy between the $N(h)$ data pairs separated by approximately the same vector h.

$$2\gamma_U(h) \text{ modeled from } \quad \frac{1}{N(h)}\sum_{j=1}^{N(h)} [u(x_j) - u(x_j + h)]^2 \tag{24a}$$

$$2\gamma_I(u; h) \text{ modeled from } \quad \frac{1}{N(h)}\sum_{j=1}^{N(h)} [i(u; x_j) - i(u; x_j + h)]^2 \tag{24b}$$

The average squared discrepancy usually increases with the interdistance $|\mathbf{h}|$. However, as opposed to the Euclidian distance $|\mathbf{h}|$, the variogram distance $2\gamma(\mathbf{h})$ is attribute-specific, and $\gamma_l(u; \mathbf{h}) \neq \gamma_l(u'; \mathbf{h})$ if $u \neq u'$. The variograms are dependent on both the modulus $|\mathbf{h}|$ and the direction of the interdistance vector $\mathbf{h}$. The variograms are said to be *anisotropic* if they depend on the direction of vector $\mathbf{h}$. Otherwise, the variograms are said to be *isotropic*. As an example, airborne pollution concentrations present greater dependence (i.e., smaller variogram values) in directions of prevailing winds.

The corresponding measure of proximity or dependence is the generalized covariance:

$$C_U(\mathbf{h}) = \text{constant} - \gamma_U(\mathbf{h}) \tag{25a}$$

$$C_l(u; \mathbf{h}) = \text{constant} - \gamma_l(u; \mathbf{h}) \tag{25b}$$

The arbitrary constant is filtered out from all subsequent utilizations of these proximity measures.

The Kriging Algorithm

Once a proximity measure is available for the attribute being estimated, a straightforward generalized linear-regression algorithm, called *kriging*, can be applied to determine the weighting system. For example, consider the estimate of the unknown indicator value $i(u; \mathbf{x})$ given by equation 26:

$$
\begin{aligned}
F(u; \mathbf{x}|(n)) &= [i(u; \mathbf{x})]^* \\
&= \sum_{j=1}^{n} a_j(u; \mathbf{x})\, i(u; \mathbf{x}_j)
\end{aligned}
\tag{26}
$$

The n weights $a_j(u; \mathbf{x})$ are given by a constrained system of normal equations, called the *ordinary kriging system* (4, 5):

$$\sum_{j'=1}^{n} a_{j'}(u; \mathbf{x}) \cdot C_l(u; \mathbf{x}_{j'} - \mathbf{x}_j) + \mu(u; \mathbf{x}) = C_l(u; \mathbf{x} - \mathbf{x}_j), \text{ for all } j = 1, \dots n \tag{27a}$$

$$\sum_{j'=1}^{n} a_{j'}(u; \mathbf{x}) = 1 \tag{27b}$$

The system described by equations 27a and 27b appears as a system of $(n+1)$ linear equations having $(n+1)$ unknowns: the n weights $a_j(u; \mathbf{x})$ and a Lagrange paramer $\mu(u; \mathbf{x})$ associated to the condition that the weights sum up to 1. A sufficient condition for the system to provide one and only one solution is that the proximity model $C_l(u; \mathbf{x})$ is positive-definite (i.e., the model is a covariance function), and that no two datum are at the exact same location (i.e., $x_j \neq x_{j'}$ for all $j \neq j'$). One such system is required for each threshold value u and for each unsampled location $\mathbf{x}$. In practice, the interval of variability of the attribute U is made discrete by K threshold values of u_k, where $k = 1, \ldots K$. Thus, at each location $\mathbf{x}$, K systems of the type given by equation 27 need to be solved to provide an assessment of the uncertainty through K discrete CDF values $F(u_k; \mathbf{x}|(n))$. CDF values for intermediary threshold values $u \in [u_k, u_{k+1}]$ can be obtained by linear interpolation:

$$F(u; \mathbf{x}|(n)) = F(u_k; \mathbf{x}|(n)) + \frac{u-u_k}{u_{k+1}-u_k} \cdot [F(u_{k+1}; \mathbf{x}|(n)) - F(u_k; \mathbf{x}|(n))] \quad (28)$$

Other nonlinear interpolation procedures might also be considered (6).

Exactitude

The ordinary kriging systems described by equation 27 provide weights such that the exactitude requirement given by equation 22 is met.

Order Relations

The K systems described by equation 27 do not ensure that the resulting K CDF values verify the order in equation 19. In practice, a piecewise linear model is fitted to the K values $F(u_k; \mathbf{x}|(n))$ so that the order relations are guaranteed. This model is used to characterize the uncertainty about the unknown. More details about order relations correction can be found in reference 7.

The Probability Kriging Algorithm

The indicator estimate in equation 23 uses only the indicator part of the information $u(\mathbf{x}_j)$ available. That estimate is enhanced if all the information available is used. A new weighted average is then defined as follows:

$$[i(u; x)]^* = \sum_{j=1}^{n} a_j(u; \mathbf{x}) \cdot i(u; \mathbf{x}_j) + \sum_{j=1}^{n} b_j(u; \mathbf{x}) \cdot u(\mathbf{x}_j) \quad (29)$$

This new estimate mixes indicator data valued either 0 or 1 and data $u(\mathbf{x}_j)$ valued in the possibly widely different measurement unit of attribute U. This scaling problem is solved by considering the rank order transforms $r(\mathbf{x}_j)$ instead of the data $u(\mathbf{x}_j)$.

If the n data $u(\mathbf{x}_j)$ are ranked in increasing order, and $r(\mathbf{x}_j) \in [1, n]$ is the rank of the datum $u(\mathbf{x}_j)$, the transformed datum $\frac{1}{n}r(\mathbf{x}_j)$ is valued between 0 or 1. The improved indicator estimate is written and is taken as a model for the uncertainty about $u(\mathbf{x})$:

$$F(u; \mathbf{x}|(n)) = [i(u; \mathbf{x})]^*$$

$$= \sum_{j=1}^{n} a_j(u; \mathbf{x}) \cdot i(u; \mathbf{x}_j) + \frac{1}{n} \sum_{j=1}^{n} b_j(u; \mathbf{x}) \cdot r(\mathbf{x}_j) \tag{30}$$

A more involved set of equations, called the *probability kriging* (PK) system, is needed to provide the ($2n$) weights, $a_j(u; \mathbf{x})$ and $b_j(u; \mathbf{x})$. More details on the PK algorithm can be found in references 7 and 8. Lemmer (6) proposed a simplification of the system described by equation 27 by considering a single common indicator covariance model $C_I(\mathbf{h})$ for all threshold values u.

The Soft Kriging Generalization

In all the preceding developments, the data available were considered "hard data", that is, without uncertainty. These data generate indicator data valued either 0 or 1.

However, in practice, data are never exactly hard if only because of measurement errors. It would be appropriate to enter actual data either under the form of constraint intervals

$$u(\mathbf{x}_j) \in [a(\mathbf{x}_j), b(\mathbf{x}_j)] \tag{31}$$

or under the form of prior probability distributions

$$\text{prob} \{U(\mathbf{x}_j) \leq u\} = F(u; \mathbf{x}_j) \in [0, 1] \tag{32}$$

For a given datum location, the indicator information can be viewed as a column of K indicator data, where one indicator datum exists for each of the K threshold values u_k. In the case of a constraint interval as in equation 31, the indicator data column is formed only outside that interval by using:

$$i(u_k; \mathbf{x}_j) = \begin{cases} 0 & \text{for all } u_k \leq a(\mathbf{x}_j) \\ \text{unknown for all } u_k \in [a(\mathbf{x}_j), b(\mathbf{x}_j)] \\ 1 & \text{for all } u_k > b(\mathbf{x}_j) \end{cases} \tag{33}$$

A "hard" datum is a particular case of a constraint interval (equation 31) having zero amplitude: $a(\mathbf{x}_j) = b(\mathbf{x}_j) = u(\mathbf{x}_j)$. In the case of local information given as a probability distribution (equation 32), the indicator datum column is filled in with values between 0 and 1 instead of equal to either 0 or 1.

$$i(u_k; \mathbf{x}_j) = F(u_k; \mathbf{x}_j) \in [0, 1] \tag{34}$$

Thus, the indicator algorithm described by equation 27 or 30 allows pooling and joint weighting of data of several types. These data types include the following:

- hard data;
- inequality constraints of the type described by equation 31, where possibly $a(\mathbf{x}_j) = -\infty$ corresponding to $u(\mathbf{x}_j) \leq b(\mathbf{x}_j)$ or $b(\mathbf{x}_j) = +\infty$ corresponding to $u(\mathbf{x}_j) > a(\mathbf{x}_j)$; and
- soft information presented under the format of a prior distribution of the type given by equation 32.

The prior distribution can stem from an actual distribution of repeated laboratory analyses or from a triangular distribution modeling expert information, which determines where in the interval $[-\infty, +\infty]$ the value $u(\mathbf{x}_j)$ is most likely to be.

The result of that indicator kriging is a probability column $F(u_k; \mathbf{x}|(n))$ valued between 0 and 1 and used as a model for the uncertainty about the unknown value $u(\mathbf{x})$.

Exactitude

If, at a location $\mathbf{x}$, prior soft information of the type described by equation 34 exists, the exactitude property of kriging entails that this prior information is restored unchanged.

$$F(u_k; \mathbf{x}|(n)) \equiv F(u_k; \mathbf{x}) \qquad \text{for all } u_k \qquad (35)$$

This equation is valid no matter what the information is at the n other locations. In other words, the process of indicator and probability kriging does not update prior information but completes it by interpolation.

In most practical cases, at every location, a minimum prior information, such as a large constraint interval (e.g., [0%, 100%]), exists for a concentration-type attribute. Consequently, at any location, the initial indicator column is at least partially filled with the following:

$$i(u_k; \mathbf{x}) = 0 \qquad \text{for all } u_k \leq 0 \qquad (36a)$$

$$i(u_k; \mathbf{x}) = 1 \qquad \text{for all } u_k > 100\% \qquad (36b)$$

The process of probability and indicator kriging does not update these prior values but completes the indicator column with probability values for those threshold values $u_k \in [0\%, 100\%]$.

Criteria for Estimation of the Unknown

Strictly speaking, the probabilistic assessment of the uncertainty about an unknown value stops at the determination of a probability distribution of the type described by equation 23. As discussed before, knowledge of that distribution suffices for the derivation of probabilities of the type described by equation 20 or 21.

The determination of a probability distribution for an unknown does not and should not require any preliminary choice of a unique estimate for $u(\mathbf{x})$. Similarly, a probability-oriented decision-making process should not require more than the assessment of the uncertainty about $u(\mathbf{x})$. Decision-making processes that require estimated values $u^*(\mathbf{x})$ are deterministic in nature.

However, traditionally the derivation of such a unique estimated value $u^*(\mathbf{x})$ is required. Because the unknown value could be any value within the interval [0%, 100%] for concentrations, making the criterion precise for selecting a unique estimated value is needed. Plausible criteria are numerous; each criterion leads to different estimates. Different uses of estimates may call for different criteria, and an "optimal for all purposes" estimate does not exist. As an example, most estimates featuring good local accuracy properties, including the kriging estimate, are smoothed in their spatial variability; thus, these estimates may prove inadequate for mapping purposes if the map is to be used to evaluate spatial variability.

An incisive layman-oriented discussion of optimality for decision making can be found in reference 9. This chapter is not the place to survey the vast array of diverse optimality criteria leading to as vast an array of estimates, all of which are optimal. The discussion in this chapter is limited to those estimates that can be derived straightforwardly from the uncertainty model $F(u; \mathbf{x}|(n))$. These estimates are a subset of the former vast array of optimal estimates and may not include the one needed for a specific application.

Loss Functions and L-Optimal Estimates

When estimating an unknown value by a single value $u^*(\mathbf{x}) = u^*$, a nonzero error $u^* - u(\mathbf{x})$ is likely to occur. Assume that the impact or loss attached to each level of error can be assessed by a function $L(u^* - u(\mathbf{x}))$. The function $L(e)$ is known (e.g., $L(e) = e^2$), but the argument $e = u^* - u(\mathbf{x})$ is unknown. However, an uncertainty model $F(u; \mathbf{x}|(n))$ that represents the distribution of the unknown $u(\mathbf{x})$ is available. Thus, the idea is to use that distribution model to determine the expected loss value:

$$E\{L(u^* - U)|(n)\} = \int_{-\infty}^{+\infty} L(u^* - u) \cdot dF(u; \mathbf{x}|(n)) \qquad (37)$$

In practice, this equation is approximated by the discrete sum

$$\approx \sum_{k=1}^{K} L(u^* - u_k') \cdot [F(u_{k+1}; \mathbf{x}|(n)) - F(u_k; \mathbf{x}|(n))] = \varphi(u^*; \mathbf{x}) \qquad (38)$$

having, for example, $u_k' = \frac{u_{k+1}+u_k}{2}$ if the attribute U interval of variability has been discretized by K threshold values u_k, where $k = 1, \dots K$. Also, usually $F(u_{K+1}; \mathbf{x}|(n)) = 1$.

The expected loss in equation 37, which is calculated from the model $F(u; \mathbf{x}|(n))$, appears as a function $\varphi(u^*; \mathbf{x})$ of the particular estimated value u^* retained. The optimal estimate for the loss function L is the value $u_L^*(\mathbf{x})$, which minimizes the expected loss $\varphi(u^*; \mathbf{x})$:

$$u_L^*(\mathbf{x}) = \text{value } u^* \text{ minimizing } \varphi(u^*; \mathbf{x}) \tag{39}$$

Derivation of the optimum $u_L^*(\mathbf{x})$ for any particular loss function L can be done through iterative calculations of the discrete sums approximating the Stieltjes integral given in equation 37; such calculations do not present any difficulty.

For some particular loss functions, the solution is straightforward. If $L(e) = e^2$ (i.e., for the least-squares criterion), the best estimate of u^* is the expected value of the probability distribution $F(u; \mathbf{x}|(n))$, which is also called the E-type estimate:

$$u_E^*(\mathbf{x}) = \int_{-\infty}^{+\infty} u \; dF(u; \mathbf{x}|(n)) \tag{40}$$

An approximation of this integral can be found in equation 45. If $L(e) = |e|$ (i.e., for the mean absolute deviation criterion), the best estimate is the median of the distribution $F(u; \mathbf{x}|(n))$, defined as

$$q_{0.5}(\mathbf{x}) = F^{-1}(0.5; \mathbf{x}|(n)) \text{ such that } F(q_{0.5}(\mathbf{x}); \mathbf{x}|(n)) = 0.5 \tag{41}$$

$$\text{if } L(e) = \begin{cases} w_1 e \text{ for } e \geq 0 \text{ (overestimation)} \\ w_2 |e| \text{ for } e < 0 \text{ (underestimation)} \end{cases} \tag{42}$$

that is, for an asymmetric linear loss function, the best estimate is the p-quantile of the distribution $F(u; \mathbf{x}|(n))$ (10):

$$q_p(\mathbf{x}) = F^{-1}(p; \mathbf{x}|(n)) \qquad \text{where } p = \frac{w_2}{w_1 + w_2} \in [0, 1] \tag{43}$$

$$\text{if } L(e) = \begin{cases} 0, \text{ for } e = 0 \\ \text{constant, otherwise} \end{cases} \tag{44}$$

The best estimate is the most plausible outcome of the distribution $F(u; \mathbf{x}|(n))$, which is the mode of the corresponding density function $f(u; \mathbf{x}|(n)) = \partial F(u; \mathbf{x}|(n))/\partial u$.

Spatial distributions of toxic chemical concentrations are usually highly skewed and would generate strongly asymmetric probability density function (PDF) models $f(u; \mathbf{x}|(n))$ for the uncertainty about an unknown. In such cases, the optimal estimates given by equations 40 and 41 are quite different, and the impact of using one criterion rather than the other can be dramatic.

Nature is rarely Gaussian, and uncontrolled spatial distributions of earth and environmental sciences attributes are almost always non-Gaussian, as opposed to distributions of repetitive measurements in a highly controlled, carefully planned laboratory experiment.

Most decision-making processes, if properly analyzed, would call for an asymmetric loss function of the type described by equation 43, whether linear or not. For example, for cleaning decision purposes, the impact of an underestimation (a hazardous area is not cleaned) is usually much higher than that of an overestimation (a safe area is cleaned unduly). Thus, for such cleaning decisions, a standard least-squares-type estimate such as equation 40 may be inappropriate. Fortunately, decision makers are aware of such shortcomings and apply various sorts of safety factors that amount to considering asymmetric loss functions. The rigorous concept of loss function and optimal L-estimates may allow greater consistency in the decision-making process once a particular and possibly subjective loss function has been chosen.

The E-Type Estimate

The E-type estimate in equation 40, although a least-squares-type estimate, is usually different from the direct ordinary kriging estimate, $u^*_{krig}(\mathbf{x})$, calculated by using the original data $u(\mathbf{x}_j)$ and the corresponding proximity measure model $C_U(\mathbf{h})$ as defined in equation 25. As opposed to the approach yielding the E-type estimate, the ordinary kriging approach does not qualify the ordinary kriging estimate for uncertainty (probability intervals) unless a Gaussian distribution for interpolation errors is assumed.

The Stieltjes integral in equation 40 defining the E-type estimate is, in practice, approximated by a discrete sum:

$$u^*_E(\mathbf{x}) \approx \sum_{k=1}^{K} u_k' \cdot [F(u_{k+1}; \mathbf{x}|(n)) - F(u_k; \mathbf{x}|(n))] \qquad (45)$$

$$\text{where } u_k' = \int_{u_k}^{u_{k+1}} u \, dF(u; \mathbf{x}|(n)) \in [u_k, u_{k+1}] \qquad (46)$$

In equation 46, u_k' is the k_{th} class mean, and u_k, where $k = 1, \dots K + 1$, are $(K+1)$ class bounds, which identify the U attribute interval of variability.

The indicator and probability kriging process provides for the probability values $F(u; \mathbf{x}|(n))$ but not for the class means u_k'. Estimation of these class means, except for the last one, usually does not pose any problem. For example, a class mean can be estimated by the equal-weighted average of the data falling into the corresponding class. Whenever the class mean u_k' relates to a narrow class amplitude $u_{k+1} - u_k$, consideration of sophisticated unequal-weighted averages is not needed.

However, for heavily skewed distributions, the last class mean may have a disproportionate influence on the estimate given by equation 45. This last class mean is highly dependent on the largest datum observed and thus on the decision to keep it, cull it, or set it back to a lower value. One conservative decision consists of using the class median for the estimate. Another solution calls for a parametric model of the distribution within that class. For example, a two-parameter log-normal model can be fitted to two robust statistics of the highest data $u(x_j) \in [u_K, u_{K+1}]$, such as u_K being the $F(u_K; x|(n))$ quantile. The median of these high data is the $[1 + F(u_K; x|(n))]/2$ quantile. The last class mean of that log-normal model is the estimate of u_K'.

The poor outlier resistance of the E-type estimate in equation 40 is shared by all mean-type estimates, including the direct ordinary kriging estimate. At least in the case of an E-type estimate, outlier sensitivity is confined to the last class. Quantile-type estimates such as given by equations 41 and 43 have much better outlier resistance and should be considered whenever their underlying optimality criteria are acceptable.

Risk and Additional Sampling Assessment

The CDF models of the type $F(u; x|(n))$, where one such function exists for each location x within a given area A, allows contouring of isopleth curves. These isopleth curves include the following:

- optimal estimate(s) $u^*(x)$ that are retained (Figures 2 and 3). These estimates can be derived independently of the uncertainty model $F(u; x|(n))$.

- probabilities that the actual unknown value $u(x)$ exceeds any given threshold (Figure 4) such as

$$\text{prob } \{U(x) > u_0|(n)\} = 1 - F(u_0; x|(n)) \tag{47}$$

- risk of a false positive:

$$\alpha(x) = \text{prob } \{U(x) \leq u_0|u^*(x) > u_0, (n)\} = F(u_0; x|(n)) \tag{48}$$

for those locations x such that $u^*(x) > u_0$ (Figure 5).

- risk of a false negative:

$$\beta(x) = \text{prob } \{U(x) > u_0|u^*(x) \leq u_0, (n)\} = 1 - F(u_0; x|(n)) \tag{49}$$

for those locations x such that $u^*(x) \leq u_0$ (Figure 6).

- p-quantiles, $q_p(\mathbf{x})$, $\mathbf{x} \in A$, where p is fixed in $[0,1]$ of the distributions $F(u; \mathbf{x}|(n))$ [i.e., contour maps of the values $q_p(\mathbf{x})$ for which the probability of exceedance is $1 - p$ (Figure 7)].

In 1982, under EPA guidance, an extensive soil sampling at a Dallas smelter site was performed. Sample locations were chosen at a pseudoregular grid spacing of 750 ft (Figure 1). The histogram of the 180 data points retained shows a mean of 447-ppm lead, a strong positive skewness and a large coefficient of variation ($\frac{\sigma}{m} = 2.3$), and a long tail: 4% of the data range from 2400 ppm to a maximum of 10,400 ppm.

Ordinary kriging that used the 180 concentration data and a corresponding covariance model was performed. The procedure yielded estimated grid values, $u^*_{krig}(\mathbf{x})$, at each node of the estimation grid. These direct-kriged estimates are contoured in Figure 2.

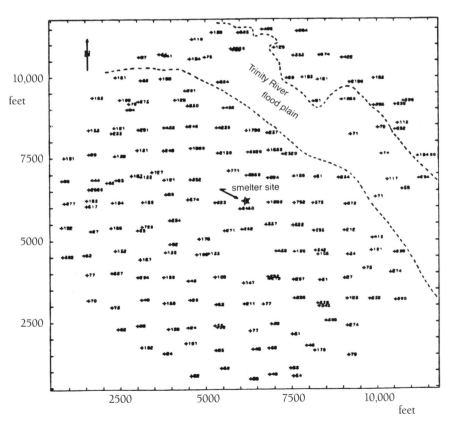

Figure 1. Posting (parts per million of Pb) of the data available over the site.

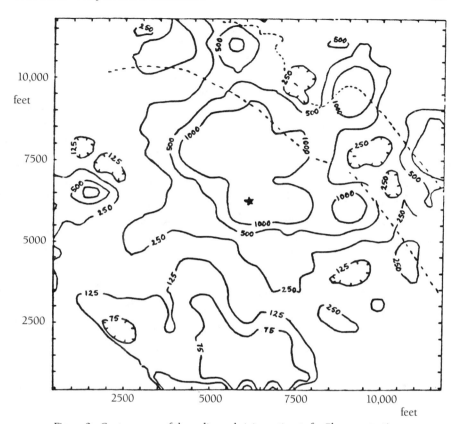

Figure 2. Contour map of the ordinary kriging estimate for Pb concentration.

A probability kriging (PK) study was performed on the same 180 data points to provide, at each grid node $\mathbf{x}$, a CDF model for the uncertainty about $u(\mathbf{x})$. The corresponding E-type estimates were retrieved by using an approximation of the type described in equation 45 and are contoured in Figure 3.

Both estimates are optimal for different criteria, although both criteria are of the least-squares type. Although different locally, both maps reveal the same large-scale structures: a high dome centered at the smelter site, a northeasterly trend of highs corresponding to the direction of prevailing winds, and a canyon of lows in the upper right corner corresponding to the flood area of the Trinity River.

Figure 4 is an isopleth map of the probability of exceedance of the 1000-ppm Pb threshold. Only those contours corresponding to a probability greater than 10% have been plotted. The airborne pollution has a clear northeasterly direction, and the presence of an outlier high on the western side of the site can be noted. This outlier spot was checked and found to be a junkyard having possible leakage from automobile batteries.

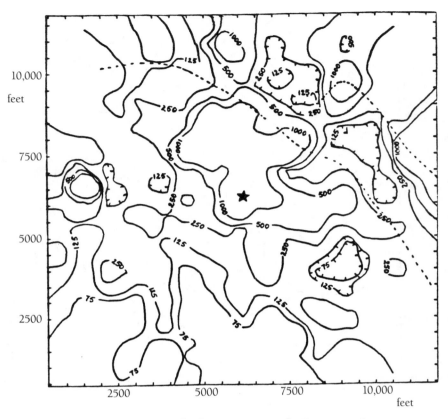

Figure 3. Contour map for the E-type estimate for Pb concentration.

Assuming that the E-type estimate has been retained to delineate the areas targeted for cleaning (e.g., those locations such as $u^*_E(\mathbf{x}) > 1000$ ppm), two questions should be answered. These questions are as follows:

- What is the risk $\alpha(\mathbf{x})$ of a false positive (i.e., the risk of declaring a location hazardous when it is not)? That risk is derived directly from the uncertainty model as in equation 48, and is contoured in Figure 5.
- What is the risk $\beta(\mathbf{x})$ of a false negative (i.e., the risk of declaring a location safe when it is not)? That second risk is also derived from the model $F(u; \mathbf{x}|(n))$ in equation 49 and is contoured in Figure 6.

The impact of the risk $\alpha(\mathbf{x})$ can be easily evaluated because it is linked to the cost of cleaning unduly. However, evaluating the impact of the risk $\beta(\mathbf{x})$ is

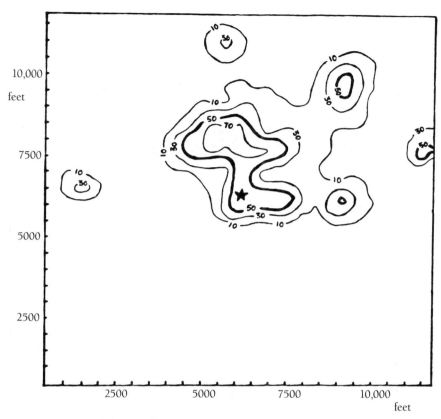

Figure 4. Isopleth map of the probability that Pb concentration exceeds 1000 ppm.

much more delicate because it involves such nonmonetary notions as health hazards and environment quality. Deciding on the threshold (u_0 = 1000 ppm for Figures 5 and 6) and on the balance of the two risks α and β is a political decision well beyond the realm of geostatistics. At best, a ranking of the areas (locations $\mathbf{x}$) targeted for cleaning can be obtained by using the uncertainty model $F(u; \mathbf{x}|(n))$.

However, the joint availability of maps such as those of Figure 3 ($u_E^*(\mathbf{x})$), Figure 5 ($\alpha(\mathbf{x})$), and Figure 6 ($\beta(\mathbf{x})$) allow for decisions concerning additional sampling. Additional sampling should be considered in areas with high misclassification risks α and β, rather than areas with high estimated values $u_E^*(\mathbf{x}) > u_0$.

For a given threshold u_0 to be performed on a given estimate, say the E-type estimate, the locations $\mathbf{x}$ such that $u_E^*(\mathbf{x}) \leq u_0$ may be ranked according to their risks $\beta(\mathbf{x})$ of a false negative. Those locations with the highest risk β are first candidates for additional sampling.

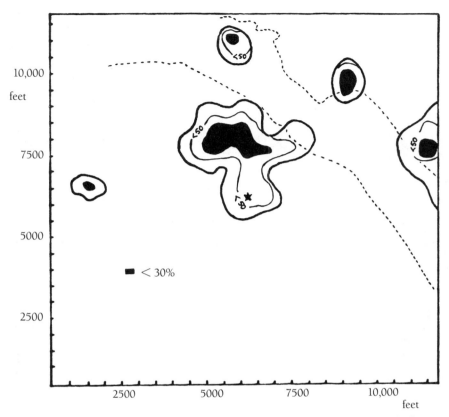

Figure 5. Isopleth map of the risk $\alpha(\mathbf{x})$ of a false positive (i.e., declaring wrongly that a location $\mathbf{x}$ is hazardous) on the basis that $u_E^*(\mathbf{x}) > u_0 = 1000$ ppm. The thick line delineates the boundary of those zones declared hazardous.

The risks α and β can be evaluated, whatever the estimate $u^*(\mathbf{x})$ retained for application of the threshold u_0.

Quantile Maps

A series of quantile maps, such as that of Figure 7, can give a fast visual appreciation of the overall reliability of any particular estimated map based on a particular estimation algorithm (e.g., the E-type estimate in the case of Figure 7).

Inversion of the CDF models $F(u; \mathbf{x}|(n))$ (e.g., equation 43) provide the p-quantile values $q_p(\mathbf{x})$. For each given value of the probability $p \in [0, 1]$, these p-quantiles can be contoured. On Figure 7, the 0.8 quantile map appears closest to the E-type estimated map; hence, the probability that the true value exceeds the E-type estimate is only 20% when averaged over the site. If that probability

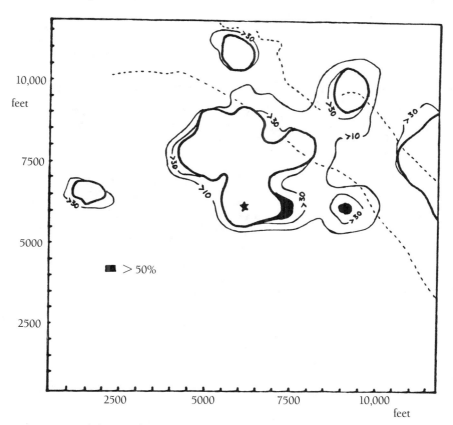

*Figure 6. Isopleth map of the risk β(**x**) of a false negative, i.e., declaring wrongly that a location **x** is safe, on the basis that $u_E^*(\mathbf{x}) \leq u_0 = 1000$ ppm. The thick line delineates the boundary of those zones declared hazardous.*

is deemed not safe enough, the E-type map may be replaced by a high p-quantile map (e.g., $p = 0.99$); the unknown true value would have, on the average, only a 1% probability to exceed the values of the 0.99-quantile map.

Unfortunately, unless the data are very numerous, the p-quantile map when p is large may depend almost totally on the model extrapolating the last calculated CDF value $F(u_K; \mathbf{x}|(n))$ towards the maximum value 1. If $q_p(\mathbf{x})$ is larger than $F(u_K; \mathbf{x}|(n))$, $q_p(\mathbf{x})$ is not data-related. The probability kriging algorithm, as any other algorithm, is no replacement for actual data, particularly those informing the tails of spatial distributions.

Ranking Areas

The decision to clean any particular subarea V within a site depends not only on the assessed distribution of the unknown concentration over that area V, but also on the assessed impact of that distribution.

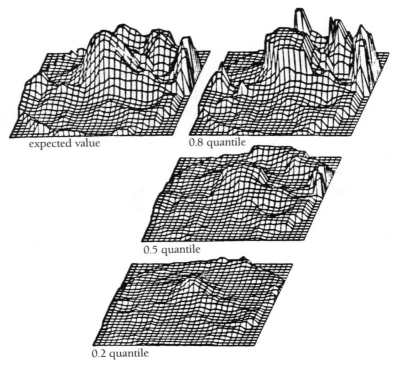

expected value   0.8 quantile

0.5 quantile

0.2 quantile

Figure 7. Perspective view of the E-type estimated parts-per-million Pb surface and three quantile maps.

Consider the impact function $h(u; x)$ of a concentration value present at location x within a particular toxic chemical site. This impact function is valued in impact units that are not necessarily dollar values. Figure 8 gives an example of such impact functions: below a tolerance level u_0, the impact is deemed nought; above that level, the impact then increases as follows:

- linearly at a low rate for uncritical locations (e.g., within fenced industrial yards), or

- at a much greater rate, or even parabolically, for critical locations such as within residential areas or schoolyards.

Because a model is available for the distribution of the unknown concentration at each location x, the total expected impact over an area A is

$$C_V = \int_V E\{h(U(x))|(n)\}dx = \int_V dx \int_0^{+\infty} h(u; x)\, dF(u; x|(n)) \qquad (50)$$

This impact can be estimated by using equation 51:

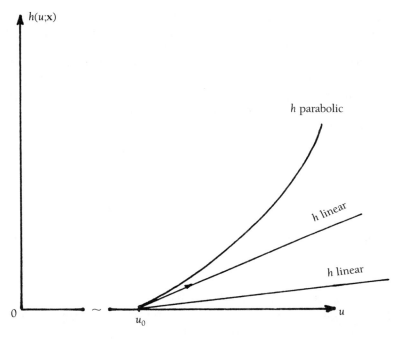

Figure 8. Examples of local impact functions.

$$C_V \approx \sum_{i=1}^{M} \sum_{k=1}^{K} h(u_k'; \mathbf{x}_i) \, [F(u_{k+1}; \mathbf{x}_i | (n)) - F(u_k; \mathbf{x}_i | (n))] \tag{51}$$

where, for example, $u_k' = \frac{u_{k+1} + u_k}{2} \in [u_k, u_{k+1}]$, and the u_ks are K thresholds discretizing the concentration interval $[0, +\infty]$. The $\mathbf{x}_i$s, where $i = 1, \dots M$, are the M locations discretizing the area V.

Areas of similar size would be ranked for their total expected impact C_V. If cleaning budget becomes available, the areas with the highest impact C_V are prime candidates.

The previous ranking is not yet an objective assessment because the impact function $h(u; \mathbf{x})$ must be given. The relative impact of a given toxic concentration on a schoolyard versus an industrial yard is very much a political and social decision.

The E-type estimate or its arithmetic average over a subarea is but a particular impact of the type given by equation 50, where the impact function is taken as arbitrarily equal to u at whatever the location $\mathbf{x}$. Selecting on the basis of the E-type estimate or, for that purpose, on the basis of any other estimate does not necessarily correspond to an objective decision process.

The linear averaging over V of the expected local impact $E\{h(U(\mathbf{x})|(n))\}$ is not necessarily a conservative choice for determining a good ranking criterion C_V. Indeed, within a schoolyard, a single spoonful of lethal concentration would be judged inadmissible. Thus, the two integral operators of equation 50 may be replaced by high p-quantile operators. For example, at each location $\mathbf{x}$ the p-quantile of the distribution of $U(\mathbf{x})|(n)$ or $h(U(\mathbf{x})|(n))$ may be retained. The p-quantile of the spatial distribution of these p-quantiles over the M locations may then be taken for ranking criterion.

Summary

The available information, whether hard data, constraint intervals, or soft information under the form of prior probability distributions, is coded as a series of indicator columns. One such indicator column exists for each location, whether a datum location or an unsampled location. The indicator and probability kriging process consists of filling the missing entries of the indicator columns with probability values. These probability values are obtained by weighting neighboring and related indicator data.

This process yields, at each location, a probability distribution for the unknown attribute. The process is nonparametric in the sense that it does not assume any particular model for these distributions. These probability distributions characterize the uncertainty about the unknowns independently of the particular estimates retained; the distributions are only information-dependent and are fully so in the sense that they are not only data-configuration-dependent, but also data-values- and data-quality-dependent.

If a loss function measuring the impact of any given magnitude of the interpolation error is available, an optimal estimate minimizing that loss can be derived. Thus, various optimal estimates may be considered depending on their particular uses. However, a single measure of uncertainty applies to all such estimates for a given information set.

In environmental applications, the availability of probability distributions for the unknowns allows contouring isopleth maps of the probability of exceedance of any given threshold, the risk α of a false positive, and the risk β of a false negative. Thus, an assessment of the need for additional sampling and a ranking of areas as candidates for cleaning are made possible.

Implementation of the indicator and probability kriging technique requires linear geostatistical tools (i.e., a variogram and ordinary kriging (normal equations) software). However, as in many statistical applications, the necessary art of approximations can only be obtained through experience.

Typical of nonparametric techniques, the elaboration through indicator and probability kriging of the distributions for the unknowns does require an appreciable amount of data. This requirement is a positive aspect because the need for additional data is clearly indicated when required for the goal at hand.

The alternatives are to acquire more data or to rely on a model that is not data-based. The general rule is to use most of the information possible through nonparametric tools, and only when that information is exhausted should parametric modeling be used.

Abbreviations and Symbols

$\alpha(\mathbf{x}) = \text{prob } \{U(\mathbf{x}) \leq u_0 | u^*(\mathbf{x}) > u_0, (n)\}$: risk of a false positive (i.e., the risk of declaring wrongly that a location $\mathbf{x}$ is polluted on the basis of a concentration estimate $u^*(\mathbf{x})$ greater than the threshold value u_0). That risk depends on the level of information (n) available.

$\beta(\mathbf{x}) = \text{prob } \{U(\mathbf{x}) > u_0 | u^*(\mathbf{x}) \leq u_0, (n)\}$: risk of a false negative (i.e., the risk of declaring wrongly that a location is safe on the basis of a concentration estimate $u^*(\mathbf{x})$ less than or equal to the threshold value u_0).

CDF: cumulative distribution function

$F(u_0 | (n)) = \text{prob } \{U \leq u_0 | (n)\}$: conditional cumulative distribution function, or the probability that the unknown u modeled by the random variable U be less than or equal to the threshold value u_0, given the information available of size (n).

$i(u_0; \mathbf{x})$: indicator values set to 1 if the attribute value $u(\mathbf{x})$ at the location $\mathbf{x}$ is less than or equal to the parameter value u_0; $i(u_0; \mathbf{x})$ is a step function with a unit value step at $u_0 = u(\mathbf{x})$.

$i^*(u_0; \mathbf{x})$: an estimate of $i(u_0; \mathbf{x})$. This estimate has values between 0 and 1, as opposed to $i(u_0; \mathbf{x})$, which is equal to either 0 or 1.

PDF: probability density function

PK: probability kriging

$\text{prob } \{U \in (a, b) | (n)\} = F(b | (n)) - F(a | (n))$: probability that the unknown value u falls in the interval $[a, b]$, where the lower bound is excluded, and the upper bound is included.

q_p: p-quantile of a cumulative distribution function $F(u)$, or the threshold value $u = q_p$ for which there is a probability $p \in [0, 1]$ for the true value to be less than or equal to it. The 1-quantile would be the maximum value that this unknown value could possibly take.

$q_p(\mathbf{x})$: p-quantile of the conditional distribution of the random variable $U(\mathbf{x})$ modeling the unknown attribute $u(\mathbf{x})$ at location $\mathbf{x}$. There is a probability p that the true value $u(\mathbf{x})$ be less than or equal to $q_p(\mathbf{x})$. This quantile $q_p(\mathbf{x})$, which is a spatial function of $\mathbf{x}$, can be contoured.

$u_E^*(\mathbf{x})$: E-type estimate of the unknown concentration at location $\mathbf{x}$. This estimate is the mean of the estimated conditional distribution $F^*(u; \mathbf{x} | (n))$ of the random variable $U(\mathbf{x})$ modeling the uncertainty about $u(\mathbf{x})$.

$\mathbf{x}$: vector of coordinates for a given location in three-dimensional space: $\mathbf{x} = (x_1, x_2, x_3)$ where, for example, x_1 is the easting, x_2 the northing, and x_3 the elevation. A function $f(\mathbf{x})$ in that three-dimensional space is a function $f(x_1, x_2, x_3)$ of the three coordinates.

Acknowledgment

Although the research described in this chapter was supported by the U.S. Environmental Protection Agency through assistance agreement CR 811893 to Leland Stanford Jr. University, this chapter has not been subjected to agency review and does not necessarily reflect the views of the agency. Therefore, no official endorsement should be inferred.

References

1. Flatman, G. In *Environmental Sampling for Hazardous Wastes*; Schweitzer, G. E.; Santolucito, J. A., Eds.; ACS Symposium Series No. 267; American Chemical Society: Washington, DC, 1984; pp 43–52.
2. Isaaks, E. H. M.Sc. Thesis, Branner Earth Sciences Library, Stanford, 1984; p 26.
3. Journel, A. G. *Proc. 19th APCOM Symp.* Society of Mining Engineers: Littleton, CO, 1986; pp 15–30.
4. Luenberger, D. L. *Optimization by Vector Space Methods*; Wiley: New York, 1969; pp 56–90.
5. Journel, A. G.; Huijbregts, Ch. *Mining Geostatistics*; Academic: London, 1978; pp 304–310.
6. Lemmer, I. C. *Math. Geol.* **1986**, *18(7)*, 589–623.
7. Sullivan, J. A. Ph.D. Thesis, available through University Microfilms: Ann Arbor, MI, 1984; pp 36–42.
8. Journel, A. G. In *Geostatistics for Natural Resources Characterization*, Part 1; Verly et al., Eds.; Reidel: Dordrecht, 1984; pp 307–335.
9. Srivastava, R. M. *CIMM*, in press, 1987.
10. Journel, A. G. In *Geostatistics for Natural Resources Characterization*, Part 1; Verly et al., Eds.; Reidel: Dordrecht, 1984; pp 261–270.
11. Journel, A. G. Indicator Approach to Toxic Chemical Sites; *Report of Project No. CR-811235-02-0*; U.S. Environmental Protection Agency: Las Vegas, 1984.

RECEIVED for review January 28, 1987. ACCEPTED May 11, 1987.

Chapter 4

Geostatistical Approaches to the Design of Sampling Regimes

George T. Flatman, Evan J. Englund, and Angelo A. Yfantis

Because of the physical and chemical laws that control the fate and transport of pollutants, field samples from a pollution plume are spatially correlated. This spatial correlation or structure is measured by directional semivariograms, which are prerequisites for the geostatistical interpolation method known as kriging. The information from the semivariograms can be used to guide the design of sampling programs, including such factors as the field sample's support (i.e., volume, shape, and compositing of subsamples) and the sampling grid design (i.e., distance between sample locations, geometry of the grid, and orientation of the grid). This chapter presents geostatistical methods, rather than subjective value judgments, to answer many of the chronic questions of field sampling.

T HE OLD ADAGE THAT A CHAIN is as strong as its weakest link implies that the prudent blacksmith will strengthen the weakest link and try to make all links equally strong. The application to quality improvement is that error variances are a chain; the analytical variance, the sampling and handling variance, and the field variance are links. The goal of quality improvement is to make the sum of the variances as small as possible, and the cost-effective way to do this is to spend more resources on the variance link that is improved most cheaply. Because of diminishing returns in variance reduction, the optimal variance to reduce is often the biggest one. The field sampling variance is often the appropriate link or variance to reduce. Variance reduction is most obviously accomplished by taking more samples, but if sampling or analytic costs are high, increasing samples may be too expensive. In many cases, the field sampling variance is economically reduced by better sampling design.

Geostatistics

The term *geostatistics* was coined by Matheron (*1*) to describe the study of "regionalized" or spatially correlated variables. In the past 20 years or so, the geostatistical literature has grown enormously, and many significant developments in theory and methodology have been presented. The practice of geostatistics has also spread from its original applications in the mining industry to such fields as soil science, forestry, meteorology, and environmental science.

The geostatistical methods described in this chapter, namely semivariograms and ordinary kriging, represent one of the approaches available to us and we selected them primarily to illustrate geostatistical concepts and their implications for sampling programs. A discussion of the pros and cons of alternate approaches, such as generalized covariance and universal kriging, is beyond the scope of this chapter. More extensive treatments of the subject can be found in references 2–5.

Random or Spatial Variables

Most field sampling plans are based on random variable statistics and assume that the sample observations are independent. However, field samples are usually spatially correlated. *Correlation* is a statistical measurement of the intuitive physical fact that samples taken close together are more similar in value than samples taken farther apart. Neglecting this correlation can make the statistics, tests, and sampling procedures of random variable statistics inappropriate; using this correlation makes the statistics, tests, and sampling procedures of spatial statistics more appropriate and powerful.

A truly random variable is completely described by its distribution of values. Samples are used to estimate this distribution and to estimate statistical descriptors such as mean, median, and standard deviation. In addition, spatial variables must be described by a measure of the correlation between each value and the values at nearby locations. Samples can be used to estimate the spatial correlation function and are frequently used to estimate localized mean values.

Localized mean estimates are often displayed in the form of isopleths or contour maps. A practical rule-of-thumb for the investigator is that if a contour map is a desired or an even plausible end product of a proposed study, geostatistical methods should be considered.

The implications for the design of a sampling program can be significant. Although random sampling is appropriate for random variables, Olea (*6*) demonstrated that the most effective sampling pattern for local estimation of spatial variables is the regular grid. Also, geostatistical studies commonly use a multiphase approach, and the first sampling phase is oriented primarily toward estimating the spatial correlation.

Semivariograms for Quantifying Spatial Correlation

One way in which spatial correlation can be measured and displayed is by a *semivariogram*, or graph of the type shown in Figure 1. The dots are the empirical semivariogram representing experimental values computed from sample data; the fitted curve is a theoretical semivariogram or an estimation of a spatial correlation function assumed to be characteristic of the sampled area. The horizontal axis, called the *lag axis*, is the distance between points in linear units such as meters or kilometers; the vertical axis, called the *gamma axis*, is the variance of differences in pollution units squared, such as parts per million squared. The experimental points are computed by averaging data grouped into distance class intervals. Variance is a function of lag. The rising nature of the points and curve follows the principle of sampling that states the variance or difference between observations increases as the distance between their locations increases.

Sill and Range of Correlation

Figure 1 is typical of many semivariograms of chemical concentrations in the environment; the rise in variance has an upper bound known as the *sill*. When the variance reaches the sill, sample locations are far enough apart to make the samples random. The distance on the lag axis at which the semivariogram's curve reaches the sill is the *range of correlation*. This distance is important to the sampling plan, the estimation of pollution over the area under investigation, and the interpolation error. The range of correlation explains a practical relationship between spatial variables and random variables; *random variables* are field samples that are farther apart than the range of correlation, and *spatial variables* are field samples closer together than the range of correlation. This range of correlation is important for choosing the correct analysis; if a classical random variable statistic is wanted, such as the mean or variance, then the sample locations must be at least the range of correlation apart. If a contour map of pollution isopleths or interpolation variance is wanted, then the closer together the sampling locations, the lower the local interpolation error. Depending on the information wanted and the spacing of the sample locations, either random or spatial variance statistical analysis can be used on field samples.

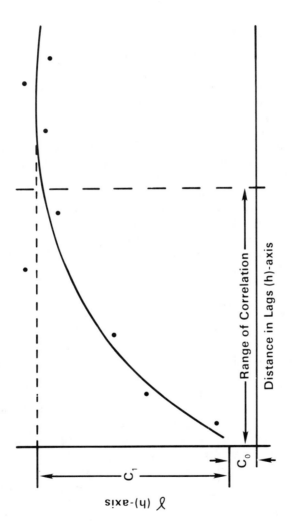

Figure 1. A typical semivariogram.

Variance Model

In Figure 1, on the vertical axis of the fitted model, the variance has two components, C_0 and C_1. The C_1 component of the variance is the measure of structural variation and has the characteristic of increasing variance between sample observations as the distance between sample locations increases. The C_0 component of the variance combines random variance factors, such as sampling and analytical error, along with any unmeasured spatial variance that may exist at distances smaller than the sampling interval; C_0 is constant for all lags. The relationship of C_0 to the need for compositing samples and the relationship of C_1 to the distance between sample locations will be discussed in a later section.

Anisotropy and Directional Semivariograms

The variance structure, as measured by the semivariogram, is often different in the range of correlation in different directions. This condition is called *anisotropy* and must be measured by directional semivariograms. Directional semivariograms are computed experimentally by grouping sample pairs into directional classes, or *windows*, as well as into distance classes. The directional ranges of correlation can change the geometry of the sampling grid and the orientation of the grid. Often, not enough preliminary data are available to compute directional semivariograms, and thus the sampling design must work with only an omnidirectional range of correlation. But an omnidirectional range of correlation and a sampling design from it are more objective and defensible than conventional random variable methods that consider only a scalar variance.

Kriging

Kriging is a linear-weighted average interpolation technique used in geostatistics to estimate unknown points or blocks from surrounding sample data. By assuming that the spatial correlation function inferred from the experimental semivariogram is representative of the points to be estimated as well as those sampled, the error variance associated with any estimate that is a linear weighted average of sample values can be computed. The kriging algorithm computes the set of sample weights that minimize the error variance.

Kriging has a number of characteristics of a desirable estimation method: sample weights can be adjusted for anisotropy; samples can be down-weighted in correlated clusters; the degree of smoothing increases as the random component (C_0) of the semivariogram model increases; and when the semivariogram model is completely random ($C_1 = 0$), the kriging estimator becomes the sample mean, as in random sample statistics.

Spatial Outliers

Spatial outliers can be found by examining a geographical plot of the data; they may fit into a random variable histogram of the data very well. In other words, a *spatial outlier* is a sample value that does not agree in magnitude with the values of its neighboring samples, especially the samples within a range of correlation. For example, a high (polluted) value in a low (background) neighborhood might be a spatial outlier but not a random variable outlier because the high value agrees with other polluted values. Once these outliers are identified, their location descriptions should be looked up in the sampling diary. If they are obviously from different sources that do not have the same correlation structure, then they should be excluded from the semivariogram evaluation. The question of whether or not to include a spatial outlier in the final local estimate of concentration must be answered on a case by case basis. This matter involves the investigator's judgment, just as the case of random variables.

Sample Support and Estimation Blocks

The *sample support* is the in situ description of the physical sample. For a sample of single core, the support is the shape and volume of the core; however, for a sample composite of many subsamples, the geometry is also a factor. The *estimation block* is the in situ volume represented by the estimated value. For simplicity, punctual kriging assumes that the input supports and output blocks are points. Naturally, the block is much larger than the support, and so the support volume estimates ought to be integrated over the block volume for a more precise (i.e., lower estimation variance) block estimate. The volume of the sample support must compromise between the need to be large for representativeness in situ and the need to be small for ease of handling and processing. All sample values must have the same support. The estimation block size should be determined by convenient working size for remediation or by the purpose of the study. For example, an appropriate remediation block might be a volume 250 ft long, 16 ft wide, and 0.5 ft thick because this amount was the minimum volume to move economically, and this shape is the cut of the bulldozer.

Primary or Semivariogram Sampling

In a multiphase sampling program using spatial statistics, the primary goal in the initial exploratory sampling is the collection of enough data to compute an empirical semivariogram and to determine the extent of the plume. These goals may conflict if limited resources are available. Widely spaced samples are needed to define extent, and closely spaced samples are often needed for semivariogram analysis. Approaches to this problem include regular grids (radial, square, or rectangular), transects, and combinations.

McBratney et al. (7) suggested transect sampling for variogram input, and this idea led to very good variograms. However, in pollution monitoring, transects alone have given very noisy variograms. This result is probably due to intrinsic noise in pollution data, which is often highly skewed and contains high coefficients of variation. A combination exploratory grid, consisting of a rectangular grid of square sampling units having an extended transect in the directions of the major axis and minor axis of the plume (8), is illustrated in Figure 2. Prior information may be used to select the best grid orientation. For example, if the plume to be investigated was made by aerial deposition from a locatable source, wind roses can be examined for wind direction and magnitude, and topographic maps can be examined for natural barriers. Only the relatively regular grid concept is important in Figure 2; the orientation is site-specific. The intense rectangular section of the square grid cell might be situated as a diamond or with the diagonals on the transects.

If the extent of the plume must be found, and funds are limited, the transect samples should be variably spaced closer together at the grid center and farther apart at the grid extremes. The purpose of this sampling is to capture the correlation structure of the plume. Inhabited areas have a high occurrence of disturbed sampling sites and local pollution from secondary sources, which are only stochastic noise to the semivariogram's calculation. Therefore, this noise

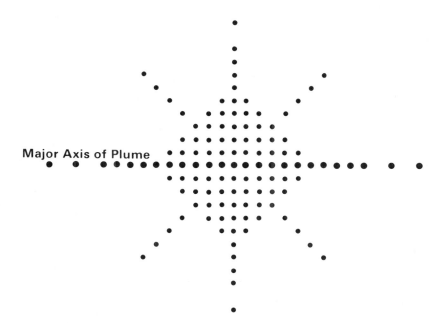

Figure 2. An exploratory grid design.

should be avoided by this sampling. The samples from the semivariogram sampling can be pooled with the secondary mapping samples if they have the same support. However, the semivariogram sampling often is the sampling that tests for the need of a larger support or compositing. If the support is changed between the sampling and the samples are pooled for analysis, then the change in support must be corrected for in the spatial analysis. The sampling team must be aware of the need to keep all samples on the same support. When the sampled locations must move from the regular grid to avoid cultural improvements or natural barriers, the spatial analysis program is corrected for this movement by the true coordinates of the new sample locations; however, no easy method is available for the program to correct for change of support. If the microvariation could be sampled and the support established before the semivariogram sampling, a complex statistical problem could be avoided in the pooling of samples for the spatial analysis.

Some samples should be taken very close together to determine the need for composite samples. This sampling can be combined with field duplicates for quality analysis and control. The smaller the coring volume, the more important compositing becomes. These microvariation samples should be taken a few multiples of the core's diameter apart. The distance between sample locations or grid unit's length needs to be estimated from the sample unit of interest (e.g., residential yard, city block, or square mile section) and the desired output unit (e.g., in remediation, the minimum area of surface soil to be removed). The optimum exploratory sampling distance is a proper fraction of these measurements, but is often determined by money available for sampling.

Secondary or Map-Making Sampling

In spatial statistics, the goal of secondary sampling is to uniformly cover the "area in question" with a density of samples sufficient to contour the plume with an acceptable error of interpolation. This sample coverage is accomplished by using the directional semivariograms to determine the orientation, shape, and size of the grid cell. Random variable statistics, where the number of samples is computed, differs from spatial statistics, where orientation, shape, and size of the grid are calculated, and the number of samples is determined from the grid.

Boundary

For secondary sampling, the extent of the sampling grid must first be chosen. The sampling grid must extend beyond the suspected plume or area in question. The area in question must be bound with sampling locations to avoid extrapolation in the estimation algorithm (kriging) for contouring. Extrapolation, which is estimating a value from data on only one side of the point, is likely to lead to unrealistically high or low values. If an action level has been set and a part of the plume has been adequately proven to be above or below the action level, then the plume need not be resampled. The sampling may be guided more by population areas or critical receptors than by the actual plume. The goals of the sampling must be written, and the areas of interest, action levels, and action areas must be defined before the optimum grid design can be made.

Compositing and Nugget

The next step in secondary sampling is choosing the sample support (9). If a residential yard is the sampling unit, then the ideal sampling process would be to take the entire yard, blend it to homogeneity, and remove the appropriate number of aliquots to meet the volume needed by the lab for analysis. However, because few residents would donate their whole yard to science, and laboratory mixing equipment such as V-blenders or ballmills cannot homogenize so large a volume, this sampling unit must be represented by a few symmetrically laid out subsamples composited together. The number of subsamples is a compromise between the size of the microvariance and the amount of time and money allowed for the digestion of the subsamples. The subsamples are laid out symmetrically because a structural or spatial correlation may occur.

The mixing of the subsamples to homogeneity is essential for compositing. If the medium is water, the task is relatively easy; for soils or sediments, the task is difficult. For media difficult to mix to homogeneity, aliquots should be taken after the mixing to make the final sample more representative. If a large nugget (e.g., random variance C_0 in Figure 1) lies on the semivariogram or if a large discrepency exists on quality analysis and quality control field duplicates, then the relative sizes of the field sampling and the laboratory analysis errors must be identified. The analysis of some pollutants has an analytical error that overwhelms the field sampling error and accounts for approximately all the semivariogram nugget. In such a case, field compositing is of little help and laboratory replicate analysis is the solution. Unfortunately, special spatial analysis is required. Yfantis and Flatman (10) derived this type of generalized spatial analysis; a computer program is available upon request.

Grid Unit Length or Distance between Sample Locations

The range of correlations, the nugget (C_0), and the sampling budget determine the *grid unit length*, or the distance between sample locations. This length determination was discussed in mathematical detail by Yfantis et al. (11). Figure 3 shows the graphs of interpolation variance as a function of the ratio of grid spacing to range of correlation for a family of semivariograms. The model variograms each have a sill ($C_1 + C_0$) equal to 100 (%). The variograms differ only in the fraction of the sill represented by the nugget component (C_0). If the

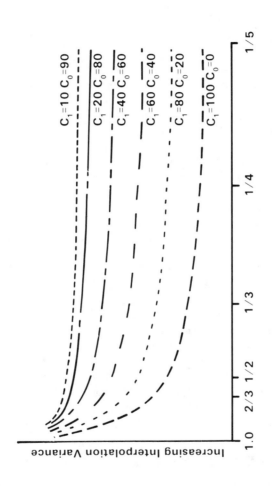

Figure 3. Diminishing returns in variance reduction for a decrease in distance between sampling locations.

semivariogram has a big nugget like the top graph of $C_1 = 10\%$ and $C_0 = 90\%$, diminishing returns set in if the sample distance is less than two-thirds of the range of correlation. For a very low nugget such as the lower two graphs, diminishing returns still set in if the sample distance is less than half of the range of correlation. The grid should be laid out with no vertices unsampled. If this design gives a too expensive sampling campaign, then the whole grid size should be adjusted, not just certain vertices unsampled as in systematic random sampling.

In the field, some vertices cannot be sampled because of man-made improvements or natural barriers, but these vertices must be sampled as reasonably close as possible, and the actual coordinates should be used in the spatial analysis program.

Grid Orientation and Shape versus Anisotropy

If the ranges of correlation are extremely different on the directional semivariograms, then the correlation structure is anisotropic. Optimum sampling patterns reflect this anisotropy. For example, the sides of a rectangular grid would be in the same ratio as the ranges of correlation for the corresponding directional semivariograms. This ratio was explained in detail by David (4), and a sampling design for logarithmic anisotropy was derived by Barns (12). Doctor and Gilbert (13) discussed sample spacing. Anisotropy is a frequent occurrence, but often the semivariogram sampling gathers too few samples to measure it. Thus, more samples may be used cost-effectively in the semivariogram sampling in order to save samples in the larger mapping sampling by identifying and taking advantage of anisotropy.

Use of the triangular grid as opposed to the rectangular grid has been discussed (6, 7, 11, 13, 14). If the nugget is large ($C_0 \gg C_1$), little is gained by the triangular grid. Also, the triangular grid makes taking advantage of anisotropy more difficult. If a triangular grid is chosen, a theodolite, which is a surveying instrument, is not needed in the field; only every other row of samples must be offset by one-half of a grid length. In practice, this action is easier than it sounds and almost as easy as the traditional square grid.

Beyond Anisotropy

Numerous additional geostatistical considerations affect environmental sampling. These considerations include spatial drift or trend, multivariate analysis, mixed or overlapping populations, concentration-dependent variances, and specification of confidence limits. A great variety of geostatistical techniques has been developed over the years to deal with these various problems, but an adequate discussion is beyond the scope of this chapter.

References

1. Matheron, G. *Econ. Geol.* **1963**, *58*, 1246–1266.
2. Flatman, G. T.; Yfantis, A. A. *Environ. Monit. Assess.* **1984**, *4*, 335–349.
3. Journel, A. G.; Huybregts, Ch. J. *Mining Geostatistics*; Academic: London and New York, 1978; pp 1–597.
4. David, M. *Geostatistical Ore Reserve Estimation*; Elsevier Scientific: Amsterdam, Oxford, and New York, 1977; pp 1–364.
5. Clark, I. *Practical Geostatistics*; Applied Science Publishers: London, 1979; p 130.
6. Olea, R. A. *Systematic Sampling of Spatial Functions*; Kansas Series on Spatial Analysis No. 7; Kansas Geological Survey. University of Kansas: Lawrence, KS, 1984; p 57.
7. McBratney, A. B.; Webster, R.; Burgess, T. M. *Comput. Geosci.* **1981**, *7(4)*, 331–365.
8. Starks, T. H.; Sparks, A. R.; Brown, K. W., submitted for publication in *Environ. Monit. Assess.*
9. Starks, T. H. *Math. Geol.* **1986**, *18(6)*, 529–537.
10. Yfantis, A. A.; Flatman, G. T., submitted for publication in *Math. Geol.*
11. Yfantis, A. A.; Flatman, G. T.; Behar, J. V. *Math. Geol.* **1987**, *19(3)*, 183–205.
12. Barns, M. G. *The Use of Kriging for Estimating the Spatial Distribution of Radionuclides and Other Spatial Phenomena*; TRAN STAT (Statistics for Environmental Studies) No. 13; Battelle Memorial Institute: Richland, WA, 1980; pp 1–22.
13. Doctor, P. G.; Gilbert, R. O. In *Selected Environmental Plutonium Research Reports of the NAEG*; White, M. G.; Dunaway, P. B., Eds.; NVO 192; Nevada Applied Ecology Group. U.S. Department of Energy: Las Vegas, NV, 1978; Vol. 2, pp 405–451.
14. Matern, B. *Medd. Statens Skogforskningsinst.* **1960**, *49(5)*, 144.

Received for review January 28, 1987. Accepted June 1, 1987.

Chapter 5

Defining Quality Assurance and Quality Control Sampling Requirements
Expert Systems as Aids

Lawrence H. Keith, M. Timothy Johnston, and David L. Lewis

Quality sampling and analysis (QSA) is a prototype expert system that helps chemists select the proper type of quality control (QC) samples and then calculates how many of each sample type are needed to meet stated confidence levels. Answers are presented either in terms of percent probability or in numbers of QC samples to be analyzed per day. A unique screen interface encourages the user to seek alternative solutions and thus to maximize the benefit of the quality assurance (QA) program.

K NOWLEDGE AND INFORMATION have a subtle but important difference. *Information* consists of facts or data obtained from study, investigation, or instruction. *Knowledge* is the fact or condition of knowing something with familiarity gained through experience or association.

A person who studies or applies information sufficiently well will become knowledgeable and perhaps ultimately become an *expert* in a subject. An *expert system* is a computer program that emulates a human expert in making decisions involving a particular subject area called a *domain*. Expert systems involving chemistry applications have primarily been confined to handling molecular structures, organic synthesis, analytical chemistry, and mathematical calculations (*1*).

The current state of affairs in the domain of quality control (QC) of sampling and analytical programs is in some disarray. Multiple definitions have

1173–6/88/0085$06.00/0 © 1988 American Chemical Society

been presented for many fundamental entities (e.g., bias, detection limits, and even QC samples), and knowledge of QC is poor. Often, professionals take heuristic approaches to QC, and the result is that much of the data collected is of unknown quality.

In years past, the U.S. Environmental Protection Agency addressed this problem by collecting and codifying available wisdom on the subject (2). Although studies such as this one have done much to unify QC techniques in sampling and analytical programs, a general lack of awareness of QC planning remains. The difficulties posed by the underlying statistical techniques and the prescriptive knowledge required to apply them still pose barriers to the wide application of well-designed QC programs by most chemists.

A small team of chemists and computer scientists at Radian recognized that the design of QC programs exhibits several characteristics of a prototype expert system. First and most importantly, human experts are involved in the selection of QC programs. The domain is complex and requires specialized education and years of experience in statistics and also sampling and analysis programs before an expert status is attained. Experts are rare and expensive, and the need for their services is great. Lastly, the knowledge concerning the selection of control strategies has a standard representation. The team recognized that in order to solve the entire problem, they must not only produce an expert system, but a system that could be widely distributed. They further stipulated that the program must run on widely available computer hardware, such as the IBM personal computer. An effort to capture the expertise of Radian's experts in the design of QC programs was commissioned, and the result was the software program presented in this chapter.

Quality sampling and analysis (QSA) is a prototype expert system that helps chemists overcome two frequently encountered QC program design problems: selection of the proper type of QC samples, and calculation of how many QC samples must be analyzed in order to meet specified confidence levels. The focus of QSA is on the design of effective QC programs, and not on the peripheral issues of method selection or control charting, which are either well-understood or addressed by other widely available computer software. Although QSA can be used for many kinds of programs, the system is especially useful for projects involving environmental sampling and analyses because of the typically low analyte concentrations and complex matrices found in these types of samples. QSA is an advanced prototype of an earlier program called QualAId (3), which was developed by using RuleMaster, a rule-induction software building tool (4).

Knowledge Engineering

The knowledge representation technique used in QSA is *learning by examples*, also known as *inductive learning*, which is an active area of research in *machine learning* (5, 6). In very simple terms, an expert "teaches" the computer by providing a set of exemplary situations, called the *training set*, and specifying the conclusion that is to be reached in each situation. The set is then converted into a classification rule that is minimally complex. The rule is often presented to humans as a *decision tree*. Figure 1 illustrates a portion of the decision tree for determining what types of QC samples are required for controlling precision.

During the development of QSA, the QC expert preferred to specify the knowledge in the decision tree format. In the case of the rule presented in Figure 1, the QC expert originally posited a much more complicated rule. When that rule was subjected to the RuleMaster induction tool, the computer produced the simpler form shown. Initially, the QC expert was skeptical of the simpler rules, but soon admitted that these rules were correct and expressed considerable enthusiasm that the computer had helped simplify the reasoning. This simplification developed by using a rule-induction algorithm is typical because humans are characteristically redundant in the formulation of rules or correlations from complex information sources.

The User Interface

The original QSA program was operated in a question and answer format, where the user was queried about needs and plans. The QSA program then used that information to calculate either the probability of achieving the user's objectives when input was in frequency of analyses, or the number of QC samples needed when input was in probability of detecting an excursion from a specified confidence level.

The problem with this approach was redundancy. Each time a parameter was changed, the program went serially through the questions and calculations for each of the necessary types of QC samples. Users found this approach confusing and time-consuming because many parameters had to be varied to see what effect the changes would have on the answers. This problem was solved by embedding the expert system in a conventional program, which conveniently partitioned the program into segments for the expert recommendations, the statistical computations, and the user interface.

The user interface was rethought entirely for the second version of QSA, which now incorporates a windowed development platform. The new design was achieved after careful thought about the presentation of information to the user and provides comprehensive support features such as on-line help and error checking.

As each question is addressed, it is presented in three forms:

1. The question is presented as a single key word within an activated portion of the screen. A reverse video area on the monitor indicates the key word for which an answer is needed.

The Control Branch - Precision

Within-day, Between-day or Both?

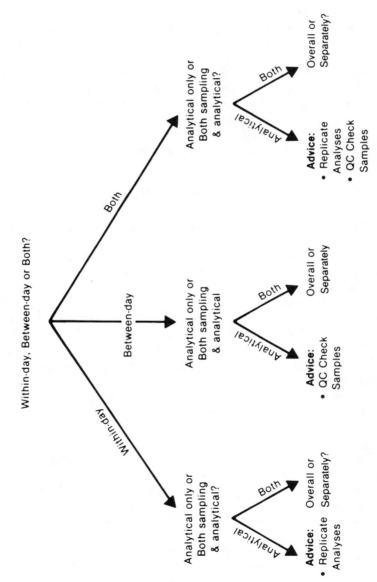

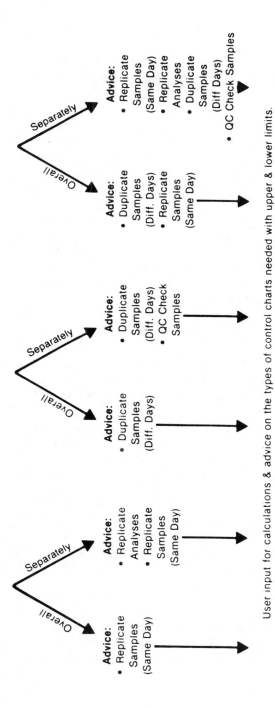

User input for calculations & advice on the types of control charts needed with upper & lower limits.

Figure 1. Part of a decision tree for selecting QC samples.

The answer is the first letter of one of the allowable responses or a number.

2. The full question is presented in a line highlighted by reverse video. If the answers consist of words, the answers are listed at the end of the question. If the answers consist of numbers, no prompts appear at the end of the question.

3. Help is instantly available through the F1 function key. This function key toggles a help screen that contains expanded definitions or explanations of what responses are expected for each question. The help is displayed in the lower portion of the screen.

The differences between the original version and the current version of QSA are most dramatic in the ease of operation of the programs. Where the first version was slow and tedious to use, the new version makes navigation through the calculations and "what if?" experiments simple. Even though the answers provided are no different, a better answer should be obtained as a result of examining more variables in the specified parameters.

Statistical Calculations

The statistical models of the sampling and analytical process assume that repeated measurements of a sample concentration will vary from the actual concentration by the sum of a systematic amount plus a random amount of variance.

The systematic variance is called *bias* and is attributable to culpable features of the sampling and analytical design. QC strategies for controlling bias attempt to detect when systematic error begins to occur and then attempt to pinpoint its cause. Some possible sources of bias include the following: faulty reporting, poor calibration, poor analytical techniques, unrepresentative sampling, sample loss or contamination, poor quality of materials and apparatus, instrument instability or nonlinearity, and inability to measure all forms of the analyte.

The random error component is related to precision. QC strategies for controlling precision must detect when the precision rises above accepted levels. Precision is also necessary for making comparisons between measured values. The three kinds of precision commonly studied include within-day, between-day, and between-laboratory precision.

QSA uses different sets of calculations for controlling bias and precision. In all cases, these calculations are based on the assumption that control charts are used to monitor bias and precision. The QSA system further assumes that the detection mechanism is an out-of-control point on a control chart. In both cases, the user may specify either the number of QC samples to analyze or the probability of detecting an excursion beyond the specified QC limits; the other possibility is computed automatically. The user can also define sample design parameters, such as the maximum length of time that can be tolerated to detect an excursion from the design parameters, number of replicates in a QC sample set, or minimum acceptable levels of bias or precision. A simplifying assumption is that the event to be detected (i.e., an excursion from acceptable bias or precision) begins to occur at the beginning of the time period window that is being monitored.

The calculations for controlling bias are based on statistical methods for detecting a shift in the mean, and the assumption inherent in the calculations is that random errors in measurements are distributed evenly about the mean. Random errors are characterized by the level of bias to detect; the relative standard deviation of the measurement error; and the maximum period of time, in days, that the problem can be allowed to persist before being detected. The *relative standard deviation* (RSD) is an estimate of the average error in the measurement due to unassignable causes and is usually expressed as a percentage of the average sample concentration. The minimum acceptable level of bias for a sampling and analysis effort, expressed as a percentage of the sample concentration, specifies when the process is considered out of control. The statistical module includes the tabled points of the cumulative standard normal distribution, which are accurate to 0.01 relative units (Z).

The calculations for controlling precision are based on the ratio of the maximum tolerable precision to the target precision, the number of samples per set, the duration of the sampling effort, and the number of sets per day. The relationship between the sample design parameters and the tolerable or target ratio also varies for each sample control strategy.

The computations take advantage of the assumption that the ratio of the sample variance to the population variance is a chi-square distribution; the points of the chi-square distribution are tabled within the calculation module for 2–10 degrees of freedom. The computation for the QC check samples must work backward from the chi-square table to find the number of analyses required per day.

The Expert System in Operation

Assume a Superfund project involving both environmental sampling and analysis is being planned. The user is concerned about controlling bias in both the laboratory and field samples. The user plans to use gas chromatography–mass spectrometry for the analyses, and the typical within-day precision of the method is estimated to be 10% RSD at levels greater than 10 times the detection limit. What kind and how many QC samples are needed?

With the aid of the QSA help functions, the user is able to clarify the project objectives and decides that bias should be controlled at both normal and low-level concentrations. The user also wants to address bias in both the sampling and analytical efforts. In other words, the user is interested in controlling bias due to low-level contamination, such as from storage and handling, and bias at normal concentration levels, such as from calibration errors or other systematic laboratory or instrument sources. Additionally, the user wants to control bias that can result from matrix effects (e.g., analyte interferences and recovery problems). Within seconds, the expert system provides the answer that laboratory blanks, field blanks, and field spikes are needed. By simply placing the cursor on the name of the QC sample type and pressing return, the user allows the program to continue and calculate how many QC samples will be needed.

This example can be continued for field spikes. The user answers each question with an optimum program within the limits defined by the samples and analytical plans. Figure 2 illustrates an example. The user would like to achieve a high probability of detecting a minimum level of 10% bias in the samples. The samples will be analyzed over a 2-week period, but the user would like to detect any bias excursions within 1 day and would like to run one field spike per day. Therefore, the user chooses frequency as the input parameter and one (1) as the frequency per day for field spike analyses. The results are calculated in less than a second and are displayed as Figure 2.

Because a probability level of 2.3% is unacceptable, the next step is to discover how many field spikes would be required to give a 95% confidence level of detecting bias under the optimum conditions. The entries are simply changed from frequency to probability, and 95% is entered for the probability level. The answer is instantly calculated to be 22 field-spiked samples per day that must be analyzed; this number is also unacceptable.

The great utility of this program is then illustrated during the next steps. By changing various parameters, the user can quickly optimize the number of field spikes that can be afforded and know precisely what has to be given up in terms of minimum levels of bias and confidence.

For example, for the parameters illustrated in Figure 2, if the probability level is changed from 95% to 90%, the number of field spikes required drops from 22 to 19. If the probability level is left at 90% and the number of days over which bias could be detected is increased from 1 to 5, the number of field blanks drops to eight per day. Next, the minimum level of detecting bias is increased from 10% to 20%; this results in a frequency of two field spikes per day. Decreasing the within-day precision level would also decrease the number of field spikes per day but that would involve a change in analytical methodology, which is not a good option in the typical example.

The number of field spikes per day is now at an acceptable level, but the user probably is not satisfied to have to wait for 5 days of analyses before a 90% probability of detecting a bias error occurs.

By changing the minimum levels of bias to detect, and varying the number of days for detection of a bias excursion, the user is quickly able to determine

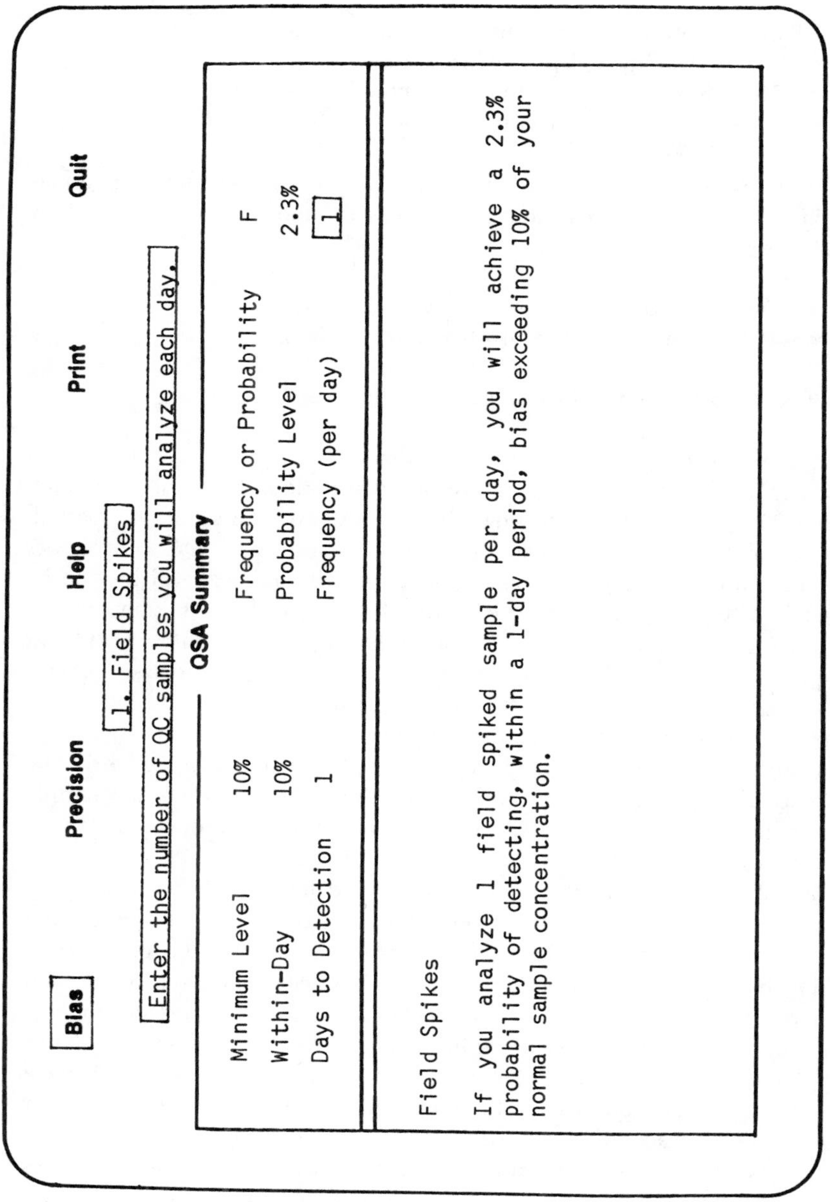

Figure 2. An example of a desired QC plan using one field spike per day provides an unacceptably low probability of detecting bias under planned sampling and analysis protocols.

that by analyzing two field-spiked samples per day, the probabilities of detecting bias exceeding 20% of the normal sample concentration are as follows:

- 94.1% probability of detecting bias within 5 days,
- 81.7% probability of detecting bias within 3 days,
- 67.8% probability of detecting bias within 2 days, and
- 43.3% probability of detecting bias within 1 day.

Furthermore, the user can calculate that the probabilities of detecting bias exceeding 30% of the normal sample concentration under the same conditions are as follows:

- >99.9% probability of detecting bias within 5 days,
- 99.9% probability of detecting bias within 3 days,
- 98.8% probability of detecting bias within 2 days, and
- 89.3% probability of detecting bias within 1 day.

In other words, although two spikes per day provide less than 50% probability for detecting a 20% bias over a 1-day period, these spikes provide almost a 90% probability of detecting a bias of 30% over the same time period. If that probability is not acceptable, then either the number of QC samples per day or the number of days until an excursion can be detected will have to be increased. With the QSA program, however, the user will always know exactly what the statistical probabilities are instead of having to guess. And, in addition to helping a user more clearly define and plan a sampling or analysis project, QSA also enables the user to know exactly what the statistical probabilities are in detecting an excursion beyond any set limits.

A potential compromise might appear as shown in Figure 3, where the probability of detecting a bias excursion exceeding 20% of the normal sample concentration within 2 days is 90% if three field-spiked samples per day are analyzed.

Basically, a similar procedure would be used for the other QC samples. Thus, in a few minutes, the user would be able to determine the most important parameters that affected the level of confidence that would be obtained in detecting bias from both laboratory and field sources. The user could then optimize the conditions that best suited the needs for the most economical conduct of the analysis. If precision is important to the study, those calculations and the required QC samples can be obtained just as quickly. An average session to completely optimize the kind and number of QC samples needed for a project only takes about 2–5 min per type of control sample.

This example shows that a quality assurance (QA) plan that simply relies on a set percentage of QC samples (e.g., 10% or 15%) to provide bias or precision statistics can result in values that may be unacceptably low. The resulting unreliable data or probability levels may be too high and therefore wasteful of

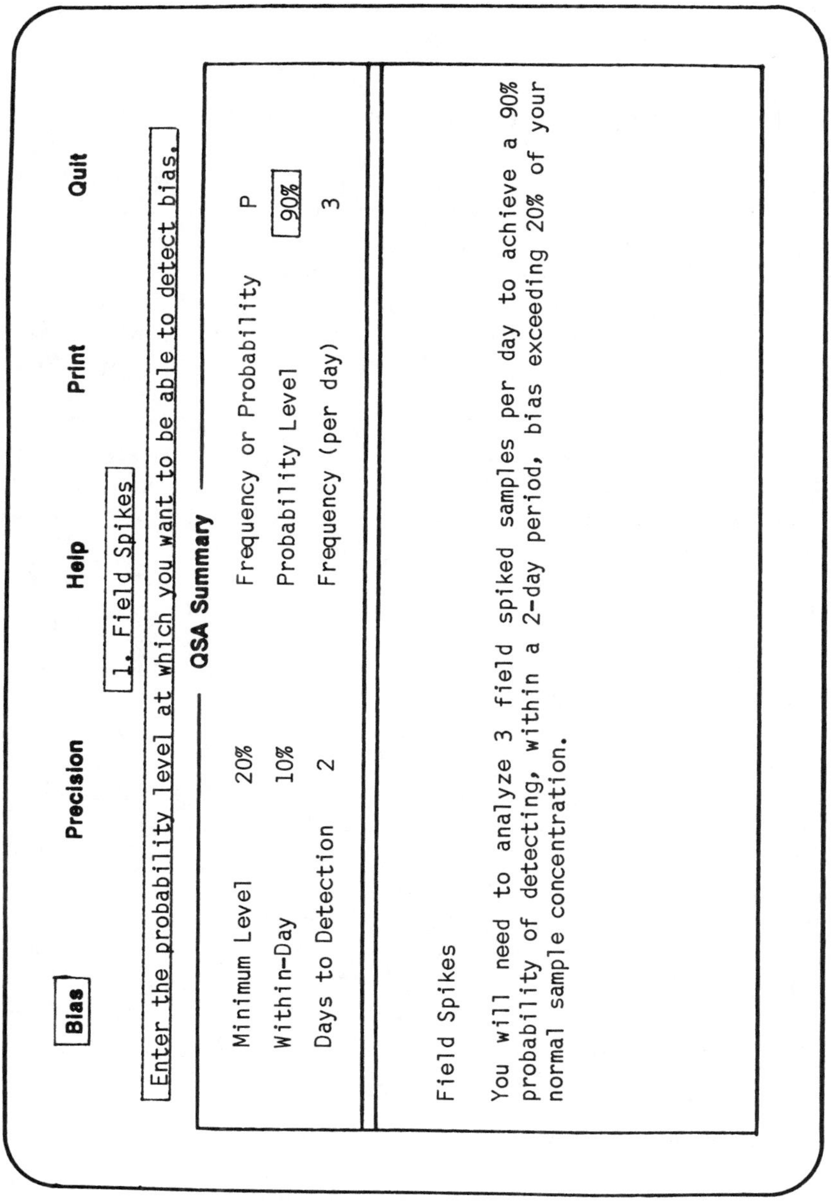

| Bias | Precision | Help | Print | Quit |

1. Field Spikes

Enter the probability level at which you want to be able to detect bias.

QSA Summary

Minimum Level	20%	Frequency or Probability	P
Within-Day	10%	Probability Level	90%
Days to Detection	2	Frequency (per day)	3

Field Spikes

You will need to analyze 3 field spiked samples per day to achieve a 90% probability of detecting, within a 2-day period, bias exceeding 20% of your normal sample concentration.

Figure 3. Example of an optimized QC plan for field spikes.

time and money in collecting, preparing, and analyzing QC samples. QA plans must consider other factors besides just the number of samples being taken and analyzed; otherwise, the wrong types or frequency of QC samples may be selected.

Directions for Further Research

The development of QSA as an aid to QC program design has provided a clear perspective of what is required of an ideal design aid. On the basis of questions and answers about the sampling and analytical effort, the program should recommend a complete QC strategy, assist the designer in setting design parameters, and assess the effectiveness of the QC program design. The program must also educate the designer, if necessary, so that he or she may properly monitor the effort as it is conducted. The ideal design aid must provide recommendations for the tools needed to effectively control the sampling and analysis program. These tools include the appropriate sampling equipment, analytical equipment, analytical method(s), and control charts to be used.

To make more effective recommendations, many of the implicit assumptions in QSA need to be relaxed, and other issues pertinent to the design of good QA programs need to be explored. Examples of these issues include

- more explicit consideration of the type of sample matrix (e.g., air, solids, and biota),
- consideration of the objectives (e.g., compliance monitoring, modeling, and rough estimates), and
- QA and QC experience level of the user.

The control strategies recommended by QSA were simplified for the purpose of producing a functioning first program. The knowledge base should be deepened to offer more specific kinds of QC. For instance, lab blanks may be separated into solvent, reagent, and system blanks.

In order to maximize the data quality, an advanced expert system should raise the level of awareness of the user about QC issues and techniques. These issues include the following:

- Standardized working definitions for key concepts must be used. For example, different interpretations of the definition of detection limit and bias can lead to different values for the same method.
- The user must be presented with a clear picture of the various control strategies and their limitations. Control charting techniques must be presented along with recommendations for using them. A user with no statistical background must be able to understand which of the charting techniques to use and must be

given enough knowledge about these techniques to allow excursions to be detected.

The novel screen interface designed for the QSA prototype is a good example of the influence that the computer interface can exert. The current version of the program encourages the user to seek alternative solutions and thus to maximize the benefit of the QA program.

The design of the user interface is a recurring issue in software development. The user interface has a tremendous impact on the utility of this expert system, as with any software program. The entire presentation of the program must be carefully considered so that the questions, answers, and intent of the program are clear to the user at all times. New interface techniques must be developed and refined so that the user's interaction with the program becomes smooth and natural.

QSA is a useful program but has even greater potential for providing much more comprehensive and sophisticated advice for a very complex technical area.

References

1. Pierce, T. H.; Hohne, B. A., Eds.; *Artificial Intelligence Applications in Chemistry*; ACS Symposium Series 306; American Chemical Society: Washington, DC, 1986.
2. Provost, L. P.; Elder, R. S. *Choosing Cost-Effective QA/QC Programs for Chemical Analysis*; U.S. Environmental Protection Agency: Cincinnati, OH, 1985; EPA Contract No. CI-68-03-2995.
3. Keith, L. H.; Stuart, J. D. In *Artificial Intelligence Applications in Chemistry*; Pierce, T. H.; Hohne, B. A., Eds.; ACS Symposium Series 306; American Chemical Society: Washington, DC, 1986.
4. Riese, C. E.; Stuart, J. D. Ibid.
5. *Machine Learning*; Michalski, R. S.; Carbonell, J. G.; Mitchell, T. M., Eds.; Tioga Publishing: Palo Alto, CA, 1983.
6. Michie, D.; Muggleton, S.; Riese, C.; Zubrick, S. In *IEEE Proceedings of First Conference on Artificial Intelligence Applications*; IEEE Computer Society Press: Silver Spring, MD, 1984.

RECEIVED for review April 1, 1987. ACCEPTED June 11, 1987.

QUALITY ASSURANCE AND
QUALITY CONTROL

Chapter 6

Defining the Accuracy, Precision, and Confidence Limits of Sample Data

John K. Taylor

The limits of uncertainty for data on samples include uncertainty due to their measurement and to the samples. Measurement uncertainty is controlled and evaluated by an appropriate quality assurance program. Sample and sampling uncertainty includes random and systematic components. The concept of quality assurance can be applied to the sampling operations to control and evaluate sampling uncertainty. The information necessary to assign statistically supported limits of confidence to sample data is discussed.

S AMPLE DATA CONTAIN A DEGREE OF UNCERTAINTY, and this uncertainty must be considered whenever the data are used. Ordinarily, measurements are made because information is needed to evaluate some property of a system so that decisions can be made concerning the system. The system or material of interest may be as small as a speck of dust or as large as the Earth. If the material of interest is small, then the sample to be analyzed—the entire object—is known with certainty. If the material of interest is very large, then only portions of the system may be examined. Clearly, some questions need to be answered. These questions include what and where to sample and how many samples will be required for a specific purpose. If, for example, the information to be found was the platinum content of the Earth, the results would be greatly influenced by the number and kinds of samples that were analyzed, and some degree of uncertainty would be associated with any conclusions drawn from the data.

Every measurement problem involving less than the entire universe of interest has some parallelisms to the example just mentioned. Researchers who ponder such matters may almost despair in making any decisions involving measurement of samples. Unfortunately, many decisions are made in ignorance or contempt of the uncertainty of the sample data. In fact, "representative samples" are often used to make decisions even though no real evidence is presented to verify that the sample represents anything other than itself. These problems could be minimized or eliminated by proper design of measurement and sampling plans and by realistic evaluation and interpretation of the data obtained by using these plans. The following discussion is presented to provide guidance in these matters.

Data Requirements

Before measurements can be made, the concept of the problem to be solved and the model to be followed for its solution must be reasonably clear (1). What needs to be measured, the levels of concern, and the permissible tolerances for the uncertainty of the data must be clearly understood. On the basis of such information, intelligent decisions can be made on how many samples are required, how to measure them, and the number of measurements needed. With controversial issues, investigators must agree on what can and should be achieved in a measurement program prior to its initiation. Otherwise, not everyone will be satisfied with the outcome.

The goals and expectations of a sampling program must be realistic and can never exceed the measurement and sample limitations. Costs and benefits must be considered in the design of almost every measurement program.

As the detection limits of methodology are approached, the number of measurements can increase by an order of magnitude (actually requiring at least nine replicates) if quantitative results are required. This increase occurs because the limit of quantitation is about 3 times the limit of detection (LOD) (2). Because both figures of merit are calculated on the basis of single measurements, a concentration level equivalent to LOD can be the limit of quantitation for the mean of nine measurements. Accordingly, making decisions on data obtained at the limits of capability of the methodology is foolish.

Sources of Uncertainty

The total variance of measurement data (s^2_{total}) can be expressed in the simplest terms as

$$s^2_{total} = s^2_{measurement} + s^2_{sample} \tag{1}$$

where $s^2_{measurement}$ and s^2_{sample} are the sampling variances due to measurement and the sample, respectively. The measurement and sampling plans and operations must be designed and executed so that the individual components may be evaluated. The possible situations that can occur are presented in Table I (3).

Table I. Measurement Situations

Situation		Significance
A.	Measurement variance	No
	Sample variance	No
B.	Measurement variance	Yes
	Sample variance	No
C.	Measurement variance	No
	Sample variance	Yes
D.	Measurement variance	Yes
	Sample variance	Yes

Situation A is confined almost exclusively to single-specimen analysis or where only semiquantitative data are required. Situation B largely pertains to the analysis of homogeneous materials. In most other cases, some degree of variability of samples is encountered. In situation C, single measurements of samples are sufficient. Unfortunately, situation D is prevalent, so both measurement and sample variability must be taken into account when evaluating sample data.

Measurement Uncertainty. Measurement uncertainty can be controlled and evaluated by an appropriate quality assurance program (*1, 4*). When properly planned and executed, the measurement variance will be known so that the number of measurements can be minimized. Otherwise, sufficient replicate measurements must be made on a sufficient number of samples to evaluate both sources of variance. Ideally, measurement uncertainty should not exceed one-third of the total uncertainty tolerance. Otherwise, replicate measurements will be needed to reduce the uncertainty to acceptable limits.

Although the quantitative uncertainty of the measured values is emphasized, the need to verify the qualitative identification of the analytes that are reported should not be overlooked. The confidence of identification must approach certainty (*3, 4*), which is achieved by a confirmation process as discussed in reference 2.

Sample Uncertainty. Sample uncertainty may contain systematic and random components arising from population and sampling considerations. Population considerations have been discussed in some detail by Provost (*5*) and will be mentioned only briefly here. Because every population has some degree of variability, a sound statistical sampling plan is necessary if the judgment of apparent differences is to be defensible. Because small differences of measured values with reference to baseline values, control areas, or norms are ordinarily of concern, the measurement of large numbers of randomly selected samples is inevitable.

Bias in samples from a population results from nonrandom sampling and from discriminatory sampling. Every population must be defined, and the act of definition biases the population, rightly or wrongly, by exclusion of certain individuals. For example, an analyte of interest may be looked for only in specific locations such as silt, and even in specific particle-size fractions. This sampling may be proper, but one must always consider how this might bias the data. Considerations such as in this example are usually based on the model employed, which should be periodically reconsidered for its adequacy.

Sampling Considerations

The sampling operations can provide both systematic and random components to the sampling uncertainty. Concerning the systematic components, the sampling equipment may not be able to experimentally realize the requirements of the model. For example, a respirable fraction of airborne particulates may be difficult to specify logically and even more so experimentally. Obviously, sampling equipment should be calibrated when factors such as flow rates, size discrimination, and temperature effects are important because errors in these factors could cause bias in sample data.

Some or all of the analyte of concern can be lost from samples because of absorption or reaction with container materials, sampling equipment, or sample-transfer lines. Deterioration arising from atmospheric contact, temperature instability, radiation effects, and interactions with other analytes can cause serious biases. The act of sample removal from its environment can disturb stable or metastable equilibria that could bias ensuing measurements. Stabilization of collected samples can be of major importance but difficult to realize. Subtle carry-over effects in sample containers and sampling lines that result from memory of previous samples can be of serious consequence.

Random components of sampling uncertainty can result from variability of all of the aforementioned sources of bias. Accordingly, considerable incentive exists to eliminate these sources of bias or, failing to do so, carefully controlling them. In addition, the sampling operations can have their own variability components due to carelessness or inability to keep all aspects in a state of statistical control.

Statistical Considerations

Systematic components of uncertainty from various sources are algebraically additive. This situation can be shown as follows:

$$B_{total} = B_1 + B_2 + ... B_n \tag{2}$$

where B_{total} is the total systematic uncertainty, and $B_1 ... B_n$ are the components of uncertainty from sources $1 ... n$, respectively. Sources of bias can be conceptually identified, but quantifying their contributions may be difficult. This statement is valid for measurement bias and especially sample bias. In many cases, all that can be done is to develop a bias "budget" and estimate the bounds of each component. Of course, corrections should be made for bias whenever possible.

The random components of sampling uncertainty from several sources, expressed as variances, are additive. Thus,

$$s^2_{sampling} = s^2_1 + s^2_2 + \ldots s^2_n \tag{3}$$

where $s^2_{sampling}$ denotes the total random components of sampling uncertainty, and $s^2_1 \ldots s^2_n$ denote components of sampling uncertainty from sources $1 \ldots n$, respectively. Identifying the sources of some, if not all of these variances, may be possible, but evaluating the individual components may be difficult and certainly time-consuming. However, evaluating the overall value by suitable replicate experiments may be possible. This evaluation may be accomplished by taking replicate samples where sample population variability is expected to be negligible (e.g., in a narrowly defined sampling area). At least seven replicate sampling operations would be required, and the excess variance over that of measurement would represent sampling variance.

Sample population and sampling variances ($s^2_{population}$ and $s^2_{sampling}$, respectively) are additive and can be represented as

$$s^2_{sample} = s^2_{sampling} + s^2_{population} \tag{4}$$

This equation demonstrates that sampling variance must be kept relatively small or else it can seriously influence measurement.

The statistical treatment of population variance was discussed by Provost (5). The total uncertainty includes the sum of the contributions arising from random and systematic sources. The random component of uncertainty is evaluated by a statistical confidence interval, and the value obtained can be reduced by replication. The systematic component of uncertainty is independent of the number of replicates. Discussion of the procedures used for the computations is beyond the scope of this chapter.

Quality Assurance of Sampling

The two aspects of quality assurance—quality control and quality assessment—can be applied to sampling as well as to measurement. Quality control includes the application of good laboratory practices, good measurement practices, and standard operations procedures especially designed for sampling. The sampling operation should be based on protocols especially developed for the specific analytical problem. Strict adherence to these protocols and sampling protocols is imperative. Sample takers must be trained to follow the protocols faithfully. All required calibrations must be made on the basis of established schedules. Special care must be devoted to sample containers and to stabilization and protection of samples. A system for assuring positive identification of samples and documentation of all sample details must be operational. Chain-of-custody procedures, whether or not required externally, are necessary if sample integrity is to be defensible, which is almost always.

Quality assessment of the sampling process depends largely on monitoring for adherence to the respective protocols. Audits on a continuing basis are the best means to accomplish this purpose. A system that uses container, field, and laboratory blanks should be part of the sampling protocol. To be meaningful, the kinds of numbers of such protocols must be individually selected for each measurement program.

Because of their critical role in the quality assessment of the sampling operation, protocols must be carefully designed, and their adequacy should be under periodic review in any monitoring program.

Preplanning is especially important because any modification of the sampling operation during a measurement program could produce a different set of samples, which may not be compatible with previous sampling operations.

Conclusion

Three basic kinds of sampling plans are in common use (3). *Intuitive sampling plans* are based upon judgment, often by technical experts, and the interpretation of data is also based on judgment. Various experts can draw different conclusions from the same data set, and choosing between them may be difficult. *Statistically based plans* provide the basis for making probabilistic conclusions that are independent of personal judgment. However, the model upon which the data are based could be open to criticism. Often, statistical plans require more samples than are feasible for various reasons. In such cases, a *hybrid plan* may be used that includes intuitive simplifying assumptions. These assumptions must be clearly stated and considered when interpreting data, or else probabilistic judgments may be made that may not be fully justified. Of course, validated and evaluated data must be used no matter which approach to sampling is followed.

In addition to these sampling plans, protocol sampling, which follows legally or contractually mandated plans, is commonly used. These plans often require a representative sample, which may or may not be specified. Although such plans may be useful in practical situations, one must remember the empiricism upon which mandated plans are often based. Such data should not be used for other than the specified applications. The practice of specifying representative samples should be discouraged whenever possible because of the virtual impossibility of demonstrating representativeness.

Sampling and the associated measurements are often made to decide on compliance with some requirement. In such cases, answers must be obtained to the following questions:

- Is the mean value of the population within acceptable limits?
- Is a specified fraction of members of the population within acceptable limits?
- Are all members of the population within acceptable limits?

Actually, only the first two questions can be answered by sampling, or rather only by statistically sampling the population. The third question cannot be answered, except by examining every individual. No matter how many individuals have been examined and found to be in compliance, noncomplying individuals may exist. As an example, if as many as 3 million individuals were found to be within acceptable limits, then there is a 5% chance that one in 1 million members could be defective.

The more one looks critically at sampling, the more one should be convinced that sampling is not a trivial exercise. Accordingly, in all but the most simple situations, all aspects of sampling should be carefully planned, and sampling experts and statistical advisors should be used as necessary if meaningful and defensible conclusions are to be realized.

Abbreviations and Symbols

B_{total}	total systematic uncertainty
LOD	limit of detection
$s^2_{population}$	sample population variance
s^2_{sample}	variance due to the sample
$s^2_{sampling}$	total random components of sampling uncertainty
s^2_{total}	total variance of measurement data

References

1. Taylor, J. K. *Anal. Chem.* **1981**, *53*, 1388A–1396A.
2. "Principles of Environmental Analysis"; American Chemical Society Committee on Environmental Improvement. *Anal. Chem.* **1983**, *55*, 2210–2218.
3. Taylor, J. K. *Trends Anal. Chem.* **1986**, *5*(5), 121–123.
4. Taylor, J. K. In *Environmental Sampling for Hazardous Wastes*; Schweitzer, Glenn E.; Santolucito, John A., Eds.; ACS Symposium Series 267; American Chemical Society: Washington, DC, 1984; pp 105–109.
5. Provost, L. P. Ibid., pp 79–97.

RECEIVED for review January 28, 1987. ACCEPTED April 6, 1987.

Chapter 7

Defining Control Sites and Blank Sample Needs

Stuart C. Black

Control sites and blanks are essential components for any monitoring or analytical program. These items are defined and guidelines for their selection are described. Some examples of their use in actual situations are presented.

CONTROLS AND BLANKS ARE ESSENTIAL COMPONENTS for experimental or for sampling and analytical studies because firm conclusions cannot be drawn from such studies unless adequate controls or blanks have been included. In one sense, *blanks* and *controls* are two names for a single process, and the name used depends on the purpose of the study. In another sense, *blank* has a specific meaning related to analytical procedures, and *control* relates to a specific type of sample, that of the environment or of a population, against which the results of a procedure are judged. Therefore, the control could be a blank. In this chapter, the common meanings of these terms are discussed, and guidance for selection and use of blanks and samples is outlined with appropriate examples.

Blank Selection

Blanks may be defined as samples expected to have negligible or unmeasurable amounts of the substance of interest. They are necessary for determining some of the uncertainty due to random errors. In any such process, the *random errors* are those errors that can be estimated by the use of standard statistical techniques. The other kind of error, which affects the total uncertainty of a process, is called *systematic error*. Systematic error cannot be estimated statistically and usually results in a consistent deviation (bias) in a final result. Discussions on the definition and use of blanks are contained in references 1 and 2.

The random errors in a sampling and analytical procedure affect the precision of a methodology because precision is estimated by the standard deviation of those errors. Because standard deviation is the square root of variance and because variance (s) can be summed where appropriate, precision can be estimated by summing the variances of the various steps. This summation can then serve as a guide for selection of blank samples. Equation 1 includes the various errors that may occur in a procedure.

$$s^2_T = s^2_a + s^2_t + s^2_s + s^2_h + s^2_p + s^2_c + s^2_m \tag{1}$$

where the subscripts identify the variances as follows: T is total, a is spatial, t is temporal, s is sampling, h is transport and storage, p is preparation, c is chemical treatment, and m is measurement variances.

Except for the first two terms in equation 1, a blank can be devised for each variance to ensure that the variance is inherent in the methodology and is not due to contamination. These blanks have specific names to identify which part of a process is being tested (1). A reagent blank is one of the most common types, and nearly every analytical procedure includes at least one, even though several other blanks can be used in analytical processes. In addition, blanks used in sample collection and preparation also have important functions. However, the reagent blank may be the most important because it can be used to estimate the lowest concentration of an analyte that can be measured, and because it also may affect the accuracy of a measurement in those cases where a blank correction is necessary (3).

The order of appearance and use of certain blanks in a sample collection and analysis program is described in this section. These blanks generally consist of distilled and deionized (DDI) water, although some are based on the analytical matrix. Methylene chloride, for example, can be used in certain organic analysis procedures.

Field Blank

This blank is used to estimate incidental or accidental contamination of a sample during the collection procedure. Capped and cleaned sample containers are taken to the sample collection site. After a sample is obtained, the collection apparatus is cleaned according to the standard operating procedure (SOP) prior to taking another sample. At that point, the collection apparatus is rinsed with DDI water, which is collected in a sample container for later analysis. If a preservative is used, then an equal amount is put into the container with the blank. As a general rule, one field blank should be allowed per sampling team per day per collection apparatus.

Transport Blank

This blank is used to estimate sample contamination from the container and preservative during transport and storage of the sample. A cleaned sample container is filled with DDI water; any preservative used in the sample is added; and then the blank is stored, shipped, and analyzed with its group of samples. This blank is more important when shipping and storage consumes several days or weeks because leaching of the material from the container can become significant. One transport blank should be allowed per day per type of sample.

Sample Preparation Blank

This type of blank is required when such methods as stirring, mixing, blending, or subsampling are used to prepare the sample prior to analysis. After a sample is processed in the apparatus, the apparatus is cleaned according to the SOP; the DDI water or the sample solvent is then processed through the apparatus and analyzed with the samples. One of these blanks should be prepared for each 20 samples processed.

Matrix Blank

This kind of blank is used to determine the presence of the analyte in the matrix when the sample matrix is not DDI water, but is instead, for example, an organic solvent. The frequency of analysis should be equivalent to that for a reagent blank. In general, the matrix blank would only be analyzed if the reagent blank analysis indicated contamination. The matrix blank would then be used to determine whether the reagents or the matrix was responsible.

Reagent Blank

This blank is one of the most important in any process. In general, the reagent blank is DDI water processed through the analytical procedure as a normal sample is processed. After use to determine the lower limit of detection, a reagent blank is analyzed for each 20 samples and analyzed whenever a new batch of reagents is used. The preferred outcome from analysis of reagent blanks is a less than detectable concentration of the analyte of interest.

Case Study of Blank Selection and Use

Several types of blanks were used in the Dallas Pb study (4) and the Love Canal monitoring program (5). In the Dallas Pb study, a field blank, a sample preparation blank (called a sample bank blank in that study), and a reagent blank were used. These blanks were prepared as indicated previously for each type (i.e., the field blank was DDI water used to rinse the soil collection equipment, the sample bank blank was a field blank used to rinse the ball-mill and sieve used to homogenize the soil samples, and the reagent blank was DDI water heated with nitric acid used to extract Pb).

None of the 76 field blanks, 77 sample bank blanks, or 148 reagent blanks had a Pb concentration that exceeded the 0.25-μg/mL detection limit for the atomic absorption spectroscopic procedure used. This concentration limit suggested that equipment-cleaning procedures were adequate and that reagents had no measurable Pb; therefore, the soil samples were the only source of detectable Pb measured by the analytical method.

Recommendation

In many analytical methods, analysis of samples can be a very expensive procedure. However, collection of blanks is easy and inexpensive if done at the time of sampling. A prudent procedure to follow then would be to collect a full suite of blank samples but analyze only the field blanks. If the field blanks indicate no problems with the method, then the other blanks could be discarded or stored as necessary. If a problem does arise, then the individual blanks can be analyzed to determine the location of the problem. Also, a reagent blank should be analyzed whenever new reagents are introduced into the method.

In those cases where an extraction device is used rather than collection of the sample itself [e.g., air filters, charcoal, ion exchangers, or a polymeric form of 2,6-diphenylene oxide (Tenax)], then the field, storage, and transport blanks should be the corresponding device. For air sampling as an example, an unused filter should be unsealed when the air filter is collected and subsequently treated as if it had been used for sampling. For many analytes, these kinds of extraction media will give positive results on analysis so that a blank correction will be necessary.

Control Selection

Control samples are basically of two types: controls used in quality control procedures to determine whether or not the analytical procedure is in control, and controls used to determine whether or not a factor of interest is present in a population under study but not in the control. The population can be a group of environmental samples, a defined group of people, or similar groups of things. The two types of controls are discussed in this section.

Calibration Control Standard

This type of control is known by several other names, such as quality control calibration standard (CCS) or calibration check standard. In most laboratory procedures, this control is a solution containing the analyte of interest at a low but measurable concentration. The precise concentration of this standard need not be known. The first sample analyzed after an instrument is calibrated is a CCS, and the result should be plotted on a control chart. Another CCS is analyzed after each 20 samples, or after each shift if fewer samples are analyzed per shift. The standard deviation of the CCSs is a measure of the instrument precision unless the CCS is analyzed as if it were a sample, in which case the CCS is a measure of the method precision.

Laboratory Control Standard

The laboratory control standard (LCS) as used herein is a certified standard, generally supplied by an outside agency. This standard is used to determine whether or not an analytical procedure is producing results comparable to other analytical laboratories. A good source of LCSs is the National Bureau of Standards, which has a variety of standard reference materials (6) containing certified concentrations of elements or compounds. The Environmental Monitoring and Support Laboratory of the U.S. Environmental Protection Agency (EPA) in Cincinnati is a useful source of organic standards in water or organic solvent solutions. An LCS should be analyzed with every batch of samples until 7–10 results are available. If those results are within the control limits specified by the program protocol, then the frequency may be reduced to one per day. However, several LCSs should be analyzed any time the analytical instrument is recalibrated. The mean value of all LCS results is a measure of the method bias.

Matrix Control

The common name for this control is *field spike*. For those sample matrices, such as sediments and sludges, where a complex mixture is liable to cause interferences in the analysis, then a field spike may be required to obtain an estimate of the magnitude of those interferences. The losses from transport, storage, treatment, and analysis can be assessed by adding a known amount of the analyte of interest to the sample during collection in the field.

Control Sites

Along with the controls used to measure the bias and precision of sampling and analysis, another very important control is the control site or control population. If the results of a study in a given area are to be judged as high, low, or insignificant, the results must be compared to the results in some other area (e.g., the control area or site). For example, if the environmental impact of a given facility, such as a waste disposal site or a coal-fired power plant, is to be assessed, then the environmental levels in the absence of that facility must be known. On the other hand, if the contribution of pollutants from an urban area to environmental pollution is to be assessed, then the contribution of pollutants from sources other than the urban area must be known.

In the laboratory, for example, a reagent blank is a control sample for the analyte of interest. But this blank has a very narrow interpretation because it sets the limit of sensitivity for an analytical methodology but provides no information on whether or not a sample contains more than "normal" amounts of an analyte. Samples from control sites are required to determine the normal analyte amounts. The sites or populations that can supply control samples can be classified as local, areal, national, or background depending on the location selected.

Local Control Site. A *local control site* is a control near in time and space to the sample of interest. The reagent blank, for example, is a type of local control. Another example is in a groundwater monitoring program, where the control site is a well upstream of a suspect facility and provides control samples against which samples from a well downstream of the facility are compared. Suspect facilities (i.e., facilities that may possibly contaminate groundwater) could be waste disposal sites or chemical plants using injection wells for disposal of waste. Factors to be considered in the selection of local control sites include the following:

- Local control sites should be upwind of the facility most of the time; a *wind rose*, which is a diagram that summarizes statistical information about the wind, should be available; and air samples should be taken only when the wind is from the sampler toward the site.

- Local control sites should be upgradient from the facility with relation to surface and groundwater flow.

- The potable water source should not be affected by site effluents.

- Travel between control site and the facility should be minimal because of problems associated with transport by vehicles.

As an example of a local control, the groundwater monitoring strategy of the EPA requires monitoring of a control well. Because groundwater is not normally a simple stream with narrow boundaries, EPA regulations require three downstream wells be monitored for comparison to the one upstream well (7). For such a monitoring system to perform adequately, the assumption must be made that at no time do the contaminants spread upstream. A much better local control would be intensive study of the background levels in the groundwater prior to construction of the facility under study. Unfortunately, background studies for the majority of suspect facilities either do not exist or the previous analyses did not include the analytes of interest. Another example of a local control is used frequently in epidemiological studies. Case–control studies are used to determine the effects of exposure to certain substances. If an unusual frequency of illness occurs among some employees, the cases would be all those employees having the same occupation in a facility. The controls would be those employees in that facility not exposed to the suspect hazardous material. The controls would also be matched on such variables as length of employment, age, sex, and time of beginning employment. A good description of this type of study is included in reference 8.

Area Control Site.   This control site is in the same area (e.g., a city or county) as the pollutant source but not adjacent to it. The factors to be considered in site selection are similar to those for local control sites. All possible effort should be made to make the sites identical except for the presence of the pollutant at the site under investigation.

An example of an area control site is the one selected in the Dallas Pb study (4). The purpose of that study was to determine whether or not the secondary lead smelters in Dallas, TX, were a cause of elevated blood-lead levels in children living near the smelters. A control site was chosen in Dallas by using the following criteria: similarity in automotive traffic, similarity in population density, and similarity in ethnic background. Water supply was the same as for the study sites, and food habits were assumed to be similar because of the similarity in ethnic background. Because automotive emissions are a source of Pb, the similarity in automotive traffic was a highly important criterion.

National Controls.   National controls can be very useful if the data are handled with care. In general, national controls tend to be very broad, or less specific, than local or area controls but can identify anomalous results. For example, if the study area data are not much different from the data in a local or area control site, then national controls can be used to suggest whether or not the data from the more local control sites may be anomalous compared to national data.

Some of the factors to consider in selecting national controls are the following:

- Similar data should be available. (National soil, water, and air monitoring programs as well as some bioassay data are available.)

- Areas should be similar. For example, data from farming areas should not be compared with data from industrial areas.

- Monitoring data should be upwind and upgradient from any possible sources such as waste disposal areas and smokestack industries.

- If possible, control data from several such areas should be chosen.

The principal problem may be the lack of adequate data, either in the type of analyte reported or in the quality of the data. Data of poor or unknown quality should not be used other than as an indicator for further data collection. If appropriate data are lacking, of course, then the use of national controls may be precluded as an option because of the expense and time required to conduct an adequate sampling program.

An example of the use of national controls is in epidemiological studies on cancer incidence or causation. National age-adjusted cancer mortality data are frequently used as the basis on which the health effects of a given exposure to pollutants are estimated (8). This type of control should be used with caution, however, because national averages tend to obscure the differences that exist in countywide or even statewide data on cancer incidence.

Background Sites. A true background site is one that has not been affected by human activities. Such sites may be difficult, if not impossible, to find because of transport processes that spread man-made contaminants throughout the biosphere. Even uninhabited locations such as desert islands, the polar regions, and continental badlands areas have been affected to some extent. For many man-made pollutants, though, carefully chosen areas can provide samples that may be considered background samples. For some naturally occurring substances, a background area may be more easily found. A good example is reported in reference 9, where the background levels of uranium in soil were determined so that the spread of contamination from a uranium processing facility could be measured.

Comparisons

In most cases, simple statistical techniques are sufficient when comparing the results of a study area or group with a control. These techniques can include a one-sided t test or an analysis of variance (ANOVA) test. However, if the results from such tests are ambiguous, plotting the two sets of data may be necessary. Bar graphs are an effective plotting technique because any differences in magnitude or distribution between the data sets become obvious. If both testing and plotting fail to show a difference, then additional sampling may be required.

Conclusion

The needs for blanks and controls are driven by the study objectives, which specify the use of data collection and the required bias and precision of the data needed to meet those objectives. In most cases, these objectives can be derived by careful study of the experimental design. Wherever in the procedure a possibility exists for introducing extraneous material into a sample collection, treatment, or analysis procedure, a blank should be devised to measure that possibility. Furthermore, if the results obtained from a sampling program need to be compared to a background, or to ambient levels, then a suite of control samples may be required to prove the presence of abnormal levels of the analyte.

The basic principal for choosing a control site is similarity as close to the study site as possible except for the presence of the contaminant of concern.

Acknowledgment

The work described in this chapter was funded in part by the U.S. Environmental Protection Agency. This chapter has not been subject to agency review and thus does not necessarily reflect the views of the agency. Therefore, no official endorsement should be inferred.

References

1. Shearer, S. D. In *Quality Assurance Practices for Health Laboratories*; Inhorn, S. L., Ed.; American Public Health Association: Washington, DC, 1978; pp 367–377.
2. *Upgrading Environmental Radiation Data*; Office of Radiation Programs. U.S. Environmental Protection Agency: Washington, DC, 1980; Chapter 5; EPA 520/1–80–012.
3. "Principles of Environmental Analysis"; American Chemical Society Committee on Environmental Improvement. *Anal. Chem.* **1983**, *55*, 2210–2218.
4. Brown, K. W.; Beckert, W. F.; Black, S. C.; Flatman, G. T.; Mullins, J. W.; Richitt, E. P.; Simon, S. J. *The Dallas Lead Monitoring Study: EMSL-LV Contribution*; U.S. Environmental Protection Agency. Environmental Monitoring Systems Laboratory: Las Vegas, NV, 1983; EPA–600/X–83–007.
5. *Environmental Monitoring at Love Canal*; Office of Research and Development. U.S. Environmental Protection Agency: Washington, DC, 1982; Vol. 1; EPA–600/4–82–030a.
6. Gills, T. E.; Hutchinson, J. M. R.; Inn, K. G. W. *Proc. 32nd Annu. Conf. Bioassay, Anal. Environ. Chem.* **1986**, 8-1.
7. U.S. Environmental Protection Agency. 40 CFR 165, Subpart F, 1983; pp 506–510.
8. Rogers, E. M. *Pathway to a Healthier Tomorrow (Understanding Epidemiology)*; Pamphlet, Health & Environmental Sciences; Dow Chemical: Midland, MI, 1980.
9. Salaymeh, S.; Kuroda, P. K. *Proc. 32nd Annu. Conf. Bioassay, Anal. Environ. Chem.* **1986**, 12-1.

RECEIVED for review January 28, 1987. ACCEPTED May 18, 1987.

Chapter 8

Assessing and Controlling
Sample Contamination

David L. Lewis

*Most environmental sampling and analytical applications offer numerous opportuni-
ties for sample contamination. For this reason, contamination is a common source of
error in environmental measurements. This chapter addresses the problem of assessing
and controlling sample contamination and the resulting measurement error.
Vulnerable points in the sample collection and analysis process as well as common
sources of contamination are discussed. The different possible effects of contamination
are also examined. Blanks are recommended as the most effective tools for assessing
and controlling contamination. Different types of blanks and their respective uses and
limitations are described. The applicability of control charts to blank measurements
and their use in assessing and controlling contamination are also discussed.*

CONTAMINATION IS A COMMON SOURCE OF ERROR in all types of environmental measurements. Most sampling and analytical schemes present
numerous opportunities for sample contamination from a variety of
sources. This chapter addresses the problem of assessing and controlling sample
contamination and the resulting measurement error. The first part of the
discussion examines the different points in the sample collection and analysis
process at which contamination is likely to occur and identifies common
sources of contamination for various measurement applications. The next
portion deals with the different possible effects of contamination. The last part
of the discussion examines the use of blanks to assess and control contamination. Different types of blanks and their respective uses are described. The
applicability of control charts to blank measurements is also discussed.

1173–6/88/0119$07.50/0 © 1988 American Chemical Society

Sources of Contamination

From an environmental sampling and analytical standpoint, *contamination* is generally understood to mean something that is inadvertently added to the sample during the sampling and analytical process. Although subsequent measurements may accurately reflect what was in the sample at the time the measurements were made, they do not give an accurate representation of the measured characteristic of the media from which the sample was taken.

Sample contamination may arise from myriad sources. To control contamination associated with a particular determination, potential sources of contamination must first be identified for the methods employed. Typically, an environmental sample may be contaminated at any of numerous points in the sample collection and analysis process. Contamination may be introduced in the field during sample collection, handling, storage, or transport to the analytical laboratory. After arrival at the laboratory, additional opportunities for contamination arise during storage, in the preparation and handling process, and in the analytical process itself. Common sources of sample contamination are summarized in Table I and discussed in greater detail in this section.

Equipment used for sample collection is a common route for introducing sample contamination in many types of environmental measurements. Sampling devices may be made of materials that contribute to sample contamination, or cross-contamination may occur as a result of improper cleaning of sampling equipment. Fetter (*1*), for example, described a variety of equipment-related sources of contamination encountered in groundwater sampling. He noted that

Table I. Potential Sources of Sample Contamination

Critical Steps in the Sampling and Analytical Process	Contamination Sources
Sample collection	Equipment and apparatus
	Handling (e.g., filtration, compositing, and aliquot taking)
	Preservatives
	Ambient contamination
	Sample containers
Sample transport and storage	Sample containers
	Cross-contamination from other samples or reagents
	Sample handling
Sample preparation	Glassware
	Reagents
	Ambient contamination
	Sample handling
Sample analysis	Syringes used for sample injections
	Carry-over and memory effects
	Glassware, equipment, and apparatus
	Reagents (e.g., carrier gases and eluents)

the first opportunity for contamination in subsurface investigations and in the installation of monitoring wells is during the process of drilling a borehole. He recommended steam cleaning of drilling rigs as a minimum measure to prevent introduction of contaminants such as gasoline, diesel fuel, hydraulic fluid, lubricating oils and greases, paint, and soil and scale from previous drilling operations. Organic-polymer-based drilling additives were cited as sources of chemical oxygen demand and total organic carbon contamination. Well-casing materials were also cited as potential sources of contaminants. These statements were supported by the work of Boettner et al. (2), which indicated that organic and organotin compounds may be leached from poly(vinyl chloride) and chlorinated poly(vinyl chloride) pipe and pipe cement.

Contamination and cross-contamination from sampling equipment is equally a problem in other types of environmental sampling. Ross (3) reported significant trace metal contamination of atmospheric precipitation samples collected in conventional polyethylene sample collectors washed only with deionized water. The highest instances of contamination were for manganese, cadmium, copper, and zinc. Rigorous acid washing of the sample collectors was shown to significantly reduce sample contamination. Collectors washed only in deionized water showed concentrations of individual elements as much as 50 times higher than corresponding acid-washed collectors. As another example, other studies (4–6) indicated that polytetrafluoroethylene (Teflon) and poly(vinyl fluoride) (Tedlar) bags can contribute significant amounts of hydrocarbon contamination to air samples collected in them.

Sample handling in the field is another potential source of sample contamination. For example, aqueous samples collected for soluble metals analysis are typically filtered and acidified in the field immediately after collection. Best et al. (7) reported significant nitrate contamination of surface water samples during field filtration. Filtration was performed with membrane filters and stainless steel filter holders that were rinsed with 5% nitric acid to minimize trace metal contaminants. Although the nitric acid rinse was followed by copious rinsing with deionized water, nitrate contamination as high as 3.5 mg/L was observed.

Acids and other chemical preservatives that may become contaminated after a period of use in the field offer another route of sample contamination during field handling (8). Sample exposure to the ambient environment should also be minimized to avoid airborne contaminants. Contamination from dry deposition of gaseous SO_2 and particulate sulfate has been cited as a source of bias in rainfall chemistry data generated during the period 1950–1980, which was prior to widespread use of collectors that open only during rain events (9). Lead and aluminum are also common airborne contaminants, especially in urban environments. Sample handling in industrial environments, often necessary during emissions testing and source characterization studies, presents even greater risks of contamination because of the considerably higher ambient levels of potential contaminants typically encountered.

Sample containers represent another major source of sample contamination. Plastic sample containers, for example, are widely recognized as a potential source of sample contamination in trace metal analyses. Moody and Lindstrom (10) examined 12 different plastic materials including conventional and linear polyethylene, polycarbonate, and several types of Teflon. They found significant levels of leachable trace elements in all of the materials examined. Linear polyethylene and the various Teflons were found to be the least contaminating bottles after cleaning. On the basis of their leaching studies, they recommended a cleaning procedure involving the use of both HCl and HNO_3, one after the other, for most trace element work.

Glass sample containers are typically used for collection of solids, sludges, and liquid wastes for organic analysis. Detergent or acid washing, followed by organic-free water rinsing and heating in an oven or muffle furnace, is recommended to minimize trace organic contamination from glass containers.

In selecting cleaning procedures for sampling equipment and sample containers, it is important to consider all of the parameters of interest. Although a given cleaning procedure may be effective for one parameter or type of analysis, it may be ineffective for another. When multiple determinations are performed on a single sample or on subsamples from a single container, a cleaning procedure may actually be a source of contamination for some analytes while minimizing contamination for others. This paradox may be caused by release of contaminants through cleaning procedures that are too rigorous or by contaminants introduced by the cleaning agent. For example, Bonoff et al. (11) reported contamination of water samples by zinc released from disposable filter cassettes following washing with 10% Ultrex nitric acid. Best et al. (7), in addition to reporting nitrate contamination introduced during field filtration, reported similar contamination resulting from the cleaning procedure used for the sample containers. During a pilot study preceding their investigation, high-density polyethylene sample bottles were cleaned by using a procedure that included a nitric acid rinse to minimize trace metal contaminants. Although the procedure used included repeated rinsings with deionized water following the acid rinse step, these rinsings were ineffective at reducing residual nitrate to acceptable levels. On the basis of the pilot study results, the cleaning procedure for field sample bottles was modified prior to subsequent field studies, and the nitric acid rinse was eliminated. Similar contamination problems have been observed in chromium analyses where chromic acid solutions were used to clean glassware and in phosphate analyses where phosphate-containing detergents were used.

Whereas sample containers represent one source of contamination during sample storage and transport, proximal storage of high- and low-level samples or samples and reagents has been identified as a potential additional source of contamination, particularly for volatile organic compounds (12). Levine et al. (13) confirmed that water samples could be contaminated by diffusion of volatile organic compounds through Teflon-lined silicone cap liners during sample shipment and storage. Although they found insignificant cross-contamination caused by storage of vials of organic-free water with saturated aqueous solutions of selected halocarbons, significant potential for contamination was observed for proximal storage of clean samples and neat reagents.

Numerous additional chances for sample contamination arise in the laboratory during sample preparation, handling, and analysis. Virtually every analytical technique presents its own special set of opportunities for sample contamination. Complex sample preparation techniques, such as those involving extraction and concentration of the analytes of interest, present proportionately more such opportunities than do simpler techniques involving fewer steps and less sample handling. Even simple, single-step preparatory techniques, however, often present opportunities for sample contamination. Bagchi and Haddad (14), for example, determined that precolumn cartridges and filter units commonly used for sample cleanup in ion chromatography can be a significant source of parts-per-billion contamination. They examined a popular, single-use cleanup cartridge and two types of filtration units designed to fit on a sample injection syringe. Their results indicated that both the cleanup cartridges and the filtration units were potential sources of sample contamination in inorganic ion analyses. Observed contaminants included lead, zinc, fluoride, chloride, nitrate, and sulfate ions.

Glassware and reagents are common sources of laboratory contamination in all types of analyses. Carry-over and memory effects from consecutive analyses of high- and low-level samples are also common to many types of instrumental methods, including gas chromatography, liquid chromatography, and many spectroscopic methods. Syringes and other devices used for sample injection are also common sources of contaminants. Sommerfeld et al. (15), for example, identified disposable plastic pipette tips as a source of iron and zinc contamination in analyses by graphite furnace atomic absorption spectroscopy. Even when the tips were acid rinsed to remove contamination, they found that the pipette tips could be recontaminated by contact with the aperture of the cool graphite tube.

Effects of Contamination

Chemical or physical properties of samples that cause errors in the measurement process are commonly known as *interferences*. Generally, two types of interferences are recognized: additive interferences and multiplicative interferences (16). *Additive interferences* are caused by sample constituents that generate a signal that adds to the analyte signal. Because they cause a change in the intercept but not the slope of the calibration curve, additive interferences have the most pronounced effect at low analyte concentrations. *Multiplicative interferences*, on the other hand, are caused by sample constituents that either increase or decrease the analyte signal by some factor without generating a signal of their own. Multiplicative interferences change the slope of the calibration curve but not the intercept.

Contamination is generally understood to mean something inadvertently added to a sample that leads to an erroneously *high* measured value. This definition, although generally implied, is not strictly correct unless we consider only contamination by the analytes of interest or by contaminants which the method employed cannot distinguish from the analytes of interest. This distinction is important because some types of contamination do not fit this general definition but instead lead to erroneously *low* measured values. These negative interferences may be multiplicative or they may be opposite in effect to additive interferences, and therefore change only the intercept of the response curve.

Multiplicative interferences are a common source of analytical error in many spectroscopic techniques, although matrix effects are a more common source of such error than contamination. Contaminants may, however, cause multiplicative interferences through adsorptive losses of the analyte of interest. These contaminants give erroneously low results. Adsorption acts as a multiplicative interference when a constant fraction of the analyte is adsorbed, regardless of analyte concentration (i.e., when relative bias is constant). When the amount of analyte is large compared to the available sites for adsorption to occur, the amount of analyte lost to adsorption tends to be constant, and relative bias decreases with increasing concentration. In such cases, adsorption causes a negative interference, opposite in effect to an additive interference. Bentonite drilling muds, for example, can absorb heavy metals in groundwater and lead to low measurements (1). Sample containers, although not usually considered contaminants in the normal sense, may also adsorb analytes from the sample and act as a negative interference.

Dilution is another example of an interference that causes the measured values to be erroneously low rather than high. This example is the special case in which the sample is contaminated by the solvent. Dilution occurs primarily in certain types of environmental sampling. Groundwater monitoring, in which monitoring wells are installed for the purpose of collecting groundwater samples, presents several opportunities for sample contamination by dilution. These opportunities may occur through addition of clean water to the well during well development, through improper well placement, through improper screening, or through cross-contamination between aquifers located at different depths (1). Many types of air sampling also present significant opportunities for sample dilution due to leaks in negative-pressure sampling systems allowing air leakage.

Regardless of the source or sources of sample contamination, the net effect is added inaccuracy in the measurement process. Like other types of measurement error, error due to contamination may be sporadic and represent special causes, or systematic and affect all measurements. Cross-contamination, such as that which often occurs during analysis when carry-over from high-level samples contaminates subsequent low-level samples, is a common source of sporadic contamination. Similarly, careless sample handling and dirty sampling equipment are often sources of sporadic contamination. Sporadic contamination most often affects the measurement process by introducing false positive results. A *false positive* is the error of concluding that an analyte is present in the media sampled when it is not. In the case of sporadic contamination where the contaminant acts as a negative interference, as in dilution or adsorption, false negatives may result. A *false negative* is the error of concluding that an analyte is not present when it is.

Contamination is a source of systematic error when the level of contamination is stable for all samples. Strictly speaking, however, stable, systematic error due to sample contamination is rare. Almost always, some element of sporadic error is associated with any source of contamination. In some cases though, the effect of this sporadic error component is small in comparison to the systematic error, or bias component. Thus, some types of contamination behave in a fashion that is primarily systematic. Systematic contamination increases the "background concentration" of the analyte of interest and thus affects the lower limit of the measurement process. Contaminated reagents are a common source of systematic contamination in many types of environmental measurements. Contaminated sample containers are another source of contamination that is often primarily systematic.

Use of Blanks To Assess and Control Contamination

The most commonly used analytical tools for assessing and controlling sample contamination are blanks. By conventional nomenclature, *blanks* are samples that do not intentionally contain the analyte of interest but in other respects have, as far as possible, the same composition as the actual samples (17). Additional descriptors, such as internal, reagent, field, solvent, and others, are used to indicate which of the various stages of the sampling and analytical process the blanks are considered to represent. Because blanks, by definition, do not intentionally contain the analyte of interest, their utility in assessing and controlling sample contamination is limited to contaminants causing additive interferences. In this regard, results for blanks are taken as a direct measure of the nonanalyte, or contaminant, signal for the corresponding samples.

Types of Blanks

Blanks play various roles in environmental measurements, depending on the analytical technique used and the goal of the blank measurements. Table II summarizes the types of blanks typically used in environmental measurements. The simplest blank, often called a *system blank* or *instrument blank*, is really not a blank at all in the sense of simulating a sample. Rather, a system blank is a

Table II. Summary of Blank Types

Common Name	Other Names	Uses	Description
		Laboratory blanks	
System blank	Instrument blank	To establish baseline response of an analytical system in the absence of a sample	Not a simulated sample but a measure of instrument or system background response
Solvent blank	Calibration blank	To detect and quantitate solvent impurities; the calibration standard corresponds to zero analyte concentration	Consists only of the solvent used to dilute the sample
Reagent blank	Method blank	To detect and quantitate contamination introduced during sample preparation and analysis	Contains all reagents used in sample preparation and analysis and is carried through the complete analytical procedure
		Field blanks	
Matched-matrix blank		To detect and quantitate contamination introduced during sample collection, handling, storage, transport, preparation, and analysis	Made to simulate the sample matrix and carried through the entire sample collection, handling, and analysis process
Sampling media blank	Trip blank	To detect contamination associated with sampling media such as filters, traps, and sample bottles	Consists of the sampling media used for sample collection
Equipment blank		To determine types of contaminants that may have been introduced through contact with sampling equipment; also to verify the effectiveness of cleaning procedures	Prepared by collecting water or solvents used to rinse sampling equipment

measure of the instrument background, or baseline, response in the absence of a sample. System blanks are often used in gas and liquid chromatographic methods to identify memory effects, or carry-over from high-concentration samples, or as a preliminary check for system contamination.

Solvent blanks are generally the next simplest type of blank and consist only of the solvent used to dilute the sample. Solvent blanks are used to identify or correct for signals produced by the solvent or by impurities in the solvent. Depending upon the analytical technique, the solvent blank may be used as a calibration blank. A calibration blank is used directly to set the instrument response to zero, or is used as one of a series of calibration standards, where the blank represents an analyte concentration of zero.

Another type of laboratory blank is the *reagent blank*. In addition to the solvent, the reagent blank contains any reagents used in the sample preparation and analysis procedure. These reagents may include color development reagents, reagents used in sample digestion steps, reagents used for pH adjustment, preservatives, or other reagents depending upon the analytical method. In methods where the solvent is the only reagent used in the sample preparation process, as in ion chromatography, for example, the composition of the reagent blank will be the same as that of the solvent blank. The distinguishing characteristic of the reagent blank, however, is that the reagent blank is carried through the complete analytical procedure in the same manner as an actual sample. This procedure should include all steps involved in sample preparation, such as cleanup, filtration, extraction, and concentration. The reagent blank thus provides a measure of contamination that may be introduced during sample preparation and analysis, whether from the reagents themselves, from glassware, or from other sources in the laboratory environment. Because it is carried through the complete analytical method, the reagent blank is also sometimes called a *method blank*.

Whereas laboratory blanks are reliable tools for assessing and controlling many types of laboratory contamination, they obviously address only a part of the overall measurement process. *Field blanks* must be used to provide information about contaminants that may be introduced during sample collection, storage, and transport. Like laboratory blanks, a number of different types of field blanks are available. The most common type of field blank, where the blank simulates the sample matrix, is sometimes called a *matched-matrix field blank*. This type of field blank is widely used in sampling involving aqueous matrices because deionized or distilled water is readily available and water blanks are easy to prepare.

Like reagent blanks, the distinguishing characteristic of matched-matrix field blanks is that they are carried through the entire sample collection and handling process so that the blank is exposed to the same potential sources of contamination as actual samples. Although exact duplication of the sample collection process for the blank sample may not always be feasible, the blank should be exposed to as many elements of the process as possible. In sampling groundwater monitoring wells, for example, preparing a blank that is exposed to the well casing and other potential contaminants of the well environment is generally not feasible. The blank sample can, however, be exposed to other elements of the collection process, such as bailers or peristaltic pumps used to obtain water samples from the well. Blanks should, of course, be treated in the same manner as other samples with regard to sample containers, field preservation, handling, and storage.

Although matched-matrix field blanks are most easily prepared for aqueous samples, they are usually prepared without too much difficulty in most air sampling applications as well. One method of preparing such blanks is by collecting samples of "clean" (e.g., breathing air or hydrocarbon-free air, as appropriate) compressed air with the normal sampling apparatus. Alternately, ambient air may be used by equipping the sampling system with an upstream purification device to remove the analytes of interest. An activated carbon canister, for example, may be used to remove trace organic compounds from ambient air to prepare air blanks for organic analyses.

Solid sampling applications typically present a greater challenge than air or water samples in preparing matched-matrix field blanks. If the media sampled is soil or soillike material, as might be encountered in characterization or remediation activities at a waste site, reagent-grade sand (i.e., silica) may be used to approximate the sample matrix. If the analytes of interest include organic compounds, the sand should be baked in an oven at 300 °C for several hours to remove any volatile species that may be present. As an alternate approach, a bulk sample may be collected from an adjacent area known to be free of the analytes of interest and used to prepare blank samples. Although this approach often provides a matrix more similar to that of the samples, the difficulty of ensuring that the blank samples are indeed free of target analytes is added. This difficulty is especially important in samples collected for metals analysis because most soils have significant naturally occurring background levels of numerous elements (*18*).

If the media sampled include solid or liquid waste materials, sludges, or slurries, a rigorous matched-matrix field blank may not be achievable. In such cases, a rough approximation is usually sufficient to identify most sources of contamination. Field blanks for these applications may be prepared by using uncontaminated soil or reagent sand, as described previously, or by preparing aqueous or solvent mixtures of soil or sand.

Another type of blank, similar in some cases to the matched-matrix blank, is the *sampling media blank*. This blank consists of the sampling media used for collection of field samples. This type of blank is used primarily in air sampling applications where filters or various types of solid adsorbents are often used in sample collection. Sampling media blanks typically involve no special preparation because the blanks are simply selected at random from the filters or traps available for sampling purposes. The blank sample media are handled and exposed to ambient conditions in the same manner as sample media used for sample collection, but no air is passed through them. After handling, sampling media blanks should be sealed in the normal manner and stored along with the other samples.

Equipment blanks are a special type of field blank used primarily as a qualitative check for contamination rather than as a quantitative measure. Equipment blanks are prepared by collecting water or solvents used to rinse sampling equipment prior to sampling. Analysis of this rinse solution then provides an indication of the types of contaminants that may have been introduced through contact with the sampling equipment. Equipment blanks may also be used to verify the effectiveness of equipment cleaning procedures and as a check for potential cross-contamination.

Nomenclature associated with blank samples is far from consistent in the literature, and distinguishing one type of blank from another is sometimes difficult except by context or by a more detailed description. Alternate names for the most common blanks are listed in Table II. Two terms frequently used ambiguously are *trip blank* and *field blank*. For example, sampling media blanks are also sometimes called trip blanks because they are typically prepared in the laboratory and act as a check for contaminants introduced during the trip from the laboratory to the field and back again. Other types of field blanks are also sometimes referred to as trip blanks. Matched-matrix blanks prepared in the laboratory and then taken to the field, for example, are sometimes called trip blanks. The term field blank is typically used in more general terms to describe any type of blank used to assess field contamination.

Regardless of the application or the type of blank used, researchers should recognize and minimize the potential for inadvertently introducing contamination during preparation of the blank or at any other point where the actual samples are not exposed to similar opportunities for contamination.

Use of Blank Results

When properly used, blanks can be extremely effective tools in assessing and controlling sample contamination and in adjusting measurement results to compensate for the effects of contamination. Used improperly, blank results can increase the variability of analytical data or be very misleading. An important part of using blanks effectively is understanding and recognizing their limitations. As mentioned previously, blanks are useful for detecting contaminants causing additive interferences but are ineffective in identifying interferences such as dilution or adsorption. Similarly, blanks cannot be used to spot noncontaminant error sources such as analyte losses due to volatilization or decomposition. Beyond these inherent limitations, the utility of blanks is determined largely by the manner in which they are used and the manner in which the results are interpreted.

Blanks serve both control and assessment functions in environmental measurements. In their control function, blanks are used to initiate corrective action when blank values above preestablished levels indicate the presence of contamination. Blanks are most often used in this control mode in laboratory operations where feedback is more nearly real-time. At the first sign of unusual contamination, analyses may be stopped until the source is identified and the contamination eliminated. If possible, affected samples may then be reanalyzed. When field blanks indicate possible contamination, resampling is usually more difficult and often impossible. Therefore, field blank data are generally used primarily for assessment rather than control. If field blank data are used for control, this control is generally accomplished only over relatively long periods of time. In their assessment role, both field and laboratory blank data may be used to define qualitative and quantitative limitations of the associated measurement data. Where appropriate, these blank data may also be used as a basis for adjusting data to compensate for background contamination. Any such adjustments, however, should be made with caution, and the average of multiple blank measurements should be used for a stable, "in control" measurement system.

Control Charts for Blanks

Whether blank data are used primarily for ongoing control or for retrospective assessment, Shewhart *control charts* (*19, 20*) provide the most effective mechanism for interpreting blank results. In the control mode, control charts can be used to detect changes in the average background contamination of a stable system. This detection is done by providing definitive limits, based on past performance, that signal when the level of contamination is greater than that which is attributable to chance causes. This signal allows corrective action to be initiated to identify and correct new or additional sources of contamination as they appear, before large numbers of samples are affected. In the assessment mode, control charts allow out-of-control periods to be easily identified so that corresponding sample data may be flagged or interpreted separately from the other data. By identifying out-of-control periods, control charts also allow more reliable estimates to be made of the average background contamination level under normal in-control periods.

Control charts have gained considerable acceptance for some types of quality control checks familiar to environmental chemists, such as recovery data for spiked samples and results for calibration check samples. Control charts have been less frequently applied, however, to results for blanks. One reason for this lack of use is that most references on development and use of control charts emphasize $\bar{X}$ and R charts, which are control charts for means and ranges, respectively. Because blanks are not commonly run in replicate, control charts for means and ranges are not usually applicable. Another problem in using control charts for blanks is dealing with "zero", or "not detected" results. This is primarily a problem in deriving the variability estimates necessary for developing control charts.

Dealing with values of zero or not detected results is easily overcome for many measurements simply by changing the reporting convention. In most spectroscopic methods, for example, even a zero analyte concentration produces a signal from which a concentration may be calculated by using the calibration function. In such cases, the calculated concentrations should be reported as calculated, even if the calculation yields a "negative concentration". In methods where a zero analyte concentration produces no measurable signal, such as in gas chromatography, where no peak is produced for integration, initial variability estimates may be developed by using results for low-level standards as discussed in the next section.

Control Charts for Individual Measurements

The problem of blanks not usually being run in replicate can be overcome by using a control chart for individual measurements. This special type of control chart is useful when no rational subgrouping scheme arises, when performance measures can only be obtained infrequently, or when the variation at any one time (within a subgroup) is insignificant relative to variation over time (between subgroups).

Although they share the same statistical basis, control charts for individuals (X charts) are different from control charts for means ($\overline{X}$ charts) and ranges (R charts) in the way the range is calculated and in the subgrouping scheme. For these reasons, individual control charts are interpreted somewhat differently than usual. In $\overline{X}$ charts, the chart reflects variability between subgroups (i.e., between means); in R charts, the chart is used to monitor variability within subgroups. In control charts for individuals, however, the range within a subgroup cannot be calculated because the subgroup size is one. Also, because individual measurements are plotted, a single chart combines all sources of variation.

The first step in preparation of an X chart for blanks is to tabulate historical data for blank measurements. This tabulation will consist of at least 20 individual results for the particular type of blank to be charted. After arranging the k results in chronological order, $k-1$ moving ranges are calculated, where the first moving range is the range between the first and second values, the second moving range is the range between the second and third values, etc. Next, the average of the moving ranges ($\overline{MR}$) is calculated, along with the average of the k measurements (X_{avg}). Before calculating the control limits for the individual values, the moving ranges are screened by first calculating the upper control limit for the moving ranges as $3.27\overline{MR}$. [The value 3.27 is the D_4 value for calculating control limits for ranges having $n = 2$, where n is the number of measurements in each subgroup (20).] Any moving ranges larger than the calculated control limit are removed, and then the average moving range is recalculated. Finally, the upper and lower control limits (UCL and LCL, respectively) for the individual values are calculated as

$$\text{UCL} = \bar{X} + 2.66\overline{\text{MR}} \tag{1}$$

$$\text{LCL} = \bar{X} - 2.66\overline{\text{MR}} \tag{2}$$

Although not tabulated in many tables of control chart factors, the value 2.66 used to calculate the control limits is the A_2 factor for calculating control limits for $\bar{X}$, where $n = 1$.

Unless the average blank value is substantially greater than zero, the LCL may be negative and thus will not be meaningful. Only the UCL can be used in these cases.

In traditional $\bar{X}$ charts, the underlying assumption is that variability within a subgroup is representative of the system variability. Control limits for $\bar{X}$ are thus derived by using the within-subgroup range to estimate the standard deviation from which the control limits are calculated. In individual control charts, the moving range between subgroups (i.e., between the individual points) is used to estimate the standard deviation. Because pairs of consecutive measurements are more likely to be affected by similar special causes than are results from different points in time, screening the moving ranges prior to calculating the control limits minimizes the contribution of these special causes. This screening prevents the control limits from being inflated by these special causes as would be the case if the standard deviation was calculated by using all the original data points.

In using this approach, the problem still arises of dealing with zero and not detected values in the blank data from which the control limits are to be calculated. In this case, the average moving range must be estimated by using alternate data. Results for low-concentration standard solutions provide the best substitute. Obviously, if the blanks of interest are, for example, matched-matrix field blanks, standards should be prepared in a similar manner by spiking the appropriate matrix with the analyte of interest. In either case, the concentration of the standard should be in the same range as the estimated detection limit (i.e., between 1 and 5 times the estimated detection limit). At this level, imprecision should be of approximately the same magnitude as that for blanks. The actual control limits are calculated by using the average moving range for the standards and the mean blank value for similar blanks.

One limitation that should be considered in using an X chart is the increased sensitivity of the limits to the distribution of the measurements. $\bar{X}$ charts are less sensitive than X charts to the distribution of individual measurements because mean values are used. Means tend to be normally distributed even when the individual values are from populations that are not normally distributed. Because contamination tends to induce positive errors in blank measurements, the resulting distribution is likely to be skewed toward positive values. If inspection of the results indicates positively skewed values, the measurement data should be transformed prior to developing the control chart, and the transformed data should be charted. A logarithmic transformation is generally most appropriate for environmental data.

Example of an X Chart for Blanks

As an example of the development and application of an X chart for blanks, consider the data in Table III. For the purpose of the example, let these data represent blank results for aqueous nitrate measurements. Assume that the results for blanks 1–20 represent historical data used to develop the example control chart, and the results for blanks 21–64 represent subsequent blank measurements. The completed control chart is shown in Figure 1.

The first step in developing the example control chart was to examine the distribution of the historical data in Table III. As shown in the frequency histogram in Figure 2, the raw results are significantly skewed as is often the case for blank data. Therefore, before proceeding further, the raw data were transformed by taking the natural logarithm of each value. This produced the transformed results listed in the third column of Table III. Figure 3 is a frequency histogram of the transformed data that shows significant improvement in the skewness of the distribution.

The next step in developing the example control chart was to calculate 19 moving ranges for the 20 chronologically ordered transformed results. These moving ranges are listed in the fourth column of Table III. The UCL for the moving range, 4.697, was obtained by multiplying the average moving range, 1.436, by 3.27. This moving range control limit was then used to screen the moving ranges prior to calculating the control limits for the blank measurements. Screening the moving ranges and removing any values exceeding the control limit prevent the control limits for the blank measurements from being inflated by values representing special causes. As indicated in Table III, the 10th moving range (blank number 11), 5.200, exceeds the moving range control limit. Therefore, this value was removed, and the average moving range was recalculated to yield a value of 1.300. Finally, upper and lower control limits for the blank measurements were calculated as the average of the transformed results (−4.273) plus and minus 2.66 times the average of the screened moving ranges (1.300), or −0.816 and −7.731, respectively.

The completed control chart for this example, shown in Figure 1, illustrates how control charts for blanks are effective tools both for ongoing control and for retrospective assessment of blank results. In the control mode, for example, out-of-control points like that for the 11th blank indicate unusual contamination from an assignable cause and should initiate corrective action to identify and eliminate the source of additional contamination. In the case of field blanks, analyzing the samples and plotting the results may not be possible until after all of the samples are collected. In such cases, control charts are still useful in

Table III. Nitrate Blank Results

Blank Number	Nitrate Concentration	ln Nitrate Concentration	Moving Range
\multicolumn{4}{c}{Historical Nitrate Blank Results}			
1	0.033	−3.411	
2	0.049	−3.016	0.395
3	0.002	−6.215	3.199
4	0.002	−6.215	0.000
5	0.008	−4.828	1.386
6	0.002	−6.215	1.386
7	0.014	−4.269	1.9468
8	0.016	−4.135	0.134
9	0.009	−4.711	0.575
10	0.009	−4.711	0.000
11	1.631	0.489	5.200[a]
12	0.063	−2.765	1.946
13	0.042	−3.170	0.405
14	0.022	−3.817	0.647
15	0.093	−2.375	1.442
16	0.022	−3.817	1.442
17	0.002	−6.215	2.398
18	0.002	−6.215	0.000
19	0.031	−3.474	2.741
20	0.004	−5.521	2.048
Average ln nitrate concentration		−4.273	
Average moving range			1.436
Moving range UCL			4.697
Average screened moving range			1.300
UCL			−0.816
LCL			−7.731
\multicolumn{4}{c}{Subsequent Nitrate Blank Results}			
21	0.006	−5.116	
22	0.028	−3.675	
23	0.021	−3.863	
24	0.020	−3.912	
25	0.031	−3.474	
26	0.002	−6.215	
27	0.002	−6.215	
28	0.026	−3.650	
29	0.040	−3.219	
30	0.726	−0.320	
31	1.111	0.105	
32	0.081	−2.513	
33	0.089	−2.419	
34	0.128	−2.056	
35	0.053	−2.937	
36	0.190	−1.661	
37	0.353	−1.041	
38	0.353	−1.041	
39	0.389	0.489	
40	0.389	−0.944	

Table III.—Continued

Blank Number	Nitrate Concentration	ln Nitrate Concentration	Moving Range

Blank Number	Nitrate Concentration	ln Nitrate Concentration	Moving Range
	Subsequent Nitrate Blank Results—Continued		
41	0.066	−2.718	
42	0.731	−0.313	
43	0.283	−1.262	
44	0.277	−1.284	
45	0.213	−1.546	
46	0.452	−0.794	
47	0.288	−1.245	
48	1.668	0.512	
49	0.051	−2.976	
50	0.056	−2.882	
51	0.253	−1.374	
52	0.054	−2.919	
53	0.097	−2.333	
54	0.672	−0.397	
55	0.221	−1.510	
56	0.206	−1.580	
57	0.293	−1.228	
58	0.128	−2.056	
59	0.431	−0.842	
60	0.180	−1.715	
61	0.108	−2.226	
62	0.052	−2.957	
63	3.586	1.277	
64	0.216	−1.532	

[a]This value exceeds the UCL for the moving range.

assessing the blank data by indicating both sporadic and systematic contamination problems and allowing the corresponding measurement data to be interpreted accordingly. The example control chart in Figure 1, for instance, shows a significant shift in background contamination during the course of the hypothetical sampling and analytical effort. Such a shift might be the result of a change in sampling or analytical procedures, a change in personnel, a new lot of sample bottles, or any one of a number of other possibilities. Identification of these types of changes in background contamination allows field sample data to be grouped and interpreted separately even if it is already too late to eliminate the new source of contamination.

Other Types of Control Charts for Blanks

X charts, like the example chart in Figure 1, are similar to $\bar{X}$ and R charts in that they are charts for individual quality characteristics. Applied to analyses of blank samples, the quality characteristic of interest is analyte concentration in the blank. Although such charts are powerful tools in detection and diagnosis

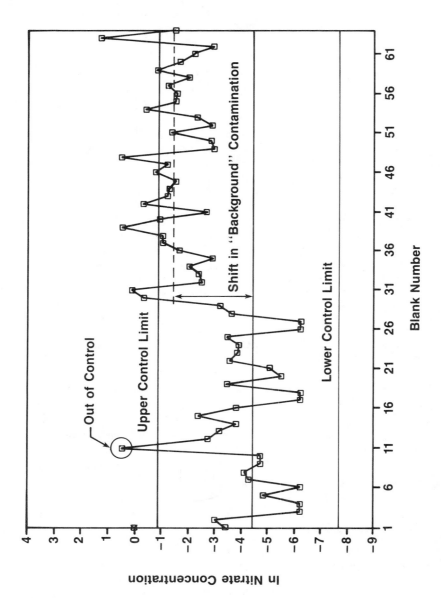

Figure 1. Example of an X chart for blank measurements.

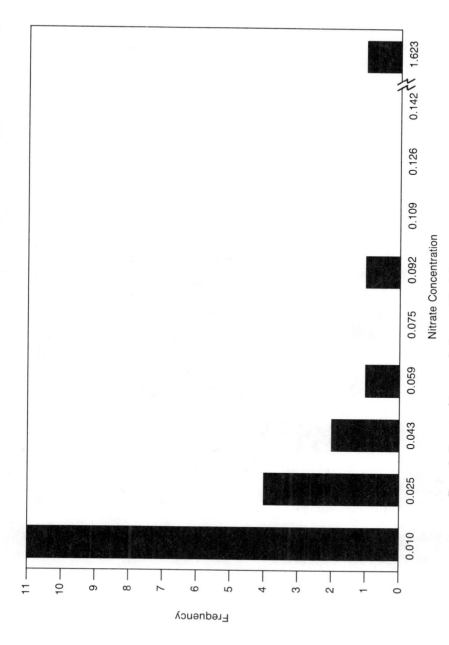

Figure 2. Frequency histogram for historical blank data prior to transformation.

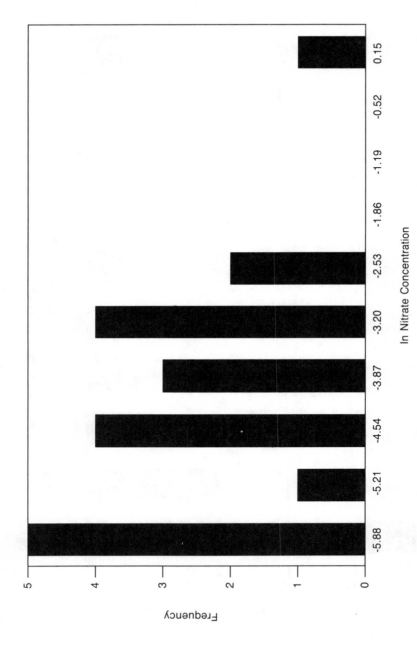

Figure 3. Frequency histogram for historical blank data after logarithmic transformation.

of quality problems in many applications, including measurement error caused by sample contamination, a separate chart is required for each quality characteristic monitored. This requirement presents a practical limitation to their applicability to measurement methods involving multiple analytes. Consider, for example, the U.S. Environmental Protection Agency's Method 624 for purgeable organic compounds. Maintaining separate control charts for each of the 31 target analytes that can be determined by this method would involve much work. In such cases, it may be worthwhile to focus primarily upon controlling contamination and giving up quantitative information about individual parameters in exchange for added ease of use.

If the user is willing to sacrifice information about average blank concentrations of individual analytes, one alternate approach that may be used involves redefining the quality characteristic of interest. For example, rather than charting the concentration of individual analytes, the total blank concentration may be used as a single measure of contamination. This single measurement may be calculated by summing individual concentrations or by summing the total signal and multiplying it by an average response factor. Although this approach permits the user to identify out-of-control periods, information is not provided for adjusting analytical data to compensate for background contamination.

Another approach, which may be appropriate when the primary emphasis is upon controlling rather than assessing contamination, is to use a control chart for nonconformities, or a c chart. Whereas X, $\bar{X}$, and R charts are control charts for variables, a *c chart* is a control chart for attributes. Rather than charting measured values of some specified quality characteristic, such as analyte concentration, c charts involve charting the number of nonconformities. Each instance of an article's deviation from specifications is a nonconformity. For blank measurements, each analyte detected in the blank or each analyte detected above a specified concentration would be considered a nonconformity. The total number of such nonconformities in each blank is the value plotted on the c chart. The center line is set at the average number of nonconformities per blank, $\bar{c}$, based on historical data. Three-sigma control limits are set at plus and minus 3 times the square root of $\bar{c}$ (i.e., $\bar{c} \pm 3(\bar{c})^{1/2}$).

Whereas the Gaussian distribution forms the basis for most commonly used control charts, c charts are based upon the *Poisson distribution*. The usefulness of c charts for a particular application thus depends upon the extent to which the Poisson distribution is an appropriate statistical model for that application. One of the required conditions for applicability of the Poisson distribution is that a large number of opportunities exist for nonconformities to occur. For this reason, c charts for blanks will generally be more applicable to methods involving large numbers of analytes than to methods involving only a few. Strict applicability of the Poisson distribution also requires that the nonconformities be independent of one another and have equal opportunities for occurrence. Because of these constraints, the use of c charts for blanks usually represents an approximation to strict theoretical applicability. Even so, the results obtained may be useful for practical purposes. Before adopting c charts for a particular application, a statistical quality control text (e.g., references 20–21) should be consulted for a detailed discussion of their use and limitations.

Assessing the Effectiveness of Control Charts for Blanks

The key role of control charts in controlling contamination is in detecting out-of-control points in an otherwise stable measurement process. Although results inside control limits do not indicate the absence of contamination, they are an indication that the effect of contamination is stable. In such a case, systematic adjustment of the measurement data using the average blank value to correct for the background contamination may be appropriate. Blank results that fall outside control limits provide a signal that some new source of contamination has entered the measurement system. Just as establishing the absence of any given analyte in a sample is analytically impossible, establishing the absence of contamination in a measurement process by analyses of blanks is also impossible. The best that can be achieved is to reduce the risk of not detecting contamination to an acceptable level.

In assessing this level of risk in using blanks to detect changes in contamination, the frequency of the blank measurements, the magnitude of change in the level of contamination that one desires to detect, and the amount of variability in the measurement system must be considered. Using three-sigma limits on the average of n measurements, the probability, P, of not detecting a bias (i.e., a change in level of contamination) of size b when the measurements are normally distributed and have a standard deviation of s (22) is

$$P = \Phi \left[\frac{3 - (n)^{\frac{1}{2}}b}{s} \right] - \Phi \left[\frac{-3 - (n)^{\frac{1}{2}}b}{s} \right] \qquad (3)$$

where Φ is the cumulative distribution function of the standard normal distribution. The probability of detecting a bias of size b in m independent tests (each based on the average of n measurements), or P_D, is

$$P_D = 1 - P^m \qquad (4)$$

where m denotes any power of P.

Consider, for example, a case in which an X chart is used to monitor blank results for a particular analyte. In this case, historical data indicate that the average blank concentration of this analyte is 3 ppb, and the standard deviation is 4 ppb. What is the probability of detecting contamination greater than 10 ppb in a single blank? Because 10 ppb represents an increase of 7 ppb above background, $b = 7$. The subgroup size for X charts is one, so $n = 1$. In this case,

$m = 1$ also because a single measurement represents only a single point on the control chart. Therefore,

$$P = \Phi\left[\frac{3 - (1)^{\frac{1}{2}}(7)}{4}\right] - \Phi\left[\frac{-3 - (1)^{\frac{1}{2}}(7)}{4}\right]$$

$$= \Phi(1.25) - \Phi(-4.75)$$

$$= 0.894 - 0.000$$

$$= 0.894$$

and

$$P_D = 1 - (0.894)$$
$$= 0.106$$

Thus, the probability of detecting an additional 7 ppb of contamination is less than 11% for a single blank analysis. If the additional contamination is from a constant source, the probability of detection improves somewhat with repeated measurements. However, because the standard deviation is relatively large compared to the added contamination, 21 measurements are required to attain a greater than 90% probability of detection. On the other hand, a shift of 20 ppb for the same measurement system would have a greater than 97% probability of being detected in a single measurement. Figure 4 illustrates this relationship by showing probabilities of detecting unusual contamination in a single blank analysis for measurement systems having standard deviations of 2, 4, 6, and 8 ppb. Contamination levels as high as 30 ppb are illustrated.

Control charts are most effective for detecting contamination when measurement variability is small relative to the level of contamination to be detected. Many measurements may be required to detect small shifts in background contamination. Also, in order for such shifts to be reliably detected even through repeated measurements, the additional contamination must be persistent. An assumption in equations 3 and 4 is that the problem persists at the same level until corrected. Although reasonable for many sources of contamination, this model is not applicable in all cases. If contamination occurs sporadically at low levels, then detecting the changes in contamination levels is much more difficult. In such cases, the only reasonable approach is to work on identifying and eliminating the source or sources of contamination.

Conclusions

Environmental sampling and analytical efforts present numerous opportunities for sample contamination from a wide variety of different sources. Regardless of

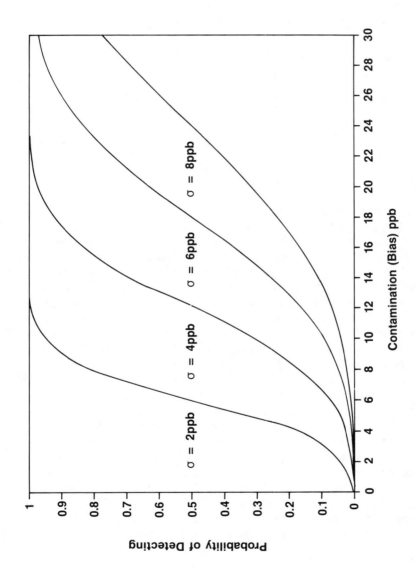

Figure 4. Example of the relationship between measurement variability and probability of detecting unusual contamination (bias) in a single blank measurement.

the source of contamination, the accuracy of the measurement process is affected. Because environmental measurements often address very low concentrations of analytes, contamination is an especially important source of potential error. To minimize error due to contamination, the potential sources of contamination must be identified and eliminated wherever possible.

Once a measurement system is established, appropriate types of blanks should be used to define background levels of contamination for the different parts of the sampling and analytical process. Blanks should also be used on an ongoing basis to assess and control contamination. In the assessment mode, the information provided by blanks may be qualitative or quantitative; blanks may be used as qualitative indicators of possible sample contamination or to derive quantitative estimates of background contamination levels. In the control mode, blanks are used to initiate corrective action when results above preestablished levels indicate unusual contamination.

Whether used primarily for assessment or for control, control charts should be used to maximize the effectiveness of blank measurements. Control charts for individual measurements, or X charts, are usually more appropriate for blanks than the more common $\bar{X}$ and R charts. Other types of control charts may be applicable or preferable under certain circumstances. Regardless of the type of control chart used, the risk of not detecting new sources of contamination in a measurement system depends upon the number of blank measurements, the magnitude of the effect of the new contamination, and the variability of the measurement system. This risk should be a primary consideration in developing the overall quality control strategy. By recognizing potential sources of contamination and using blanks to detect changes in background levels, reducing or correcting for contamination is generally possible, and the associated measurement biases can thus be reduced to acceptable levels.

Abbreviations and Symbols

A_2	factor for determining three-sigma control limits for $\bar{X}$ charts by using the average range, R
b	size of bias
$\bar{c}$	average number of nonconformities per blank
Φ	cumulative distribution function
D_4	factor for determining the upper three-sigma control limit for R charts by using the average range, R
k	number of results
LCL	lower control limit
m	number of independent tests
$\overline{MR}$	average of the moving ranges
P	probability of not detecting a bias
P_D	probability of detecting a bias

s	standard deviation
UCL	upper control limit
X_{avg}	average of k measurements

References

1. Fetter, C. W., Jr. *Ground Water Monit. Rev.* **1983**, *3*, 60–64.
2. Boettner, E. A.; Gall, G. L.; Hollingsworth, Z.; Aquino, R. *Organic and Organotin Compounds Leached from PVC and CPVC Pipe*; U.S. Environmental Protection Agency. U.S. Government Printing Office: Washington, DC, 1981; EPA 600/1-81-062.
3. Ross, H. B. *Atmos. Environ.* **1986**, *20*, 401–405.
4. Cox, R. D.; McDevitt, M. A.; Lee, K. W.; Tannahill, G. K. *Environ. Sci. Technol.* **1982**, *16*, 57–61.
5. Seila, R. L.; Lonneman, W. A.; Meeks, S. A. *J. Environ. Sci. Health* **1976**, *A11*, 121.
6. Kopczynski, S. L.; Lonneman, W. A.; Winfield, T.; Seila, R. *J. Air Pollut. Control Assoc.* **1975**, *25*, 251.
7. Best, M. D.; Drouse, S. K.; Creelman, L. W.; Chaloud, D. T. *National Surface Water Survey, Eastern Lake Survey (Phase I—Synoptic Chemistry) Quality Assurance Report*; U.S. Environmental Protection Agency, in press; EPA 600/4-86/011.
8. *Handbook for Analytical Quality Control in Water and Wastewater Laboratories*; Office of Research and Development. U.S. Environmental Protection Agency: Cincinnati, OH, 1979; EPA 600/4-79-019.
9. Fowler, D.; Cape, J. N. *Atmos. Environ.* **1984**, *18*, 183–189.
10. Moody, J. R.; Lindstrom, R. M. *Anal. Chem.* **1977**, *49*, 2264–2267.
11. Bonoff, M. B.; Filbin, G. J.; Gremillion, P. T.; Kinsman, J.; Mudre, J.; Schriner, C. "Field Operations Report"; Living Lakes, Inc. Aquatic Liming and Fish Restoration Demonstration Program; Living Lakes, Inc.: Washington, DC, 1987.
12. *Fed. Regist.* **1984**, *49*, 29–39; Appendix A to 40 CFR Part 136.
13. Levine, S. P.; Puskar, M. A.; Dymerski, P. P.; Warner, B. J.; Friedman, C. S. *Environ. Sci. Technol.* **1983**, *17*, 125–127.
14. Bagchi, R.; Haddad, P. R. *J. Chromatogr.* **1986**, *351*, 541–547.
15. Sommerfeld, M. R.; Love, T. D.; Olsen, R. D. *At. Absorpt. Newsl.* **1975**, *14*, 31–32.
16. *Trace Analysis: Spectroscopic Methods for Elements*; Winefordner, J. D., Ed.; Wiley: New York, 1976; Chapter 2.
17. Commission on Spectrochemical and Other Optical Procedures for Analysis. *Spectrochim. Acta, Part B* **1978**, *33B*, 248–269.
18. Shacklette, H. T.; Borngen, J. G. *Element Concentrations in Soils and Other Surficial Materials in the Conterminous United States;* U.S. Geological Survey Professional Paper 1270; U.S. Geological Survey: Reston, VA, 1984.
19. Shewhart, W. A. *The Economic Control of Quality of Manufactured Product*; Van Nostrand: New York, 1931.
20. Grant, E. L.; Leavenworth, R. S. *Statistical Quality Control*, 5th ed.; McGraw–Hill: New York, 1980.
21. Burr, I. W. *Statistical Quality Control Methods*; Marcel Dekker: New York, 1976.
22. Provost, L. P.; Elder, R. S. *Choosing Cost Effective QA/QC Programs for Chemical Analysis*; Office of Research and Development. U.S. Environmental Protection Agency: Cincinnati, OH, 1985; EPA 600/S4-85/056.

Received for review February 13, 1987. Accepted May 8, 1987.

Chapter 9

Storage and Preservation of Environmental Samples

M. P. Maskarinec and R. L. Moody

The chemical analysis of environmental samples is complicated by the lack of a definitive data base establishing scientifically valid preanalytical holding times for such samples. This study is designed to establish the preanalytical holding times for volatile organic compounds and explosives in water and soil under various storage conditions. Procedures are also being developed to reproducibly prepare the large number of samples necessary to create such a data base. These sample preparation procedures and the data base will also make producing materials for reference and quality assurance purposes possible.

ANALYTICAL METHODS FOR ENVIRONMENTAL SPECIMENS require that samples be transported to sophisticated facilities prior to analysis. As a result, considerable time may elapse between the collection of the sample and the initiation of the analytical procedure. In addition, an environmental sample is sometimes necessarily stored for an extended period of time while decisions are made as to what procedures are to be used and what determinations are to be carried out. Unfortunately, no definitive studies have been completed on the length of time during which a sample can be stored without compromising the data generated by the analysis. One major reason for this lack of data is the extreme difficulty involved in accounting for all of the storage variables, analytes, methods, and sample matrices within a single experimental design. The purposes of the study described in this chapter are to define the time frame in which samples can be expected to remain stable, and to recommend storage conditions that maximize this holding time without undue expense. The study has been necessarily limited to compounds for which considerable sampling and analytical cost savings would accrue from extending the holding times. The

1173–6/88/0145$06.00/0 © 1988 American Chemical Society

analytes of interest in this study include two classes of compounds: 19 volatile organic compounds and four explosive (semivolatile) compounds.

The development of procedures for the precise fortification of volatile organic compounds into both water and soil is necessary for any stability study on these compounds. The degree to which differences in stability can be identified is related directly to the statistical variability in the analysis of the compounds in the individual aliquots. This variability is a combination of the variability associated with the fortification technique and the variability of the analytical methodology. The variability of the fortification technique can be determined by replicating a number of individual samples and can be separated from the analytical variability by the repeated analysis of aliquots of an individual sample. When developing techniques for the fortification of volatile organic compounds into environmental matrices, controlling the conditions expected to influence the stability of the compounds during the fortification and aliquot steps becomes important. These considerations apply not only to volatile organic compounds but to any analytes of interest.

Development of the preanalytical holding times for environmental samples requires consideration of the sample matrix; the level of contamination; the storage condition; and the factors governing loss of sample integrity, including biological action, chemical action, volatilization, and photodegradation. Furthermore, the maintenance of sample integrity for a single analyte determination differs from that of a multianalyte determination, where the integrity is governed by the least stable analyte. Because many of the currently popular analytical methodologies require significant economic outlays, ensuring maximum data reliability is crucial. If the sample has been compromised, useful data may still be acquired if a data base is established that provides an estimate of the degree of compromise. Such a data base is now being generated. The study described in this chapter was designed to provide for some of these needs.

Experimental Design

This study was designed to accommodate as many of the highest priority variables as possible within budgetary limits. The factors considered are shown in Table I.

Choice of Factor Range or Value

Sample Type. Virtually all samples acquired in environmental settings are either soils or aqueous liquids. These sample types are the highest priority samples from the point of view of this study.

Level of Contamination. The major problem expected is the availability of samples containing a wide range of contaminants. The use of analyte spikes will

Table I. Factors Considered in the Storage and Preservation Study

Factor	Range or Value
Sample type	soil or liquid
Level of contamination	blank, low, and high levels
Storage conditions	
soil samples (explosives)	-20, 4, and 25 °C
soil samples (volatile organic compounds)	-70, -20, and 4 °C
water samples (explosives)	4 and 25 °C, extract at 4 °C
water samples (volatile organic compounds)	4 and 25 °C, extract at 25 °C
Time intervals	0, 3, 7, 14, 28, 56, 112, and 365 days
Concentration levels	5–10 and 100–500 times the detection limit

facilitate the selection of sample matrices and ensure intercomparability of the data generated from matrix to matrix. From an analytical standpoint, the use of surrogate spikes serves effectively as a standard addition procedure.

Storage Conditions. The choice of storage conditions is dictated by practicality as well as previous knowledge on the stability of contaminants. Soil samples can probably be frozen, although aqueous liquid samples probably should not be frozen. The preparation of an extract is expected to substantially increase the holding time of aqueous liquid samples but might not be as useful for soil samples, where storage at 70 °C will likely inhibit degradation. For volatile organic compounds, the extract is a trap resulting from purge and trap extraction and is stored at room temperature; for explosives, the extract is a sorbent cartridge [styrene–divinylbenzene copolymer resin (XAD–4)] stored at 4 °C.

Time Intervals. The time intervals have been chosen on a logarithmic basis because decay curves are expected to be primarily logarithmic. However, on a practical basis, the intervals chosen must bracket those intervals that have current wide acceptance.

Thus, the number of subsamples at time zero that would be generated from a single sample or sampling point, assuming quadruplicate aliquots, would be calculated as follows: level × storage condition × storage time × replicates, or $2 \times 3 \times 8 \times 4 = 192$. The number of blanks would be calculated as follows: time × storage conditions, or $8 \times 3 = 24$.

Concentration Levels. The analytes of interest include volatile organic compounds and explosives. Because of the relatively large numbers of analyses required for each sample, the components of the spike solution must be chosen by using the following criteria:

    1. The analytical procedure to be followed should be a U.S. Environmental Protection Agency (EPA) relevant procedure using

instrumentation widely available and having documented precision and accuracy.

2. The analytes should represent a range of related components for purposes of extrapolation to other compounds on the current hazardous substances list.

3. The analytes should be selected on the basis of sensitivity to a variety of storage conditions.

On the basis of these considerations, the analytes and spiking concentrations proposed are listed in Table II.

Data Analysis

The treatment of the data assumes that the expected concentration, $E[C]$, of the analytes of interest will decrease with time (t) as an exponential function:

$$E[C] = A \exp(-Bt) \tag{1}$$

where A is a preexponential constant, B is a decay constant, $A > 0$, $B > 0$, and $t = 0 \ldots 365$ days.

The parameters A and B can be estimated either by least squares (e.g., $\ln C = \ln A - Bt + \text{error}$) or by nonlinear least squares (e.g., $C = A \exp(-Bt) + \text{error}$). The decay constants are indexed as B_{ijk}, where i = ith sample, and $i = 1, 2, 3 \ldots n$; j = jth spike level, and $j = 1$ or 2; and k = kth storage condition, and $k = 1, 2,$ or 3.

The hypothesis tests are then presented as follows:

A. no storage conditions effects

H_0: $B_{ij1} = B_{ij2} = B_{ij3}$
H_1: $B_{ijk} \neq B_{ijk'}$ for some $k \neq k'$

B. no contaminant level effects

H_0: $B_{i1k} = B_{i2k}$
H_1: $B_{ijk} \neq B_{ij'k}$ for some $j \neq j'$

C. no sample matrix effects

H_0: $B_{1jk} = B_{2jk} = B_{3jk} \ldots$
H_1: $B_{ijk} \neq B_{i'jk}$ for some $i \neq i'$

where H_0 is the null hypothesis, and H_1 is the alternative hypothesis.

Table II. Analytes and Spiking Concentrations

	Water (μg/L)		Soil (μg/kg)	
Compound	Low Level	High Level	Low Level	High Level
Volatile Organic Compounds[a]				
Bromomethane	50	500	250	2500
Chloroethane	50	500	250	2500
Methylene chloride[b]	50	500	250	2500
1,1-Dichloroethylene	50	500	250	2500
1,1-Dichloroethane	50	500	250	2500
Chloroform	50	500	250	2500
Carbon tetrachloride	50	500	250	2500
1,2-Dichloropropane	50	500	250	2500
Trichloroethylene	50	500	250	2500
Benzene	50	500	250	2500
1,1,2-Trichloroethane	50	500	250	2500
Bromoform	50	500	250	2500
1,1,2,2-Tetrachloroethane	50	500	250	2500
Tetrachloroethylene[b]	50	500	250	2500
Toluene	50	500	250	2500
Chlorobenzene	50	500	250	2500
Ethylbenzene	50	500	250	2500
Styrene	50	500	250	2500
Oxylene	50	500	250	2500
Explosives				
HMX	100	2000	2000	20,000
RDX	50	1000	1000	10,000
TNT	50	1000	1000	10,000
2,4-DNT	50	1000	1000	10,000

[a] The compounds listed as volatile organic compounds constitute EPA's volatile mixtures VI and VII.
[b] These compounds were prepared and spiked from commercially available compounds.

Thus, by the use of these hypothesis tests, extrapolating the maximum holding time allowed for a particular analysis on either of these two sample matrices is possible from these data.

Analytical Methods

The volatile organic compounds are analyzed by using EPA Method 8240 (1). A total of 19 compounds were added to the matrices, including those in EPA's volatile mixtures VI and VII (2). Methylene chloride and tetrachloroethylene were also added to the matrices. Samples were introduced by using the purge and trap technique with all appropriate internal standards and surrogates. In the case of the extract samples, the sample was purged initially onto the traps with the surrogates, and the analysis included a second purge to introduce the internal standards followed by thermal desorption. Soil samples were analyzed with a needle sparger modified to accommodate a 40-mL vial (i.e., the sample container).

The explosives include cyclotetramethylenetetranitramine (HMX), cyclo-trimethylenetrinitramine (RDX), trinitrotoluene (TNT), and 2,4-dinitrotoluene (DNT). These explosives were obtained from the U.S. Army Toxic and Hazardous Materials Agency Standard Analytical Reference Materials (USATHAMA SARMS) program. All these compounds are readily determined by high-performance liquid chromatography with UV detection (automated). The column used was octadecylsilane (Zorbax C–18), 250 $\times$ 4.6-mm inner diameter. For these four compounds, the mobile phase consisted of 25% methanol, 25% acetonitrile, and 50% water (constant volume). The detector was an absorbance detector (Waters Associates model 440) operated at 254 nm. Quantitation was by external standardization.

The sorbent tubes used for collection of the explosives were styrene–divinylbenzene copolymer (XAD–4) cartridges (SKC, Inc.). About 1 mL of water was passed through each cartridge, and the cartridges were stored at the conditions stated in Table I. Desorption was accomplished with diethyl ether, and the analysis was carried out as described previously.

Sample Description

Three water samples and three soil samples were chosen for the initial study. The water samples were as follows: distilled-in-glass grade (Burdick and Jackson), groundwater acquired from one of the newly drilled Resource Conservation and Recovery Act wells at Oak Ridge National Laboratory (ORNL), and surface water acquired from First Creek on the ORNL plant site. The soil samples were as follows: THAMA reference soil, which contained no semivolatile organic compounds; clay soil, which was obtained from the 0800 area (cartographer's designation) of ORNL and had some biological activity; and clay-loam soil, which was rich in microbial activity toward organic compounds. The source of the clay-loam soil was the area around a hazardous landfill in Mississippi.

The data gathered with this limited sample set is not adequate for complete characterization of the preanalytical holding times. Additional samples will be identified after the initial data are gathered. This procedure allows a reduction in the number of storage times for the additional samples and an increase in the number of samples that can be studied.

Certain storage conditions and spiking procedures will allow the mainte-nance of sample integrity over long periods of time. As these conditions are identified, sample aliquots spiked and held by these procedures will become candidates for use as reference materials for performance evaluation and methods development. With additional bulk samples, an assortment of reference materials or procedures for the preparation of quality control samples can be provided to interested parties.

Sample Storage

As soon as the spiked samples were prepared, they were stored under the appropriate conditions. A laboratory was set up to provide refrigerated storage for the appropriate samples. This room had two refrigerators operating at each refrigerated storage temperature to provide backup. Each refrigerator unit was equipped with a temperature trace recorder and an alarm system that was connected to the operations control center at ORNL. An alarm condition is received at the operations control center if any of the refrigerator units experiences a temperature excursion outside the acceptable range. The operator on duty at the time then follows written instructions as to how to correct the situation. These instructions may include transferring the affected samples to the working unit of the pair, if necessary. The temperature recorder traces are kept on file if needed for quality assurance purposes.

Results and Discussion

Characterization of the Sample Matrices

Water 1 used in this study was distilled-in-glass grade (Burdick and Jackson), the characteristics of which are well-known. Waters 2 and 3 for this study were well water and surface water samples, respectively, taken from the ORNL reservation area. These water samples were analyzed for total organic carbon, total dissolved solids, pH, conductivity, anions, volatile organic compounds, and metals by inductively coupled plasma emission spectroscopy. Both the well and surface waters were slightly basic and had a pH of 7.34 and 7.74, respectively. The well water exhibited a higher conductivity (841.5 μS) than did the surface water (322.3 μS). This result was not surprising because the well water contained an overall higher concentration of ions than the surface water. Also, the total dissolved solids of the well water (543 mg/L) was higher than that of the surface water (213 mg/L). The total organic carbon content for well water was 2.0 μg/mL and for surface water was 1.7 μg/mL. Both well and surface water contained SO_4^{2-}, Cl^-, B, Ba, Ca, Mg, Na, Si, and Sr at a range of concentrations. The groundwater concentration was generally 2–4 times higher than the surface water. In addition, the groundwater contained Mn, V, and Zn. The values for Mn and Zn were at least 30 times more concentrated (compared to the detection limits) than in the surface water. Volatile organic compound analyses were performed; the well water contained 9.5 μg/L of *trans*-1,2-dichloroethene and 11.1 μg/L of tetrachloroethene. The surface water contained 6.4 μg/L of chloroform. Chemical analyses have been performed on the USATHAMA reference soil by the University of Maryland Department of Agriculture and Department of Agronomy. These analyses are documented in a report from the University of Maryland (3).

Sample Preparation Procedures

One of the primary difficulties in implementing a study on holding times and storage conditions is obtaining a large number of sample aliquots containing the analytes of interest at an acceptable precision. Although accuracy of the procedure is also important, the absolute concentration of the analytes is of lesser significance. Of course, when developing the proper techniques for producing these aliquots, the scientist must pay attention to the perceived limitations of stability. Two ways are used to approach this situation: the sample can be fortified and then an aliquot can be taken, or aliquots can be individually fortified. The advantage of taking aliquots from the fortified bulk sample is the elimination of systematic errors resulting from individual fortification. The advantage of individual fortification is that no opportunity arises for change of the sample during the taking of the aliquot. Depending on the analytes and matrix of interest, one procedure or the other may be preferred.

For the case of volatile organic compounds in water, where the sample is stored in a 40-mL vial with a polytetrafluoroethylene-faced (Teflon-faced) septum and no headspace, both approaches were investigated. Individual vials were filled with water and fortified. A large volume of water was placed in a poly(vinyl fluoride) (Tedlar) gas sampling bag and fortified, and aliquots were taken. The individual vials were found to exhibit systematic variability (i.e., an occasional vial would contain concentrations consistently below or above the target concentration). This result is unacceptable. Vials filled from the bag, which maintains a zero headspace condition during the procedure, exhibited no such behavior and provided relative standard deviations of less than 4% for virtually all compounds. The sample in the bag was determined to be stable for at least 3 days under refrigeration. This amount of time allowed all the aliquots to be taken. Furthermore, this technique provides excellent reproducibility from sample to sample so that concentration effects are not observed.

The major difficulty of determining the preanalytical holding time for soil samples containing volatile organic compounds is achieving adequate recoveries and reproducibilities for the fortification procedure. In this case, bulk fortification is not useful because excessive losses occur during homogenization and aliquot removal. Therefore, a technique was developed that combines the soil aliquot with a portion of fortified water prepared as previously described. The major advantage of this procedure is the use of poly(vinyl fluoride) (Tedlar) bags for the preparation of the fortified solution. Because of the possibility of further losses during the transfer of the sample to the analytical system, a modified polytetrafluoroethylene (Teflon) fitting was constructed to couple the volatile organic analysis vial containing the sample directly to the purge manifold during analysis. These two steps should minimize headspace losses for analysis of these samples and thus improve both the recoveries and reproducibilities. By using this technique, the precision of the analysis of individual vials was less than 7% relative standard deviation for all compounds. In summary, the procedure involves the use of air-dried soil, which is activated with respect to microbial systems by addition of a small amount of water (ca. 40% water saturation) 3 days prior to the introduction of the volatile organic compounds. The volatile organic compounds are then added in a volume of water sufficient to achieve 80% saturation.

The procedures with respect to explosives (i.e., semivolatile organic compounds) are more straightforward. These analytes do not undergo the same rate of change as do the volatile compounds. For water samples, addition of the compounds in a methanolic solution and subsequent aliquot removal provides excellent reproducibility. For soil samples, individual aliquots that are analyzed without subsampling are used. Again, the problem of soil homogeneity overrides the systematic error problem.

Storage Effects

Several possible causes of compromising the integrity of samples containing volatile organic compounds are known. Losses can occur by mass transfer of the compounds through the septum cap of the vial. These losses will be greater for the most volatile compounds as well as for the least soluble compounds. No headspace should be left in the vial because diffusion in the gas phase is 5 orders of magnitude faster than in the liquid phase. Losses can also occur as a result of microbial degradation. Temperature is an important factor in the magnitude of these losses. Losses can also occur as a result of chemical reactions such as oxidation or dehydrohalogenation. Temperature is also an important effect in this case.

Results of this study to date indicate that if the vial is properly filled and capped, no losses will occur by diffusion through or around the cap. For example, no volatilization and loss of the very volatile chloromethane was observed after 56 days. In fact, the only significant changes in the composition of the sample have been the dehydrochlorination of 1,1,2,2-tetrachloroethane to trichloroethylene, and of trichloroethane to dichloroethylene. The transformation to trichloroethylene occurred within 3 days at room temperature but took longer than 28 days at 4 °C. The transformation to dichloroethylene occurred after 28 days at room temperature and had not occurred after 56 days at 4 °C. With respect to the samples stored on the traps, no significant changes had occurred after 56 days. Because the traps are far less likely to exhibit matrix variability, this procedure may represent the best long-term storage condition for volatile organic compounds. No quantitative or qualitative differences were found between the high- and low-level samples. No contamination of the blanks has been observed, even though these blanks are stored with the high-level samples. This observation is a further indication of the integrity of the septum caps. No data are available yet on the stability of the soil samples.

With respect to the explosive compounds, matrix effects predominate. Samples with biological activity cause a rapid degradation of the TNT, and all samples stored at room temperature show logarithmic decreases in the TNT concentration with time. Apparently, both chemical and biological effects occur, although the biological effects predominate. The DNT, although much more stable than TNT, shows a similar effect. For a biologically active sample, all TNT is lost after 28 days at room temperature, while 20% of the DNT still remains. At refrigerator temperature, 80% of the TNT was lost, but only 40% of the DNT was lost. Although denitrification of the TNT may create additional DNT, this possibility did not appear to be the case. Denitrification would be expected to result in equal amounts of 2,4- and 2,6-DNT, and no 2,6-DNT was detected. More likely, oxidation of the TNT occurred, and this situation can probably be remedied by preservation. Also, a concentration effect appeared. This effect was that the low-level samples degrade faster than the high-level samples. This degradation may be a function of the microbial population. Again, no degradation has been observed on the sorbent traps. This observation indicates the utility of sorbent sampling as a preservation technique.

Abbreviations and Symbols

A	preexponential constant
B	decay constant
2,4-DNT	2,4-dinitrotoluene
$E[C]$	expected concentration
H_0	null hypothesis
H_1	alternative hypothesis
HMX	cyclotetramethylenetetranitramine
i	sample number
ICP	inductively coupled plasma emission (spectroscopy)
j	spike level
k	storage condition
ORNL	Oak Ridge National Laboratory
RDX	cyclotrimethylenetrinitramine
SARMS	Standard Analytical Reference Materials
t	time
TNT	trinitrotoluene
USATHAMA	U.S. Army Toxic and Hazardous Materials Agency

Acknowledgments

The research in this chapter was sponsored by the U.S. Environmental Protection Agency (EPA) under interagency agreement no. 17441744A1, EPA no. DW89932002010; by the Department of Defense, U.S. Army Toxic and Hazardous Materials Agency under interagency agreement U.S. Department of Energy (DOE) no. 17431743A1; and by the U.S. Navy under interagency agreement DOE no. 17431743A1, Navy no. NO537A86P000004 under Martin Marietta Energy Systems, Inc., contract DEAC0584OR21400 with the DOE.

References

1. *Test Methods for Evaluating Solid Waste*; U.S. Environmental Protection Agency: Washington, DC, 1982; SW–846.
2. U.S. Environmental Protection Agency. *Qual. Assur. Newslett.* **1984**, 6(3).
3. U.S. Army Toxic and Hazardous Materials Agency. Report to U.S. Army Armament Research and Development Command (USAARRADCOM) Product Assurance Directorate; University of Maryland College of Agriculture: College Park, MD, 1981.

RECEIVED for review January 28, 1987. ACCEPTED May 11, 1987.

Chapter 10

Evaluating and Presenting Quality Assurance Sampling Data

Franklin Smith, Shrikant Kulkarni, Lawrence E. Myers,
and Michael J. Messner

Objectives of this chapter are to (1) present procedures for evaluating quality assurance (QA) information for sampling as presented in a project sampling plan or a quality assurance project plan (QAPjP), and (2) discuss techniques for presenting the resulting QA sampling data in final project reports. Specifically, five data quality indicators (DQIs) including precision, bias, representativeness, completeness, and comparability are discussed. Each of these five DQIs is sensitive to the manner in which sampling is carried out. Total variability of monitoring data is treated in terms of three sources of error: sampling design, sampling implementation, and analysis. Preferred data quality assessment procedures for quantifying or estimating the DQIs for each source of error are described where applicable.

A MEASUREMENT PROGRAM SHOULD HAVE built-in quality control checks, applied by project personnel, to monitor and control data quality at specific points in the measurement process. This chapter focuses on quality assurance (QA) procedures used to assess and document data quality for each of the sources of error applicable to the subject measurement process. The objectives of this chapter are to discuss procedures for evaluating QA sampling information presented in a sampling plan or in a quality assurance project plan (QAPjP) as required by the U.S. Environmental Protection Agency (EPA) (1), and to present techniques for reporting QA sampling data in project reports or computerized data bases as a means of documenting data quality. Specifically,

1173–6/88/0157$06.00/0 © 1988 American Chemical Society

this chapter will discuss the five data quality indicators (DQIs) required in EPA's interim guidelines for preparing QAPjPs (1): precision, bias, representativeness, completeness, and comparability. Each DQI is sensitive to the manner in which sampling is carried out.

In this chapter, variability in environmental sampling and analysis results is treated in terms of three individual sources of error (2). Sampling is divided into design and implementation phases. The design phase of sampling addresses where and when to sample and how many samples or portions to collect (3). The implementation phase of sampling is concerned with how the individual samples or portions will be collected, processed, packaged, transported, and stored prior to analysis. The analytical phase is the third source of error in the measurement process. This phase is generally the most thoroughly estimated and best controlled source of error. Each source of error may have systematic and random components. For this chapter, the *measurement process* includes all three sources of error, and the *measurement system* consists of the implementation and analysis phases only.

Other classifications of the measurement process are possible. Sample preparation and extraction, which might be regarded as part of the implementation phase, are important for gas chromatography–mass spectrometry (GC–MS) procedures but might be negligible or nonexistent for other measurement techniques. Sample workup, which is so critical to GC–MS analysis, could be treated as a fourth phase of the measurement process between the implementation and analysis phases.

Although an estimate of the overall uncertainty in the measurement data is the primary need of the data user, a knowledge of the magnitude and characteristics (i.e, systematic or random) of the individual sources of error is necessary in order to design a cost-effective sampling plan or QAPjP. Thus, the major sources of error should receive the main emphasis in designing a sampling plan or QAPjP. All three sources of error may not be important for some measurement processes. For example, quantification or estimation of all three sources of error is important when the measurement process involves estimating a population parameter from a statistical sample, and the sample collection and analysis phases are separated in space and time. The other extreme is represented by in situ monitoring with a continuous analyzer where the value of the measurement at a point in space and time is of interest; thus, the design and implementation phases of sampling should have zero or very small levels of error.

Evaluating Quality Assurance Procedures for Collecting Sampling Data

To be evaluated as satisfactory, a QAPjP must, among other things, adequately address the five DQIs as defined in guidelines issued by the EPA's quality assurance management staff (QAMS) (*1*). Preferred procedures for assessing each indicator are discussed. These procedures may not be appropriate for some environmental measurement activities because of technology or resource constraints.

Taken together, the objectives given in the QAPjP for the DQIs should provide evidence that the program's *data quality objectives* (DQOs) will be realized. DQOs are statements of the levels of uncertainty that a decision maker is willing to accept in the decisions made on the basis of the measurement data (*4*). The DQOs are based on the intended use of the data.

Precision

According to EPA's Office of Research and Development,

> Precision *[emphasis added]* is a measure of mutual agreement among individual measurements of the same property, usually under prescribed similar conditions. Precision is best expressed in terms of standard deviation. Various measures of precision exist, depending upon the prescribed similar conditions (*1*).

The relationship of the total variance to the components of variance (*2*) is

$$\sigma^2_{t} = \sigma^2_{d} + \sigma^2_{i} + \sigma^t_{a} \tag{1}$$

where σ^2_{t}, σ^2_{d}, σ^2_{i}, and σ^2_{a} are the total variance, the design phase of sampling variance, the implementation phase of sampling variance, and the analytical variance, respectively.

For measurement processes where probability sampling is involved, for example, precision of the process (σ_{t}) is estimated by the standard deviation of the individual measurements. No direct means is available to estimate the design-phase precision (σ_{d}) or implementation-phase precision (σ_{i}). Measurement system precision ($\sigma^2_{i} + \sigma^2_{a})^{\frac{1}{2}}$, however, can be estimated by the collection and analysis of collocated samples (*see* next paragraph), and analytical precision (σ_{a}) can be estimated by analysis of laboratory-replicated samples. Thus, with estimates of σ^2_{a} and ($\sigma^2_{i} + \sigma^2_{a}$), an estimate of σ_{i} can be calculated (*2*). An estimate of σ_{d} can be calculated if estimates of σ_{i}, σ_{a}, and σ_{t} are available by using the relationship in equation 1.

The sampling plan or QAPjP should state QA objectives for σ_{t}, ($\sigma^2_{i} + \sigma^2_{a})^{\frac{1}{2}}$, and σ_{a}. Also, procedures to be used for deriving estimates for each of these measures of precision should be discussed. For example, the preferred procedure for estimating the precision of the measurement system is through the collection and analysis of collocated samples. *Collocated samples* are two or more portions collected at the same point in time and space so as to be considered identical (*5*). This procedure will quantify combined random errors for the implementation and analysis phases of the measurement process. Replicating or splitting a sample any time after collection does not include all sources of error of the total measurement system; thus, a poor, underestimate of precision may result.

Procedures for handling collocated samples should be included in the sampling plan or QAPjP. To obtain an accurate estimate of precision, the collocated samples should be separated after collection and allowed to pass through the shipment, handling, and analysis phases at time intervals similar to that experienced by the individual study samples. If the collocated samples are not separated in time, they may vary in the same manner and at the same rate as they pass through the rest of the measurement process and the change will not be detected. Thus, an optimistic estimate of precision will result.

Subdividing one or both of the collocated samples just prior to analysis provides an estimate of analytical precision. Subtraction of the analytical variance from the measurement system variance $[(\sigma^2_i + \sigma^2_a) - \sigma^2_a = \sigma^2_i]$, as previously discussed, provides an estimate of precision of the implementation phase of sampling.

The sampling plan or QAPjP should clearly state the type of precision being used and describe the conditions under which the data quality assessments are to be made. For example, intralaboratory precision is the variation associated with a single laboratory or organization and is usually referred to as *repeatability* (6, 7). Intralaboratory precision should be described as short-term (e.g., within-day) or long-term (e.g., across-days) precision according to how the collocated or replicated samples are passed through the implementation and analysis phases of the measurement process. Interlaboratory precision, or *reproducibility* (6, 7), is the variation associated with two or more laboratories or organizations using the same measurement method.

If the nature of the sample acquisition procedure prevents the assessment of the precision of the entire measurement system, replicated samples used to assess precision should be selected to incorporate as much of the measurement system as possible. Data quality assessments should be made by using the same measurement method at concentration levels typical of the range observed in routine analysis.

The design of the data quality assessment efforts in the sampling plan or QAPjP must address such factors as the data quality needs (i.e., the DQOs), the precision of the measurement system, the number of field samples to be collected and analyzed, and the number and frequency of data quality assessment samples to be used.

For batches having a large number of samples, a fixed frequency for replicate measurements (such as one sample in 10 or 20) is recommended. For small batches, more frequent repetition may be desirable to ensure that sufficient data are available to assess precision. Alternatively, data quality assessment results from multiple batches with a common matrix, analyzed by the same measurement system, may be combined. If the environmental measurements normally produce a high percentage of results below the method detection limit, samples for replicate measurement should be selected from those containing measurable levels of analyte. Where this selection is impractical, such as with complex multianalyte methods, matrix spikes or matrix spike duplicate samples should be prepared so that spiked concentrations are within the range of interest (8). Recovery and precision are then calculated for each matrix spike or matrix spike duplicate.

Bias

According to EPA's Office of Research and Development,

> Bias *[emphasis added]* is the degree of agreement of a measurement (or an average of measurements of the same thing), X, having an accepted reference or true value, T, usually expressed as the difference between the two values, $X - T$, or the difference as a percentage of the reference or true value, $100(X - T)/T$, and sometimes expressed as a ratio, X/T (1).

The relationship of total process bias to the biases of the individual sources of error is as follows:

$$B_t = B_d + B_i + B_a \qquad (2)$$

where B_t is the bias of the total process, B_d is the bias in the design phase of sampling, B_i is the bias in the implementation phase of sampling, and B_a is the analytical bias.

For measurement studies where probability sampling is used, the level of bias in the final estimate is a function of the adequacy of the sampling plan design (9-11). The bias level probably cannot be quantified but only estimated on the basis of assumptions. As was the case of precision, this part of the sampling plan should be evaluated by a qualified statistician.

Like precision, bias of the implementation phase of sampling cannot be estimated directly. Combined implementation and analytical bias ($B_i + B_a$) is estimated by using collected samples spiked in the field. (The systematic component of error associated with sample collection is not included in this estimate.) That is, a field sample is subdivided in the field, at least one fraction is spiked with a known quantity of the target analyte, and each fraction is analyzed. The percent recovery of the spike is calculated, and by combining several such results, an average percent recovery or bias (i.e., average percent recovery $-$ 100%) for all process steps after collection through analysis is obtained. Likewise, laboratory spikes are used to estimate bias of analysis (B_a). Then, $B_i = (B_i + B_a) - B_a$. Estimates of bias should, where possible, be based on externally (i.e., external to the project staff and equipment) and independently prepared spiking or spiked materials or externally and independently conducted analyses.

A documented spiking protocol and consistency in following that protocol are important elements in obtaining a meaningful data quality assessment and should be included in the sampling plan or QAPjP. Spikes should be added at different concentration levels to cover the range of expected target analyte concentrations. For some measurement systems (e.g., continuous analyzers used to measure pollutants in ambient air), the spiking of samples is not practical, and assessments are made with appropriate blind reference materials. Ideally, spiking materials or reference materials should be introduced into the samples at the collection site so that the bias assessment includes any losses caused by handling, preservation, and storage of the collected samples. If the matrix type or the measurement system prevents such practices, bias assessments should be made for as large a component of the measurement system as possible.

A representative sample of the sample lot or batch is selected for spiking, and the selected samples are subdivided and analyzed, one portion without spiking and one portion after spiking, in order to measure recovery. The spiking frequency will depend upon the data quality needs of the program, the bias and precision of the measurement system, the size of the sample lot, and other considerations. To properly assess the bias for a small sample lot, a relatively high percentage of the samples might have to be spiked. Where the method performance for multiple sample lots of similar matrix types is expected to be equivalent, combining information so that fewer spikes are required in each lot may be possible.

For certain multianalyte methods, such as EPA Method 608 for organochlorine pesticides and polychlorinated biphenyls in water, bias assessments are complicated by mutual interference between certain analytes that prevent all of the analytes from being spiked into a single sample. For such methods, lower spiking frequencies can be employed for analytes that are seldom or never found.

Representativeness

According to EPA's Office of Research and Development,

> Representativeness [emphasis added] expresses the degree to which data accurately and precisely represent a characteristic of a population, parameter variations at a sampling point, a process condition, or an environmental condition (1).

The QAPjP or sampling plan should provide clear instructions for where and when to sample and how many samples to collect.

For estimating an average concentration over some region, representativeness of a sample is assured by random sampling from the target population. Therefore, the preferred approach is to use a probability sample designed by statisticians working with scientists experienced in the sampling and measurement processes. These individuals should also review the QAPjP. If possible, sampling should be done in stages so that data from one stage can be used in designing the next stage.

For estimating a maximum concentration over the same region, scientific judgment is commonly applied to choose sampling locations at or near the maximum. If possible, these locations should be confirmed as having the desired property by additional random sampling from the region.

In some cases, the sampling instructions are given in guidelines or regulations. In such cases, these guidelines or regulations should be discussed in the QAPjP with a complete reference provided. In other cases, sampling procedures may be justified by historical information on the region or population.

Completeness

"*Completeness* [italics added] is a measure of the amount of valid data obtained from a measurement system compared to the amount that was expected to be obtained under correct normal conditions" (1).

With the present emphasis on DQOs and QA objectives for DQIs in QAPjPs, completeness could also be defined as follows: completeness = 100 × valid data collected or data needed for the DQO.

By using either of these definitions, a QA objective for completeness is required in the QAPjP. The basis for the stated objective (e.g., 90% completeness) and the consequences of failure to achieve this objective should be discussed. For example, the stated objective may be required to ensure that a satisfactory level of power is realized for an experiment designed for hypothesis testing, or to ensure that a probability interval of desired width is achieved in an experiment designed for parameter estimation.

The phase or phases of the measurement process where data loss is most likely to occur should be pointed out with alternative procedures specified, if necessary, to meet the completeness objective. Examples of possible data losses include, but are not limited to, specific sampling sites being inaccessible at the time of data collection, breakage or spilling of sample during handling or shipping, instrument failure (e.g., continuous analyzers), and holding time being exceeded before analysis.

Comparability

"*Comparability* [italics added] expresses the confidence with which one data set can be compared to another" (1).

For example, to show comparability of data sets from different monitoring programs, the following topics should be addressed:

1. how the sampling stations in each network are sited, so that each network had the same probability of collecting a representative sample (for example, were the sites selected to detect maximum values or values representative of an area or volume?);

2. how the same observables (analytes) measured in each program are reported in the same units and corrected to the same standard conditions or, if . not, provide a mathematical or gravimetric relation between observables;

3. how the procedures and methods for collection and analysis of any particular observable are the same between networks, or provide a mathematical relation;

4. how the data quality is established and how there will be sufficient documentation of audits and DQIs to determine data adequacy; and

5. how, in terms of accuracy and precision, data for one observable, measured by one method or equivalent methods, can be combined or compared in a statistically defensible manner among programs.

The first topic establishes equivalency of objectives; the second and third topics address equivalency of measurement data among programs, and the fourth and fifth topics show equivalency of data quality among programs. If siting, collection techniques, and measurement procedures and methods are equivalent for two data sets, then proceeding with a determination of the equivalence of data quality for a given observable is appropriate.

Presenting Quality Assurance Sampling Data

Where and how to present QA data are always questions of concern. First, the decision maker using the measurement data should have an estimate of its uncertainty. This estimate could be in the form of probabilities of a false positive or false negative in hypothesis testing, a confidence interval around the estimated population parameter in parameter estimations, or a probability interval for individual measurements. This uncertainty estimate requires that at least precision and bias estimates or some combination thereof accompany the final results of the research or monitoring study. That is, the uncertainty information should be documented where the project's final result appears— either in a hard copy project report or in a computerized data base.

Also, a future user of the measurement data for such purposes as model validation should have a complete discussion of the QA and quality control (QC) program including calculated data quality assessment results. This discussion could be a QA and QC section in the project report or an independent QA and QC document to accompany the project report or to be maintained as a reference for computerized data bases. In such a section or document, each of the five DQIs should be addressed, and for precision, bias, and completeness, each of the three sources of error should be discussed. This error source discussion allows the data user to perform an error analysis for application.

Precision

Three precision estimates may be obtained from the QA program. First, the precision of the measurement process is estimated by the standard deviation of the actual reported values. Second, the combined precision of the implementation and analytical phases is estimated from analysis of collocated or field-replicated samples. Third, the precision of analysis is estimated from analysis of laboratory-replicated samples or repeated measurements of field samples spiked in the laboratory with the target analyte. Reporting these three precision estimates provides measurement data users with the precision information necessary to assess the quality of the data for their particular needs. Future users of the measurement systems should know the system's performance capabilities and major sources of error.

Preferred indices for precision include standard deviation and relative standard deviation. These indices, if standardized, would aid in assessing comparability across programs or studies and in improving understanding in the scientific community. Procedures for presenting precision estimates in order of preference are as follows:

- Precision should be presented as a function of the measured value across the applicable range. Ideally, the presentation could be a graph containing the actual data points, the mathematical relationship providing the best-fit curve, and confidence intervals about the best-fit curve.

- Tables should be presented containing data quality assessment data points and regression equation coefficients when appropriate.

- Calculated values should be presented of the standard deviation at discrete measured values that cover the applicable range.

Precision estimates should be supported with certain descriptive information. This information should include, but not be limited to, labels identifying interlaboratory or intralaboratory precision, the components of the process included in the estimate, and any outlier identification schemes used for the data quality assessment data points. Interlaboratory precision should be further described as short- or long-term, single- or multiple-analyst, single- or multiple-instrument, and single- or multiple-calibration.

Bias

Two bias terms can result from the QA program. These are bias of the combined implementation and analysis phases based on the analysis of field spiked samples or reference materials, and the analytical bias obtained from repeated analyses of field samples or reference materials. Bias of the design phase of sampling can only be estimated. This estimate made with statistical tools or scientific judgment as appropriate should be stated along with any underlying assumptions.

The preferred measure of bias is the difference between the average measured value and the true value. Percent recovery (100% bias) is also frequently used in environmental measurement programs.

Procedures for presenting bias information in their preferred order are as follows:

- Bias should be presented as a function of the true value over the applicable range. This presentation could be a graph containing the actual data quality assessment data points, the best-fit curve, confidence intervals about the best-fit curve, and the regression equation for the best-fit curve when appropriate.

- A table of actual data quality assessment data points and the regression equation should be presented.

- Bias values calculated at discrete measured values covering the applicable range should be presented.

Bias data should be accompanied with certain supporting information. This information should include, but not necessarily be limited to, a description of how and under what conditions the bias data were collected, and the number of data points involved.

Representativeness

As stated previously, representativeness of a statistical sample from a population or a discrete portion collected from a process or source cannot be quantified because "truth" is not known. Representativeness can be estimated through application of statistical or scientific approaches supported by the actual measurement data.

A description of the design and implementation phases of the sampling plan as actually carried out during the program with appropriate photographs (e.g., of the sampling site configuration) and drawings with an assessment of the resulting representativeness should be provided in the QA and QC report.

Completeness

For in situ measurement programs such as ambient air monitoring, where a sample (e.g., a 1-h average) missed cannot be replaced, a straightforward calculation (1) is used to arrive at a numerical value for completeness. For other sampling and analysis activities (e.g., where discrete samples are collected and shipped to a laboratory for analysis), replacing lost samples at a later time may be possible, or more samples than the sampling plan requires may be collected originally to ensure that the required level of completeness is satisfied.

In either situation, missing data should be explained as to how and where in the measurement process the data were lost. For example, all missing data may have resulted from inability to collect the samples in the field, from loss or breakage during shipment, or from failure to analyze before allowable storage time was exceeded. This knowledge is important in designing similar collection and analysis plans in the future or for improving the efficiency of an ongoing sampling program. Reporting how the lost data were distributed across time and space, when appropriate, is equally important. For example, was the lost data randomly scattered over all sampling sites, at one site but random in time, or over all sites because of one instrument failure at one site?

Some recommendations have been made that completeness be used in the planning phase of a program and discussed in terms of the consequences of incompleteness. For example, what is the effect of incompleteness on the probability interval about an estimated parameter or on the probability of making a wrong decision in hypothesis testing?

Comparability

Reasons should be given for why this data set is comparable to an existing data set. These reasons would include, but not necessarily be limited to, the following:

- comparison of siting criteria,
- observables measured,
- sampling and analysis protocol,
- comparison of QA and QC programs,
- comparison of precision and bias estimates,
- units of reporting, and
- correction of measured values to standard conditions (e.g., temperature and pressure).

Abbreviations and Symbols

σ_a	analytical variance
σ_d	design phase of sampling variance

σ_i	implementation phase of sampling variance
σ_t	total measurement process variance
B_a	analytical bias
B_d	design phase of sampling bias
B_i	implementation phase of sampling bias
B_t	total measurement process bias
DQI	data quality indicator
DQO	data quality objective
GC–MS	gas chromatography–mass spectrometry
QA	quality assurance
QAPjP	quality assurance project plan
QC	quality control
T	true value
X	measured value

References

1. *Interim Guidelines and Specifications for Preparing Quality Assurance Project Plans*; Office of Monitoring Systems and Quality Assurance. Office of Research and Development. U.S. Environmental Protection Agency: Washington, DC, 1980; QAMS–005/80.
2. Box, G. E. P.; Hunter, W. G.; Hunter, J. S. *Statistics for Experiments*; Wiley: New York, 1978; p 571.
3. Horwitz, W. *Nomenclature of Sampling in Analytical Chemistry*; Center for Food Safety and Applied Nutrition, Food and Drug Administration: Washington, DC, 1986.
4. Johnson, G. L.; Blacker, S. Presented at the 1987 EPA/ACPA Symposium on Measurement of Toxic and Related Air Pollutants, Research Triangle Park, NC, May 1987.
5. *Calculation of Precision, Bias, and Method Detection Limit for Chemical and Physical Measurements*; Quality Assurance Management and Special Studies Staff. Office of Monitoring Systems and Quality Assurance. Office of Research and Development. U.S. Environmental Protection Agency: Washington, DC, 1984; Chapter 5.
6. Mandel, J. In *Interlaboratory Testing Techniques, Chemical Division Technical Supplement*; American Society for Quality Control: Milwaukee, WI, 1978; pp 40–48.
7. Smith, Woolcott. In *Statistics in the Environmental Sciences*; Gertz, S. M.; London, M. D., Eds.; ASTM Special Technical Publication 845; American Society for Testing and Materials: Philadelphia, PA, 1984; pp 90–97.
8. Mandel, J. *ASTM Standard. News* 1977, *56*, 17–20.
9. Horwitz, W. *Sampling for Nonstatisticians*; presented at the Association of Official Analytical Chemists Spring Workshop, Indianapolis, IN, April 1983.
10. Grant, D. M. *Water Eng. Manage.* 1982, *129*(7), 28–32.
11. Kratochvil, B.; Taylor, J. K. *Anal. Chem.* 1981, *53*(8), 924–938A.

RECEIVED for review January 28, 1987. ACCEPTED May 18, 1987.

SAMPLING WATERS

Chapter 11

Sampling Waters
The Impact of Sample Variability on Planning and Confidence Levels

U. M. Cowgill

This chapter addresses generic difficulties associated with water sampling: problems of sampling related to the physical state of the sample (e.g., ice, snow, and dew), how these disconcerting situations are resolved or circumvented, and how these considerations affect proposed sampling protocols and desired confidence limits of subsequently obtained chemical results. Types of contamination expected from sampling tools composed of various materials, and sorption and leaching of compounds from such materials are considered. In addition, examples of variation in chemical composition are presented in relation to replication, sampling frequency, and sampling location in various water bodies. Lakes, rivers, groundwater, extensive ice development, stratigraphy encountered in dew depositions, and variations found during a 3-h rainstorm are among the examples shown.

S AMPLING IN ENVIRONMENTAL ANALYSIS is akin to dishwashing in analytical chemistry. If the dishes are dirty, all analytical activity thereafter is for nought. Similarly, nothing is gained from the chemical endeavor if the samples are not representative of the environment from which they are taken. The first portion of this chapter is devoted to details demanding attention if proper samples are to be gathered. The next portion concerns specific types of samples and associated problems to be considered in sample collection. The remainder of the chapter addresses how these various topics affect proposed sampling protocols and desired confidence limits of subsequently obtained chemical results. The majority of the examples are drawn from inorganic

1173–6/88/0171$06.00/0 © 1988 American Chemical Society

chemistry because few studies have been devoted to the variability of trace organic compounds in relation to sampling. Further discussion of organic compound sampling may be found in the works of Keith (1-3). A good general review of sampling for chemical analysis may be found in the publications of Kratochvil and co-workers (4, 5).

Problems Associated with Sampling

Much of the material presented in this chapter was gathered in the process of trying to discover the best way to sample particular bodies of water or water in a physical state other than the liquid. Such data are not usually published. But this kind of problem does reflect the kind of cogitation that is a necessary preamble to any extensive and successful research effort. The Linsley Pond Study (North Branford, CT) (*Arch. Hydrobiol.*, 1970–1978) involved a very lengthy preliminary planning stage wherein proper sampling could be identified only by careful chemical analysis. Sampling had to be adequate for precise measurement of major as well as minor constituents. Such requirements had to apply not only to water but to ice, snow, rain, mud, plants, and microscopic biota. Most extensive geochemical studies involve detailed comprehensive preliminary work. Many of the examples used to illustrate sampling pitfalls originated from those various preliminary studies carried out on a variety of water bodies. Much of the unpublished work referred to in this chapter was generated in various places and over several decades; therefore, these data will be referred to by date rather than by place.

The purpose of sampling dictates the nature of the sample to be gathered, the equipment used to acquire the sample, the size of the sample, and the frequency of sampling. Generally, the purpose of water sampling is to acquire a representative sample of the body of water being studied. Once such a representative sample has been obtained, the investigator may be interested only in studying its general chemical composition and may therefore have no interest in seasonal changes. Thus, the necessity of frequent sampling is limited.

Contamination from Sampling Equipment

Table I shows the type of contaminants contributed to water samples by materials used in sampling tools and monitoring-well construction. Poly(vinyl chloride)-threaded (PVC-threaded) or poly(vinyl chloride)-cemented (PVC-cemented) joints may have contaminants that leach into contiguous water. Such contaminants may be removed below the limits of detection in distilled water by steam cleaning. The source of the steam should be distilled water. Five separate steam washes will remove undesirable substances (Cowgill, U. M., *Symposium on Field Methods for Ground Water Contamination Studies and Their*

Table I. Contaminants Contributed to Water Samples by Materials
Used in Sampling Devices or Well Casings

Material	Contaminants Contributed Prior to Steam Cleaning
PVC-threaded joints	Chloroform
PVC-cemented joints	Methyl ethyl ketone, toluene, acetone, methylene chloride, benzene, ethyl acetate, tetrahydrofuran, cyclohexanone, three organic Sn compounds, and vinyl chloride
Polytetrafluoroethylene (Teflon)	Nothing detectable
Polypropylene or polyethylene	Plasticizers and phthalates
FRE	Nothing detectable
Stainless steel	Cr, Fe, Ni, and Mo
Glass	B and Si

SOURCE: Adapted from Cowgill, U. M., *Symposium on Field Methods for Ground Water Contamination Studies and Their Standardization*, American Society for Testing and Materials, in press, and Cowgill, U. M., unpublished data, Dow Chemical Company, 1985.

Standardization, American Society for Testing and Materials, in press). Similarly, steam cleaning of fiberglass-reinforced epoxy material (FRE), polypropylene, polyethylene, and glass will substantially reduce the quantities of substances that leach out of these materials into the contiguous water (Cowgill, U. M., Dow Chemical Company, unpublished data, 1985). As a result of this experience, I recommend that no sampling devices or well casings should be used without prior, thorough, steam cleaning.

Two incidents involving the use of PVC casing with solvent-welded joints are noteworthy. Many contracting firms have installed wells employing PVC-cemented joints around industrial plant sites. Substances originating from that cement have been detected in groundwater a decade after installation (Cowgill, U. M., Dow Chemical Company, unpublished data, 1985). Another incident occurred when the Midland, MI, water treatment plant changed their method of water treatment. When this finished water was initially transported to the Dow Chemical Company's Aquatic Toxicology Laboratory, a quarter mile of cast-iron pipe was destroyed and replaced with PVC pipe containing cemented joints. After 84 days, 1.3 mg/L of methylene chloride, 3 μg/L of chloroform, and 6 μg/L of tetrahydrofuran were detected in the water despite the fact that 7,257,600 gal of water (pH 8) had passed through this quarter mile of pipe.

Additionally, the chemical contribution to water transported through soldered pipes cannot be ignored. Tin and lead are the most common contaminants encountered under such circumstances. Water containing high quantities of calcium tends to extract the lead preferentially. Tin, however, continues to be removed in small amounts for decades (6).

Sorption and Leaching of Contaminants
by Sampling Tool Materials

Barcelona et al. (7, 8) published a study comparing the maximum sorption of dilute (400-μg/L) halogenated hydrocarbon mixtures in water by various plastics. PVC exhibited a maximum sorption of 622 μg/m^2, and polytetrafluoroethylene (PTFE) (Teflon) sorbed only 237 μg/m^2. Hunkin (Hunkin, G. G., Sr., Ground Water Sampling Inc., personal communication, 1985–1986) employed similar but not identical procedures outlined by Barcelona et al. (7) and found that the maximum sorption by FRE was 203 μg/m^2. This figure was obtained by exposing 6 g of FRE particles for 72 h to 400 μg/L of individually halogenated hydrocarbons similar to those used elsewhere (7). Barcelona et al. (7, 8) did not use exposure times in excess of 60 min. The data revealing the actual loss of halocarbons from solution are quite interesting. In 72 h, the sorption on the FRE sample was 20.25% (Hunkin, G. G., Sr., Ground Water Sampling Inc., personal communication, 1985), and sorption values of PTFE and PVC in 1 h were 38% and 98%, respectively. These data were calculated on the basis of the halocarbon concentration in the fluid expressed in terms of sample surface area. When the data are expressed in terms of organic concentration loss, the percent sorption is slightly lower. The presence of organic carbon has an effect on sorption that apparently is material-dependent (8). The depletion of halocarbons from the solution is more dependent on the type of material than on the tubing diameter (8). However, when a constant flow rate is used, losses are more likely to increase rather than decrease as the tubing diameter becomes larger (7, 8). Clearly, sorption is a function of the mass as well as the surface.

This discussion shows that the use of PTFE or PVC compromises the value of total organic carbon as an indicator of pollution. This result has not been found for FRE (Hunkin, G. G., Sr., Ground Water Sampling Inc., personal communication, 1986). Thus, growing evidence suggests that thermoplastic materials sorb many priority pollutants efficiently. Furthermore, this sorption may be so effective that detection of undesirable compounds in water may be delayed when synthetic sampling tools are suspended in wells and have long contact with well water, or when sampling devices used for continuing sampling of surface water studies are made of thermoplastic materials. Barcelona et al. (7, 8) showed that PTFE tubing brought about an adsorptive loss of chlorinated hydrocarbons of 21% in less than 1 h. Similar losses have been observed for FRE in 72 h (Hunkin, G. G., Sr., Ground Water Sampling Inc., personal communication, 1986).

Miller (9) noted that the extent of adsorption of metals at low concentrations on container walls is determined by metal concentration, pH of the sample, length of contact with the container, sample and container chemical composition, and the presence of dissolved organic carbon as well as complexing agents (10). Robertson (11), in his elemental study of seawater, reported that adsorption of In, Sc, and Ag could be significantly reduced when the sample was stored in glass or polyethylene. This reduction occurs by adjusting the pH of the sample to 1.5 with HCl. Robertson (*12*) also pointed out that PVC contained Zn, Fe, Sb, and Cu; that polyethylene contained Sb; and that these metals could leach from such materials into the contiguous water. Boettner et al. (*13*) reported the leaching of alkyltin and organic pipe cement from PVC and chlorinated PVC into contiguous water at pH 5 and 37 °C. Dibutyltin dichloride and dimethyltin dichloride were also detected in this study. Miller (*9*) noted that adsorption and subsequent leaching of Pb were more rapid when Cr(VI) was present in the solution along with six volatile organic compounds than when Pb and Cr(VI) were alone in solution. The Cr(VI) did not adsorb or leach from PVC, polyethylene, or polypropylene. The intent of this discussion is not to engage in a lengthy review but only to point out that contaminant leaching from sample tools is quite complex and requires serious attention by those involved in sampling water for trace quantities.

Replication

The number of replications required to characterize a water body is determined by the purpose of sampling. The purpose falls into two general categories: (1) sampling for the purpose of description (i.e., the chemical composition of the water body), and (2) sampling for the purpose of monitoring (i.e., the substances to be monitored have already been identified). Replication in the second category must be described in terms of acceptable arithmetic means, standard deviations, and confidence limits. In an area to be monitored, descriptive sampling is necessary not only to discover what is present but to obtain chemical background information as well.

Before proceeding further, replication should be defined. If three aliquots are analyzed from a 1-L sample, and if three 1-L samples are collected and an aliquot is removed from each liter and analyzed, then the results of these two approaches will differ. The first approach is a good way to check the instrument and ascertain that it is functioning properly. However, these data are biased and nonrandom in a statistical sense from the standpoint of the water body, but not from the standpoint of evaluating the entire analytical technique. The second approach is true *replication*. Any variation among the three aliquots gathered from three separate samples is the experimental sampling error. This type of replication provides the measure of sampling precision. Furthermore, the built-in randomization ensures the validity of this measure of precision. In addition, replication of this sort will help to illuminate gross errors in analytical measurements. The first approach could not be used in this manner.

The first aproach (three aliquots per liter) to sampling will usually result in a smaller percent coefficient of variation than the second approach (*14*), and may occasionally exhibit a smaller standard deviation. However, the second approach will provide a result more representative of the body of water being studied at the time of sampling. Because the second approach results in the ultimate objective of all water sampling, it is the most likely approach to achieve needed results over time. Table II illustrates this point with some results from Linsley Pond in North Branford, CT (Cowgill, U. M., unpublished data, 1965–1980).

An examination of the data shown in Table II reveals that the percent coefficient of variation is always lower when three aliquots per liter are analyzed than when one subsample is analyzed from each of three 1-L samples. This study also shows that the concentrations are neither consistently higher nor lower but that as the concentration of the substance of interest declines, the variation between the two sampling regimens increases. Significant differences are encountered between the two systems. The second approach, however, does not indicate the true value or even how close to the true value either analysis may be. This point is discussed further by McBean and Rovers (14).

Another approach to the problem is to gather larger samples. Table III shows the results of this endeavor. All the samples were gathered from each side of a boat with a Van Dorn bottle outfitted with a graduated-steel tape (Cowgill, U. M., unpublished data, 1965–1980). [Further discussion on available types of Van Dorn bottles may be found in reference 15.] No significant difference arose between 20- and 30-L samples taken from either side of the boat. In addition, the percent coefficient of variation becomes progressively smaller. The data shown from either side of the boat involving 20- and 30-L samples are, statistically speaking, no different from the results obtained by subsampling each of three 1-L samples replicated 10 times. Results obtained from either side of the boat contain too much discrepancy in the smaller volume samples. To illustrate the problems of replication, the worst possible case has been selected, that of the euphotic zone, or the zone of a productive lake in which much of the living activity takes place. In addition, the depth varies by 2.5 m. The conclusion is that replicated 20-L samples provide reasonable results regardless of which side of the boat the samples are collected from or how small the concentrations of the substance of interest may be. Another point is that each body of water has a "personality of its own", and this kind of descriptive sampling needs to be

Table II. Chemical Results Obtained from Various Sampling
Techniques on Linsley Pond

Number of Subsamples Replicated 10 Times	Ca (mg/L)	Mg (mg/L)	Sr (µg/L)	P (µg/L)	Pb (µg/L)
Three aliquots per liter	53 (4.5)	30 (8.3)	130 (20)	20.4 (5.1)	2.4 (0.5)
One aliquot from each of three 1-L samples	29.7 (5.0)	11.8 (3.8)	64 (20)	72 (30)	13 (4)
χ^2, P<	0.02	0.005	0.001	0.001	0.01

NOTE: The depth sampled was from the surface to 2.5 m below the surface. The numbers in parentheses denote the standard deviation.

Table III. Chemical Results Obtained by Collecting Samples
from the North and South Side of the Boat

Sample Size	Ca (mg/L)	Mg (mg/L)	Sr (µg/L)	P (µg/L)	Pb (µg/L)
		South Side			
Three 2-L samples	55	56.2	155	100	5
	(10.0)	(10.0)	(25)	(25)	(3.0)
Three 5-L samples	40	22	120	110	10
	(11.0)	(10.0)	(21)	(25)	(3.1)
Three 20-L samples	30	12	75	75	15
	(3.0)	(2.1)	(10)	(10)	(2.4)
Three 30-L samples	31	14	70	80	18
	(2.8)	(2.1)	(10)	(8)	(2.1)
		North Side			
Three 2-L samples	76	30	95	135	22
	(16.0)	(10.0)	(15)	(17)	(4.5)
Three 5-L samples	34	18	104	85	6
	(7.0)	(5.0)	(15)	(8)	(5.5)
Three 20-L samples	29.8	11	68	72	13
	(4.1)	(2.0)	(9)	(9)	(3.0)
Three 30-L samples	31.4	15	72	78	15
	(3.8)	(2.4)	(11)	(9.3)	(2.8)

NOTE: The numbers in parentheses denote the standard deviation. The sampling stratum is the same as that in Table II.

carried out to characterize the chemical composition of the body of water properly.

In the case of sampling for monitoring purposes, the substances of interest have been identified, and their background values (e.g., mean, range, and variance) have been ascertained. The data are assumed to have been examined for normality (i.e., the data exhibit a normal distribution over time). In this case, the investigator may wish to specify the conditions of variability that will be tolerated. Thus, the estimated standard deviation may be set within a certain percentage of its true value at a particular confidence level. Or, the investigator may specify an accuracy of a sample mean by specifying the percent coefficient of variation that will be tolerated at a particular confidence limit. Further discussion on methods used to estimate the optimum amount of replication to characterize such water may be found in Berg's book (*16*). The volume of the sample in monitoring regimes is partly determined by the substance being monitored as well as its expected concentration. The volume should always be of sufficient size to avoid the north–south side of the boat dilemma.

Frequency of Sampling

When engaging in descriptive sampling, the frequency of sampling will be determined by the purpose of sampling. In the case where seasonal variation is of interest, weekly sampling will be necessary. To illustrate the extent of variation, Table IV depicts variation in some elements in Linsley Pond over a 1-year period of study (17). Explanations for these oscillations have been described in detail elsewhere (17, 18); however, these variations represent the period of human disturbance, the chemical composition of the ice, the chemical composition of the density current, the thawing of the ice, the periods of vegetative decay, the periods of vegetative growth, and the algal bloom.

When the purpose is monitoring for some group of substances, the frequency of sampling is often specified by permits or other types of regulatory action. In the absence of regulatory criteria, sampling frequency may be based on historical data illustrating the variation of some variable of interest through some specified period of time. The rule of thumb is that the length of the record should be at least 10 times as long as the longest period of interest. Thus, if the longest period of interest is 1 year, then 10 years of data would be required. Berg (16) provided a detailed discussion of this approach.

Types of Blanks

The types of blanks required for proper quality assurance may be described as equipment blanks, field blanks, and sampling blanks. *Equipment blanks* are obtained as follows: Prior to departure from the laboratory, all equipment should be soaked in the best grade of double-distilled (glass) water available. This soaking water should then be stored in appropriate glass vessels for analysis. *Field blanks* are obtained as follows: Prior to the collection of a sample, all equipment used to collect that sample and that will have contact with that sample must be soaked in the best grade of double-distilled (glass) water available. This water should then be stored in appropriate glass vessels for analysis. *Sampling blanks* are obtained as follows: During the period that a sample is being gathered, bottles containing the best grade of distilled water available should be exposed to the air. This exposure should extend from the beginning to the end of sampling. At the termination of sampling, the bottles containing blanks should be sealed, labeled, and placed in 4 °C chests for eventual analysis.

These three types of blanks may become of paramount importance in the event that erratic results are obtained from sample analysis. These blanks, on analysis, may identify unsuspected contaminants associated with distilled water purity, improper dishwashing procedures, contamination problems associated with travel, and latent air contaminants that may have been sorbed by the samples during collection. This last problem has been encountered most frequently when groundwater samples were involved. The sampling blank was found to contain substances that in the past had been absent from groundwater samples.

Table IV. Striking Weekly Changes in Some Elemental Concentrations in Various Strata in Linsley Pond Observed during a 53-Week Study

Date	Strata (m)	Ti (kg)	Fe (kg)	Mg (kg)	Sr (kg)	Pb (kg)	Br (kg)	Mn (kg)	Hg (g)
Sept. 14, 1965	0–14.5	—	—	5400 (100)	29 (2)	5.8 (1)	—	—	—
Sept. 20, 1965	0–14.5	—	—	8300 (100)	49 (5)	10.5 (2)	—	—	—
Nov. 1, 1966	8–11	—	—	—	—	—	—	580 (11)	—
Nov. 8, 1966	8–11	—	—	—	—	—	—	40 (4)	—
Feb. 6, 1966	0–14.5	—	—	—	—	—	—	—	—
Feb. 13, 1966	0–14.5	3 (0.2)	180 (10)	—	—	—	300 (5)	—	200 (15)
Feb. 20, 1966	0–14.5	26 (2)	1100 (50)	—	—	—	750 (10)	—	900 (90)
June 26, 1966	0–2.5	—	—	—	—	—	—	—	198 (10)
July 3, 1966	0–2.5	—	—	—	—	—	—	—	42 (5)

NOTE: The quantities in each stratum represent the total amount found in that layer. The numbers in parentheses denote the standard deviation.

Nature of Samples and Problems Associated with Sample Collection

This section will be devoted to problems associated either with the type of water body being studied (e.g., lakes, rivers, streams, oceans, or aquifers) or the physical state of the sample (e.g., snow, ice, rain, fog, or dew).

In many water bodies, the stratum from which the sample is gathered is of utmost importance from the standpoint of chemical composition. Data presented in Table V illustrate this point.

In lakes shallower than 5 m, sufficient wind action exists so that mixing can be assumed to occur. Therefore, neither chemical nor thermal stratification is likely to be encountered. In lakes deeper than 5 m, chemical stratification as well as thermal stratification may be found. The examples used in Table V for Linsley Pond represent the height of the density current (February 20, 1966), the spring homothermal period (March 13 and 20, 1966) and the period of eutrophication (anoxic mud surface) during the summer months (July and August 1966). Cedar Lake, CT, is not known to stratify. Thus, when sampling such water bodies, something should be known about the morphometry and morphology of lakes, reservoirs, and estuaries prior to developing any kind of sampling regimen.

Rapidly flowing rivers that are shallow fail to exhibit any sort of consistent, recognizable chemical stratification. The example presented for this type of river is the Jordan River, which was sampled in the summer of 1963 in the northern Jordan Valley of Israel (6). A rapidly flowing river of considerable depth may undergo chemical stratification unaccompanied by any pronounced thermal stratification. The Ohio River at Pittsburgh, PA, was examined over an 8-year period (Cowgill, U. M., University of Pittsburgh, unpublished data, 1970–1978).

The data presented in Table V show that the type of monitoring to be carried out in such water bodies should determine the stratum from which a sample is to be collected, and that the position should be strictly adhered to for the entire period of sampling.

Similar types of problems have been encountered in groundwater sampling, especially when much of the well casing has been screened. Data presented in Table VI were gathered sporadically over a period of 14 years from a 33-m-deep well that had 5 m of the casing screened. The original purpose of sampling this abandoned but not closed drinking water well was to see if measuring any chemical interchange between the groundwater and nearby lakes was possible. Thus, prior to all sampling, a minimum of 10 bore volumes was removed. Each bore volume was sampled for temperature, pH, and conductivity. When these three variables approached stability after freshwater had entered the well casing as noted by a sudden drop in temperature, sampling at various depths within the screened area was carried out. The closed bailer (19) used in this study was 2 m long.

Table V. Variation in Chemical Composition in Relation to the Sampling Position in Various Water Bodies

Date	Strata (m)	Mn (µg/L)	Fe (µg/L)	P (µg/L)	Ca (mg/L)	Ti (µg/L)	K (mg/L)	Na (mg/L)
		Linsley Pond, CT (stratified)						
Feb. 20, 1966	0–2.5	300 (35)	400 (38)	38 (4)	27 (3)	—	—	—
	8–11.0	1000 (83)	5000 (450)	75 (8)	30 (3)	—	—	—
Mean for March 13, 1966 and March 20, 1966	0–2.5	250 (24)	480 (50)	59 (6)	28 (3)	—	—	—
	8–11	2000 (100)	900 (83)	64 (6)	29 (3)	—	—	—
Mean for July and August	0–2.5	400 (43)	300 (35)	55 (6)	31.3 (4)	—	—	—
	11–14	6000 (550)	2700 (100)	264 (20)	34.0 (2)	—	—	—
		Cedar Lake, CT (unstratified)						
April 17, 1966	0–2.5	—	—	32 (3)	—	4.1 (0.1)	—	—
	2.5–5.0	—	—	34 (3)	—	4.2 (0.8)	—	—
July 17, 1966	0–2.5	—	—	15 (2)	—	2.2 (0.4)	—	—
	2.5–5.0	—	—	16 (2)	—	2.1 (0.5)	—	—
Aug. 7, 1966	0–2.5	—	—	16 (2)	—	2.0 (0.3)	—	—
	2.5–5.0	—	—	15 (2)	—	1.9 (0.2)	—	—

Continued on next page.

Table V.—Continued

Date	Strata (m)	Mn (µg/L)	Fe (µg/L)	P (µg/L)	Ca (mg/L)	Ti (µg/L)	K (mg/L)	Na (mg/L)
Jordan River, Israel								
June–August, 1963 (weekly)	0–0.5	—	934 (88)	—	51.3 (5)	—	0.82 (0.1)	3.5 (0.5)
	2–2.5	—	950 (85)	—	52.0 (5)	—	0.88 (0.1)	3.8 (0.5)
Ohio River, Pittsburgh, PA								
1970–1978 (biweekly)	0–0.5	—	1.2[a] (0.2)[a]	—	55 (6)	—	2.1 (0.1)	6.4 (1.0)
	8–10	—	4.3[a] (0.4)[a]	—	80 (7)	—	4.2 (0.3)	10.5 (1.5)

NOTE: The numbers in parentheses denote the standard deviation.
[a] These values for Fe are expressed in milligrams per liter.
SOURCE: Adapted from references 6, 17, and 18, and U. M. Cowgill, unpublished data, 1970–1978.

Table VI. Variation in Chemical Composition in Relation to Sampling Strata
in a 33-m Drinking Water Well Containing a 15-m Screen

Strata (m)	Mo ($\times 10^{-2}$ µg/L)	V ($\times 10^{-4}$ µg/L)	Bi ($\times 10^{-2}$ µg/L)	Hg (µg/L)	Be ($\times 10^{-2}$ µg/L)
19–21	15 (2.5)	0.8 (0.2)	8.5 (1.7)	1.6 (0.4)	0.7 (0.2)
24–26	10 (2.0)	30 (4.0)	4.5 (1.0)	18.3 (1.7)	18 (3.0)
29–31	28 (3.5)	1115 (97.3)	18.2 (2.0)	3.0 (0.5)	11 (1.0)

NOTE: The numbers in parentheses denote the standard deviation.

Most of the variability is reasonably consistent within the depth sampled (i.e., no percent coefficient of variation is greater than 28.5), yet the variation between depths is statistically significant in most cases. The cause of this variation is natural. The substrata vary chemically. Careful sampling, which avoided any excessive disturbance of the recharged water column, showed consistent variation over time and was later confirmed to originate from the substrata by sampling the various substrata and leaching them with natural water.

Position of sample collection is also important in the collection of ice and snow. During the 53 weeks of the Linsley Pond Study, the pond was ice-covered from January 16 through March 6. During that time, the thickness of the ice varied from 5 to 25 cm. In the open water, the variation in chemical composition was statistically significant in the vertical direction and varied little within an ice stratum in the horizontal direction. To obtain enough material to make chemical analysis precise enough, ice was gathered weekly so that on melting, the ice provided a volume of 250–300 L of water. Table VII shows the variation in chemical composition of ice in relation to depth of ice. It should be noted that there is a poverty of data on the chemical composition of ice from lakes and ponds. Maksimovich and Yashchenko (20) noted that the mineral content of ice in lakes and ponds varied between 22 and 90 mg/L. The sources of the chemical composition of ice are the chemical composition of the surface water from which ice forms; entrapped dust that contributes to the concentration of Fe, Ti, and Mo (17, 18); entrapped plankton; and the rate at which formation occurs.

The elements Ca, Si, Al, P, Ba, Sr, and Mn are concentrated in the ice in reference to the surface centimeter of water, but this concentration–depth relationship is not the case for Fe and Ti. How ice is to be sampled is largely dependent upon the purpose of sampling. If the concern is the geochemical effect of melting on the receiving water, then adequate results can be obtained by bulk sampling in a number of localities on the lake. If the interest is how the composition of the surface water is related to that of the ice in contact with it,

Table VII. Variation in Chemical Composition of Ice in Relation to the Thickness of the Ice, and Chemical Composition of Water in Relation to Depth of Water

Thickness or Depth (cm)	Ca	Si	Al	Fe	P	Ba	Sr	Mn	Ti
				Ice Thickness					
0–5	29,000 (1500)	30,000 (3000)	800 (80)	600 (55)	38 (4)	67 (6)	62 (6)	200 (20)	18 (2)
5–10	27,500 (2000)	33,000 (3000)	600 (65)	500 (60)	42 (4)	46 (6)	59 (6)	150 (15)	22 (2)
10–15	28,000 (2000)	32,000 (3000)	800 (80)	550 (60)	50 (5)	67 (6)	60 (6)	180 (20)	18 (2)
15–20	28,000 (1000)	25,000 (1500)	500 (55)	430 (40)	40 (4)	65 (6)	60 (6)	100 (10)	12 (1)
				Water Depth					
0–1	1800 (150)	4800 (300)	180 (15)	600 (60)	9 (1)	8 (1)	3 (0.5)	32 (3)	18 (2)

NOTE: All values are in micrograms per liter. Numbers in parentheses denote the standard deviation.

then the ice in a series of strata should be sampled and the ice samples'
composition compared with that of the surface water.

The increase of Ca, Si, Al, Fe, Ba, Sr, Mn, and Ti in the surface ice is in
part related to entrapped atmospheric dust. The chemical composition of ice
varies with the depth of the ice more than with the horizontal component, may
vary significantly from day to day, and varies significantly on a weekly basis
(17, 18) during the period of ice cover.

The chemical composition of snow is highly variable. In severe winters,
snow can contribute substantially to the elemental composition of a water body.
Ideally, snow should be sampled at the time of snowfall because its composition
changes considerably on standing. However, if snow is sampled immediately
upon deposition, conditions tend to reflect only various chemical changes in the
atmosphere. The chemical composition during a snowfall changes in much the
same way as during rainfall. This situation will be discussed later in this section.
If snow hasn't occurred for some time, the initial snowfall tends to be, from a
chemical viewpoint, quite concentrated. As the snowfall progresses, the
composition is diluted until, by the end of a long snowfall, the composition
becomes a rather poor grade of distilled water containing only a few elements
(Cl, Br, I, S, B, Na, K, Ca, Mg, and N) that can be detected with ease. Thus, if
sampling is done shortly after a 0.5-m snowfall, a distinct chemical variation will
be encountered between the snow that fell initially and the snow that recently
fell. This discussion refers to snowfalls within 100 km of industrialized areas.
High-altitude snowfalls are chemically much more dilute. This summary on the
chemical (elemental) composition of snow is the result of sporadic investigations
that have occurred in lake basins at various altitudes in diverse places (Cowgill,
U. M., unpublished data, 1963–1981).

The chemical composition of rain is highly variable and the comments just
made about snow also apply. Table VIII shows the chemical results of a 3-h
rainfall collected in March 1978 on the roof of a 12-story building in Pittsburgh,
PA. The rain was collected in polytetrafluoroethylene (Teflon) rain collectors that
were covered with very fine nylon netting to prevent the entrance of soot in the
samples. Collections were made at 10-min intervals. The most pertinent data are
shown in the table.

These data show that the rain has been reasonably successful in cleaning
the atmosphere of its contaminants with the exception of sulfate and total
nitrogen. Pittsburgh is a limestone region, and certainly some of the Ca and Mg
in rain originates from limestone dust, but the remainder of the contribution
arises from the local cement industry. The major point is that reports of the
chemical composition of a particular rainfall are meaningless if the aliquot of
the rainfall analyzed is not timed and dated. As may be observed from
examination of the data in Table VIII, the composition varies with time.
Reporting the average composition of a rainfall is meaningless because 70 min
of rain having a pH of 2.4 will damage vegetation and buildings. The fact that
the final pH approached that of normal rain would not be useful if the purpose

Table VIII. Variation in Chemical Composition of a 3-h Rainstorm in Pittsburgh, PA, in March 1978, and Variation in Chemical Composition of Uncontaminated Rain

Time (min)	pH	Cl	SO$_4$	B	Na	K	Mg	Ca	N
Pittsburgh Rainstorm									
0–10	1.2	10.1 (2)	29.3 (4)	0.18 (0.03)	5.8 (1)	0.61 (0.1)	2.75 (0.5)	3.05 (0.6)	4.7 (1)
10–20	1.5	8.8 (2)	25.8 (4)	0.15 (0.03)	5.2 (1)	0.54 (0.1)	2.40 (0.5)	2.68 (0.5)	4.2 (1)
50–70	2.4	6.5 (1)	18.8 (4)	0.12 (0.02)	3.7 (1)	0.39 (0.1)	1.77 (0.5)	1.90 (0.5)	3.0 (0.5)
120–150	3.9	4.6 (1)	12.0 (3)	0.08 (0.01)	3.1 (0.05)	0.28 (0.1)	1.00 (0.2)	1.40 (0.5)	1.8 (0.5)
170–180	5.2	4.2 (1)	7.6 (2)	0.08 (0.01)	2.4 (0.05)	0.25 (0.1)	0.38 (0.1)	0.25 (0.1)	0.86 (0.1)
Uncontaminated Rain									
—	—	5.1	1.7	—	3.1	0.2	0.3	0.2	<0.009

NOTE: All values are in milligrams per liter. The numbers in parentheses denote the standard deviation.
SOURCE: Adapted from reference 21 for uncontaminated rain data only.

of collection is to monitor and avoid damage to living things and structures of civilization. Further discussion on adequate precipitation sampling can be found elsewhere (22–24).

Great interest has been expressed recently on the chemical composition of fog and dew. With the possible exception of the first monolayer of seawater, no physical state of water is more difficult to sample than fog and dew. Dew was collected with a kind of glass vacuum pump from three arbitrarily selected strata of a grass field. A total of 10 acres was collected on an early summer morning (Cowgill, U. M., Yale University, unpublished data, 1968). These 10 acres provided 1 L of dew from each of the three strata. Three replicates of each strata were analyzed. Table IX illustrates some of the problems encountered with dew collection and analyses. The samples had to be concentrated over P_2O_5 to detect K, Cl, S, and N with ease. The biggest single problem is that reproducibility is rather poor in the sense that the percent coefficient of variation is much greater

Table IX. Chemical Composition of Three Strata of Dew Extracted from a Grass Field

Date	Strata	K	Cl	SO_4	N
July 15, 1968	top	40	100	1200	30
		(20)	(50)	(500)	(16)
	middle	37	575	985	39
		(20)	(300)	(510)	(17)
	bottom	79	1000	1962	56
		(35)	(500)	(990)	(30)
July 25, 1968	top	65	975	1500	75
		(30)	(500)	(700)	(40)
	middle	69	1000	1783	95
		(31)	(500)	(800)	(48)
	bottom	54	1400	2000	98
		(25)	(700)	(990)	(50)
Aug. 15, 1968	top	74	1000	1700	85
		(35)	(500)	(810)	(42)
	middle	55	1000	1400	85
		(25)	(500)	(750)	(42)
	bottom	100	930	975	78
		(42)	(410)	(500)	(40)
Aug. 29, 1968	top	82	1500	1565	85
		(40)	(750)	(700)	(42)
	middle	80	1600	1511	90
		(40)	(800)	(711)	(42)
	bottom	55	755	800	44
		(22)	(400)	(410)	(20)

NOTE: The grass field consisted of *Phleum pratense* L. (Timothy) and *Bromus inermus* L. (Bromegrass). All values are in micrograms per liter. The numbers in parentheses denote the standard deviation.

than would normally be acceptable. The percent coefficient of variation did not improve noticeably with time.

Some variation among the strata can be seen, although the bottom strata are probably influenced by some upward movement from the soil to the dew. These samples were collected at the same time in the morning (5:00 a.m.). However, a great deal more data need to be collected before any conclusions can be reached concerning the chemical composition of dew. The Cl, S, and N probably originated from the atmosphere, and the K originated from the grass. Since these data were gathered, various synthetic materials that absorb many times their weight have become available. The use of such material to collect data from a predetermined number of stations might provide a successful alternative to the vacuum cleaner approach to the sampling of dew.

A similar approach has been used to collect fog. Unfortunately, samples collected simultaneously 200 m apart provided a 200% spread on elements such as S, N, and Ca. Until duplicate samples can be collected simultaneously and upon analysis provide some reasonable degree of concordance, the chemical composition of fog will remain elusive.

Finally, the oceans present sampling problems not unlike those of stratified lakes. The precautions that have been noted under that heading should be adhered to in marine sampling. The chemical composition of near-shore waters is chemically far more variable than any composition thus far encountered in freshwater bodies. Some work (25, 26) is in progress to estimate atmospheric contributions to ocean waters. Further discussion may be found in Broecker and Peng (27).

Planning and Desired Confidence Level of Chemical Results

Before engaging in a research project in which the accuracy will be dependent upon proper sampling, the purpose of sampling must be clearly identified. Once this purpose has been identified, the items to be studied must be selected, and the concentration range expected for these items must be described. At this point, sampling tools that do not contain and will not sorb the items of interest must be procured. The degree of replication must be discussed not only in the light of the desired precision but also with the view of the confidence limits within which the desired data should fall. Frequency of sampling will largely be determined by the purpose of sampling. Finally, background data must be gathered from the water body to be studied. Once such data have been collected, more information will be available about the items of interest; their concentration range; their seasonality; and their mobility within the system to be studied, and an appropriate research effort can be described, with proper sampling becoming the cornerstone of the effort.

References

1. Keith, L. A. *Environ. Sci. Technol.* **1981**, *15*, 156–162.
2. Keith, L. A. *Advances in the Identification and Analysis of Organic Pollutants in Water*; Ann Arbor Press: Ann Arbor, MI, 1981; Vol. 1, pp 3–479.
3. Keith, L. A. *Advances in the Identification and Analysis of Organic Pollutants in Water*; Ann Arbor Press: Ann Arbor, MI, 1981; Vol. 2, pp 481–1170.
4. Kratochvil, B. G.; Taylor, J. K. "A Survey of the Recent Literature on Sampling for Chemical Analysis (Final Report)"; NBS Technical Note 1153; National Bureau of Standards: Washington, DC, 1982.
5. Kratochvil, B. G.; Wallace, D.; Taylor, J. K. *Anal. Chem.* **1984**, *56*, 113R–119R.
6. Cowgill, U. M. *Int. Rev. Gesamten Hydrobiol.* **1980**, *65*, 379–409.
7. Barcelona, M. J.; Helfrich, J. A.; Garske, E. E. *Anal. Chem.* **1985**, *57*, 460–464.
8. Barcelona, M. J.; Helfrich, J. A.; Garske, E. E. *Anal. Chem.* **1985**, *57*, 2752.
9. Miller, G. D. *Proc. Nat. Symp. Aquifer Restor. Ground Water Monit. 2nd* **1982**, 236.
10. Massee, R.; Maessen, F. J. M. J.; DeGoeij, J. J. M. *Anal. Chim. Acta* **1980**, *127*, 181–193.
11. Robertson, D. E. *Anal. Chim. Acta* **1968**, *42*, 533–536.
12. Robertson, D. E. *Anal. Chim. Acta* **1968**, *40*, 1067–1072.
13. Boettner, E. A.; Ball, G. L.; Hollingsworth, Z.; Aquino, R. *Organic and Organotin Compounds Leached from PVC and CPVC Pipe*; U.S. Environmental Protection Agency. U.S. Government Printing Office: Washington, DC, 1981; EPA-600/1-81-062.
14. McBean, E. A.; Rovers, F. A. *Ground Water Monit. Rev.* **1985**, *5*, 61–64.
15. Lind, O. T. *Handbook of Common Methods in Limnology*; C. V. Mosby: St. Louis, MO, 1974; pp 26–31.
16. Berg, E. L. *Handbook for Sampling and Sample Preservation of Water and Wastewater*; U.S. Environmental Protection Agency. U.S. Government Printing Office: Washington, DC, 1982; EPA-600/4-82-029.
17. Cowgill, U. M. *Arch. Hydrobiol.* **1970**, *68*, 1–95.
18. Cowgill, U. M. *Arch. Hydrobiol.* **1976**, *78*, 279–309.
19. Gibb, J. P.; Schuller, R. M.; Griffin, R. A. *Proceedings of the Sixth Annual Symposium on the Disposal of Hazardous Wastes, Chicago, Illinois, March 17-20, 1980*; U.S. Environmental Protection Agency: Cincinnati, OH, 1980; pp 31–38; EPA-600/9-80-010.
20. Maksimovich, G. A.; Yashchenko, R. V. *Zh. Khim. Kii* **1963**, Abstract 21 E 51.
21. Gorham, E. *Geochim. Cosmochim. Acta* **1955**, *7*, 231–239.
22. Asman, W. A. H.; Jonker, P. J. In *Deposition of Atmospheric Pollutants*; Georgic, H. W.; Pankrath, J., Eds.; D. Reidel: Amsterdam, 1982; pp 115–123.
23. Campbell, S.; Scott, H. In *Quality Assurance for Environmental Measurements*; ASTM Special Technical Publication 867; Taylor, J. K.; Stanley, T. W., Eds.; American Society for Testing and Materials: Philadelphia, PA, 1985; pp 272–283.
24. Raynor, G. S.; Hayes, J. V. In *Sampling and Analysis of Rain*; ASTM Special Technical Publication 823; Campbell, S. A., Ed.; American Society for Testing and Materials: Philadelphia, PA, 1983; pp 50–60.
25. Green, D. R. *National Research Council of Canada Publication No. 16565*; Marine Analytical Chemistry Standards Program; National Research Council: Halifax, Canada, 1978.
26. Duce, R. A. *Special Environmental Report 12*; WMO No. 504; World Meteorological Organization: Geneva, Switzerland, 1979.
27. Broecker, W. S.; Peng, T-H. *Tracers in the Sea*; Lamont–Doherty Geological Observatory. Columbia University: New York, 1982. (For Atlases containing chemical data [GEOSECS], see pp 670–672.)

RECEIVED for review January 28, 1987. ACCEPTED March 27, 1987.

Chapter 12

Assessment of Measurement Uncertainty

Designs for Two Heteroscedastic Error Components

Walter S. Liggett

Specification of quality assurance samples is discussed for sampling and measurement error having two independent, heteroscedastic components. Each component is heteroscedastic because its standard deviation depends on concentration. This error model is appropriate for environmental studies in which the samples are measured in several batches and the sample concentrations vary widely. The design considered specifies that a duplicate of a routine sample and two reference samples be included in each batch. Some batches have reference samples with the same concentration; others do not. The adequacy of this design depends on the number of batches, on the relative sizes of the within-batch and among-batch error components, and on the concentrations in both the samples of interest and in the quality assurance samples.

U SE OF REFERENCE AND DUPLICATE SAMPLES to estimate the standard deviation of the measurement error is the subject of this chapter. Estimation of the error standard deviation is only one aspect of error assessment. In some studies, limits on sample contamination and the sensitivity of the sampling and measurement procedures are the most important aspects. These aspects are important if the primary objective of the study is the detection of a toxic chemical at concentrations above a health-effects level. In other studies, the detection of correctable problems with the sampling and measurement procedures is the most important aspect. Estimation of the error standard deviation is particularly important in studies based on comparison of concentrations at different times or different locations. Such comparisons occur, for example, in the long-term monitoring of supposedly similar locations.

Estimation must be based on an assumed probability model of the measurement error. Various models might be considered. In the most elementary model, each measurement is accompanied by an independent realization of the error, and the realizations are normally distributed with zero means and equal standard deviations. The model considered in this chapter is more general in two ways. First, the measurement error is assumed to consist of two statistically independent components, a within-batch component and an among-batch component. The batch structure is caused by the division of the samples into batches for processing. Second, the standard deviation of each component of the measurement error is allowed to depend on concentration. In this sense, the error components are *heteroscedastic*.

The assumed error model is intended to reflect physical sources of variation in the sampling and measurement procedures. Examples of error sources include incomplete mixing in the subsampling procedure, sample contamination during handling, sample degradation during transport, and variation in the analytical instrument. The amount of variation contributed to a measurement by each of these sources will usually depend on the concentration of the analyte of interest in the sample. Investigators often assume that the standard deviation of the measurement error is proportional to concentration (i.e., the relative standard deviation is constant). Although this assumption is often better than the assumption of constant standard deviation, an assumption more general than either of these is usually needed if the range of concentrations in the samples is large. Because a large range of concentrations is typical of many environmental studies, this chapter is based on a more general assumption about the dependence on concentration.

Consider an environmental study in which samples from several locations are repeatedly collected and analyzed for a chemical constituent of interest. If the study involves a large number of samples or a long period of time over which the samples are collected, then the investigator cannot avoid an error model that contains multiple components. Either a large number of samples or a long period of time forces the investigator to process the samples in batches. Inevitably, errors in measurements from the same batch are more similar than errors in measurements from different batches. Often, samples in the same batch are subjected to the same physical environment and are measured under the same calibration of the instrument, for example. Thus, measurements on samples processed in batches must be considered to have at least two error components, a within-batch component and an among-batch component.

Studies that involve multiple heteroscedastic error components similar to those discussed should be designed so that the dependence of standard deviation on concentration can be estimated for each component. This estimation cannot be accomplished without some assumptions about the dependence of the component standard deviations on concentration. As discussed in the next section, this chapter is based on simple parametric models of the dependence on concentration. In this case, estimation of the parameters in these models is part of the estimation of the measurement error properties. More general models than the ones considered here might be adopted at the risk of a substantial increase in the number of quality assurance samples required.

To assess the sampling and measurement error, an investigator might introduce various types of quality assurance samples. In this chapter, estimation of the error properties is based on two types. The first type is a duplicate sample collected coincidentally with the routine sample taken at some location. In the Eastern Lake Survey portion of the National Surface Water Survey (*1*), these pairs of samples are called *routine-duplicate pairs*. If they are processed in the same batch, they are useful in assessing the within-batch error (*2*). The second type is a prepared reference sample inserted into the sample processing procedure as close to the beginning as possible. Such reference samples are called *audit samples* in the Eastern Lake Survey. The Eastern Lake Survey also included other types of quality assurance samples aimed just at the error of the analytical laboratory.

In the choice of quality assurance samples, a judgment must be made on what sources of error each reflects. Most sources of error in the sampling procedure cannot be assessed through reference samples. Duplicate samples provide some information on the sampling procedure but cannot be used to assess the among-batch component of the sampling error. The sampling error would have an among-batch component if, for example, the concentration in the material collected were to depend on temperature and moisture content and thus on the season when the sampling was done. In this chapter, the routine-duplicate pairs are assumed to be subject to the same sources of error as the reference samples. This assumption seems plausible for lake sampling such as the sampling in the Eastern Lake Survey. This assumption is suspect in the case of sampling bulk materials because collection of a duplicate sample involves a subsampling error that cannot be made part of the handling of reference samples. Thus, the designs discussed in this chapter must be generalized for the case of bulk materials.

On the basis of the assumed error model and the feasibility of various types of quality assurance samples, the quality assurance samples for an environmental study can be specified. The specification should be judged first on whether the error properties that are part of the error model can be estimated. For example, separation of the error components requires routine-duplicate pairs or duplicate reference samples in some batches. Estimation of the dependence of the standard deviation on concentration requires quality assurance samples at various concentrations. If estimation of the error properties is possible, then the adequacy of the estimates for the purpose of the study becomes the issue. The case in which this judgment is simplest is the one in which the study involves comparisons between measurements that might plausibly differ because of measurement error alone. In this case, the quality assurance samples have to provide estimates of the error properties that are good enough to allow the measurement error to be distinguished from deviations due to other causes. Examples of this case are discussed in the section on required sample sizes.

This chapter is organized as follows: In the next section, the model for the measurement error is discussed. In the following section, the allocation of the quality assurance samples is specified. Then, estimation of the parameters in the error model is detailed. Next, on the basis of some simple study protocols, various choices for concentrations of the quality assurance samples are explored. In the final section, possibilities for adapting alternative statistical methods from the literature are considered.

Measurement Error Model

The error model consists of two independent heteroscedastic error components, the within-batch component and the among-batch component. All the measurements are assumed to be made in one laboratory so that there is no need to consider an interlaboratory error component. Each measurement x made on a sample having concentration μ is given by

$$x = \mu + h(\mu)\epsilon_a + g(\mu)\epsilon_w \tag{1}$$

where the random variables ϵ_a and ϵ_w are independent and normal and have mean 0 and variance 1, and the functions of μ, $g(\mu)$ and $h(\mu)$ give the standard deviations of the within-batch and among-batch error components, respectively, as functions of concentration. Although containing only two components, equation 1 applies when there are several sources of error if the contribution of each source is normal and has mean zero.

The parametric forms assumed for $g(\mu)$ and $h(\mu)$ are given by equations 2 and 3, respectively.

$$g(\mu) = \sigma(\mu + \tau)^{1-\lambda} \tag{2}$$

$$h(\mu) = (\eta_0 + \eta_1\mu)(\mu + \tau)^{1-\lambda} \tag{3}$$

where the parameters σ, λ, τ, η_0, and η_1 are unknown prior to the measurement of the samples and must be estimated as part of the error assessment. For cases in which the component standard deviations increase with concentration, these forms will often provide an adequate representation. The case in which each component has constant standard deviation is given by $\lambda = 1$ and $\eta_1 = 0$. The case in which each component has constant relative standard deviation is given by $\lambda = 0$, $\tau = 0$, and $\eta_1 = 0$. The parameter τ is needed because even if the within-batch error component has constant relative standard deviation for high concentration, the standard deviation will generally not go to zero with decreasing concentration. The factor $(\eta_0 + \eta_1\mu)$ in $h(\mu)$ may have to be generalized in some cases. Generalizing this factor and the corresponding methods discussed in this chapter is not difficult.

The particular parameterization shown in equations 2 and 3 is a consequence of our approach to elimination of the dependence of the within-batch error on the concentration. Our approach is the power transformation (3)

$$y = \frac{(x + \tau)^\lambda - 1}{\lambda} \qquad \text{for } \lambda \neq 0 \tag{4a}$$

$$y = \log (x + \tau) \qquad \text{for } \lambda = 0 \tag{4b}$$

where y is the transformed measurement of concentration x. Assume that $g(\mu)/(\mu + \tau)$ and $h(\mu)/(\mu + \tau)$ are much less than 1 for the true value of (λ, τ). For values of (λ, τ) close to the true value, the result of the power transformation applied to x can be approximated by

$$y = \frac{(\mu + \tau)^\lambda - 1}{\lambda} + h(\mu)(\mu + \tau)^{\lambda-1}\epsilon_a + g(\mu)(\mu + \tau)^{\lambda-1}\epsilon_w \tag{5}$$

This approximation is the Taylor series expansion about $g(\mu)\epsilon_w = 0$ and $h(\mu)\epsilon_a = 0$. This approximation shows that for the true value of (λ, τ), the within-batch standard deviation after transformation has the value σ, and the among-batch standard deviation after transformation is given by $\eta_0 + \eta_1\mu$. In general, a power transformation induces dependence between independent error components. This dependence can be ignored if $g(\mu)/(\mu + \tau)$ and $h(\mu)/(\mu + \tau)$ are small as assumed. In the approximation given by equation 5, the two error components are additive and independent after transformation. The effect of the transformation is to change the dependence on concentration.

Multibatch Design for Error Assessment

Of the many ways to allocate quality assurance samples to the batches of routine samples, only one will be considered in this chapter: the collection of one duplicate for every batch processed and the insertion of two reference samples in every batch processed. In some batches, the reference samples have the same concentration, and in other batches, the reference samples have different concentrations. Two reference samples with the same concentration contribute effectively to the estimation of the properties of the within-batch error. Two reference samples having different concentrations contribute effectively to the estimation of η_1, the parameter that gives the dependence of the among-batch error on concentration.

This allocation is comparable to the allocation used in the Eastern Lake Survey. If blank samples are considered as reference samples, then the Eastern Lake Survey had three reference samples: two blanks and another reference sample as well as a routine-duplicate pair in each batch. Unfortunately, although blanks may be useful in some ways, they may not be useful in the estimation of error properties. Blanks are easy to prepare and are very helpful in detecting contamination of the samples. However, blanks have a concentration that is at an extreme and thus may not provide information useful in the assessment of measurements of nonzero concentrations. Laboratories will often report concentrations for blank samples as none detected. Even if a value is reported, the value may not reflect instrument behavior for nonzero concentrations. For example, for an instrument that has a search for peak location and peak width in its processing software, the behavior at zero concentration may differ from the behavior at low concentrations (4). Thus, reference samples with concentrations above the instrument detection limit are often of more help in estimating error properties.

The allocation of quality assurance samples considered in this chapter seems worthy of consideration for reasons other than the performance achieved in estimation of error properties. This design seems desirable because of its simplicity and because it spreads the quality assurance samples throughout the study. In a large study, this property is important because in the sequential processing and analysis of a large number of batches, changes in the probability properties of the error may occur. If the quality assurance samples are spread throughout the study, some types of changes can be accommodated by separation of the measurement error assessment into two parts: one for batches analyzed before the change and one for batches analyzed after the change. For example, in the Eastern Lake Survey, the original procedure for preparing aliquots for nitrate analysis introduced nitrate contamination. This contamination was discovered about halfway through the survey, and new measurements on unfiltered samples held for a long period had to be substituted. Thus, for the nitrate measurements in this survey, the sampling and measurement error had to be assessed separately for each period.

For the purposes of this chapter, only six samples in each batch need to be considered. Let t index the batches, where $t = 1, 2, \ldots N$. The measurements for the locations of interest are denoted by x_{1t} and x_{2t}. Although in a batch, more samples could be from locations of interest, the examples in this chapter only require notation for two such locations. The measurements on the routine-duplicate pair of samples are denoted x_{3t} and x_{4t}. The measurement on the first reference sample, which is of type q_1, is denoted by x_{5t}. The measurement on the second reference sample, which is of type q_2, is denoted by x_{6t}. The symbols q_1 and q_2 denote functions of t that indicate the reference sample type. We do not consider the case in which the routine-duplicate pair is drawn from one of the locations of interest.

Estimation of the Error Properties

From the measurements on the quality assurance samples, x_{3t}, x_{4t}, x_{5t}, and x_{6t}, the properties of the error can be estimated. To assess the within-batch error, the pairs of measurements made on nominally identical samples are used. These samples are either from the same location or reference samples of the same type. These pairs are used to choose the proper transformation from the family given in equation 4. This choice is made by transforming the measurements in each pair and then testing the transformed measurements for constant within-batch standard deviation. A confidence region for the transformation parameters (λ, τ) is given by the set of transformations for which the hypothesis of constant within-batch standard deviation is not rejected.

Let the number of sample pairs, which is the number of routine-duplicate pairs plus the number of batches with reference samples of the same type, be denoted by p. Let a_i be the average of the transformed measurements on the ith pair, and let d_i be the difference between the transformed measurements on the ith pair. Denoting the transformed measurements by y_{jt}, where $j = 3, 4, 5,$ or 6, leads to

$$a_i = \frac{y_{3t} + y_{4t}}{2} \quad \text{or if} \quad q_1 = q_2, \, a_i = \frac{y_{5t} + y_{6t}}{2} \tag{6a}$$

$$d_i = y_{3t} - y_{4t} \quad \text{or if} \quad q_1 = q_2, \, d_i = y_{5t} - y_{6t} \tag{6b}$$

Let the index i be chosen so that the averages and differences are ordered in increasing a_i (i.e., so that $a_1 \leq a_2 \leq \ldots \leq a_p$). Equation 5 shows that d_i involves only the within-batch error and that the standard deviation of d_i is $\sqrt{2g(\mu)}(\mu + \tau)^{\lambda-1}$.

If the transformation applied to the measurements is the one that gives within-batch error with constant standard deviation, then the differences d_i are independent of the averages, and sorting the differences on the averages does not alter the independence of the differences from each other. This fact is the basis of our statistical test for constant standard deviation. Consider the sequence $(d_{2k-1})^2 + (d_{2k})^2$, where $k = 1, 2, \ldots m$, and where m is the integer part of $p/2$, denoted $[p/2]$. (The symbol p denotes the number of nominally identical samples.) If the transformation applied is the proper one, then this sequence consists of independent identically distributed chi-squared variates, each having two degrees of freedom. On the other hand, if the transformation is not the proper one, then the members of this sequence will vary not only because of randomness but also because the standard deviation of d_i is not constant.

$$S_j = \frac{\sum_{k \leq j} (d_{2k-1})^2 + (d_{2k})^2}{\sum_{k \leq m} (d_{2k-1})^2 + (d_{2k})^2} \quad \text{where } j = 1, 2, \ldots m - 1 \tag{7}$$

The symbol S_j denotes a normalized partial sum that would equal j/m if the sequence $(d_{2k-1})^2 + (d_{2k})^2$ were constant. As discussed by Durbin (5), if the d_i values are independent and normal with constant standard deviation, then the S_j values are distributed as the order statistics from a sample of $m - 1$

independent observations from a uniform (0, 1) distribution. In other words, the S_j values have the same probability properties as $m - 1$ independent uniform (0, 1) variates that have been arranged in increasing order. This hypothesis can be tested by using, as test statistics,

$$c^+ = \max_j [S_j - (j/m)] \tag{8a}$$

$$c^- = \max_j [(j/m) - S_j] \tag{8b}$$

Thus, c^+ and c^- can be used as statistics for the choice of transformation. Alternatively, the test statistic

$$z = \frac{\Sigma S_j - [(m - 1)/2]}{[(m - 1)/12]^{\frac{1}{2}}} \tag{9}$$

might be compared with an appropriate percentile of the normal distribution (6). Thus, z is an alternative statistic for the choice of transformation. Equations 8 and 9 each lead to a confidence region. These regions consist of the values of (λ, τ) for which the statistic is less than the critical value.

The Monte Carlo experiments discussed in this chapter are based on the second of these interval estimates for (λ, τ). In a thorough statistical analysis, a point estimate of (λ, τ) is also needed so that transformed measurements can be examined. Various point estimates of the transformation might be suggested. A reasonable choice is the (λ, τ) that minimizes the larger of c^+ and c^-.

Transformed measurements on the reference samples can now be used to obtain point estimates for the among-batch error. In doing this, the error made in estimating (λ, τ) is ignored. Consider the sum of and the difference between reference-sample measurements from the same batch. From equations 2, 3, and 5, equations 10a–d were obtained.

$$\text{var } (y_{5t} + y_{6t}) = \left\{ 2[\eta_0 + \eta_1 \mu(q_1)] \right\}^2 + 2\sigma^2 \qquad \text{if } q_1 = q_2 \tag{10a}$$

$$\text{var } (y_{5t} - y_{6t}) = 2\sigma^2 \qquad \text{if } q_1 = q_2 \tag{10b}$$

$$\text{var } (y_{5t} + y_{6t}) = \left\{ 2\eta_0 + \eta_1[\mu(q_1) + \mu(q_2)] \right\}^2 + 2\sigma^2 \qquad \text{if } q_1 \neq q_2 \tag{10c}$$

$$\text{var } (y_{5t} - y_{6t}) = \left\{ \eta_1[\mu(q_1) - \mu(q_2)] \right\}^2 + 2\sigma^2 \qquad \text{if } q_1 \neq q_2 \tag{10d}$$

where var is the variance, and $\mu(q_1)$ and $\mu(q_2)$ are the concentrations of the two reference samples. The variances on the left sides of these equations can be estimated from the reference samples and then used to estimate η_0 and η_1. Estimation of the variance of the within-batch error σ^2 also involves the differences between the routine-duplicate pairs. The discussion in Davidian and

Carroll (7) suggests various approaches to estimating η_0 and η_1. One approach is to subtract the estimate of $2\sigma^2$ from the estimates of the variances on the left sides and then take square roots. From these square roots, the parameters η_0 and η_1 can be estimated by weighted linear regression. Weighting is important because the variability of a variance estimate increases with the true variance.

For the case in which η_0 is much greater than $\eta_1\mu(q_1)$ and $\eta_1\mu(q_2)$, equation 10d provides a reasonable estimate of $\eta_1{}^2$. The reason is that var(y_{5t} − y_{6t}) does not involve the large variance component associated with η_0. From equation 10, for the case $q_1 \neq q_2$,

$$\eta_1{}^2 = \frac{\text{var }(y_{5t} - y_{6t}) - 2\sigma^2}{[\mu(q_1) - \mu(q_2)]^2} \tag{11}$$

When using this equation for estimation, the difference between the estimate of var(y_{5t} − y_{6t}) and the estimate of $2\sigma^2$ may be negative. The usual approach to this problem is to replace a negative difference by 0.

In the course of the foregoing analysis, problems with the assumed error model might be discovered. For example, the power transformation might not be appropriate in the sense that equation 2 does not hold for any value of (λ, τ), or the routine-duplicate pairs might reflect error sources that do not influence the reference samples. If the test for homogeneity of the transformed within-batch error rejects homogeneity for every (λ, τ), then the investigator should conclude that one of these conditions holds. A graph of S_j versus j might be instructive. For example, if the differences that correspond to the reference samples are smaller than the other differences, then the investigator may conclude that the second condition holds. Problems with the applicability of the error model might seriously affect study conclusions if the problems are not discovered until the data analysis stage of the investigation. If problems with the error model can be anticipated, then preliminary experimental work to discover the possibilities should be performed, or quality assurance samples appropriate to the most complicated case envisioned should be used.

Required Sample Sizes

Because more error characteristics must be estimated, (λ, τ) and (η_0, η_1) in addition to σ^2, more quality assurance samples are needed, one would expect, than in the simple case in which only a single standard deviation is to be estimated. The question is how many. In this section, a partial answer to this question is provided through Monte Carlo experiments. Two hypothesis-testing situations are considered for which the probability can be obtained of detecting a real difference. This probability, which depends on the size of the real difference, is the power of the test. This probability is studied not only as a function of sample size but also as a function of the concentrations of the quality assurance samples because these concentrations determine how well error properties can be estimated. Of course, the statistical analysis that accompanies an environmental study is much more intricate than a single hypothesis test.

To investigate various designs for the quality assurance samples, situations should be considered that are as simple as possible without being misleading. Two situations are considered in this section: one in which the null hypothesis is that the concentrations at the two locations are equal, and the other in which the null hypothesis is that the concentrations at the two locations are different but predictable from an environmental model. The first situation involves only the within-batch error. The second situation involves the among-batch error because of the dependence of the among-batch error on concentration. For concreteness, imagine that the two locations are two lakes located close enough together that the natural variation due to the weather is the same for each lake and that the samples are taken from the center of each lake so that natural mixing averages other sources of local variation. For two such lakes, the hypothesis that they have equal concentrations might be plausible, and deviations from this hypothesis might be of interest. In this case, the first situation applies. More likely, two such lakes will have different mean levels and seasonal behavior, and deviations from the repetitive seasonal behavior might be of interest. In this case, the second situation applies.

Situation 1

Four examples of the first situation will be considered. The four examples differ in the concentration assigned to the first location of interest, μ_1, and in the concentrations assigned to the reference samples. As shown in Table I, the first and second examples are based on $\mu_1 = K \exp(2)$, where $K = 4$, and the third and fourth examples are based on $\mu_1 = K \exp(4)$. Two types of reference samples are specified. In half of the batches, the two reference samples are of different

Table I. Proportion of Exceedances of the Critical Value

μ_1	Reference Samples		Sit.	Ex.	N	Proportion of Exceedances 1	2	3
K exp(2)	K exp(−1)	K exp(5)	1	1	16	0.80	0.82	0.79
K exp(2)	K exp(−1)	K exp(5)	1	1	32	0.62	0.63	0.69
K exp(2)	K exp(1.5)	K exp(2.5)	1	2	16	0.95	0.92	0.96
K exp(2)	K exp(1.5)	K exp(2.5)	1	2	32	0.89	0.83	0.86
K exp(4)	K exp(−1)	K exp(5)	1	3	16	0.28	0.34	0.43
K exp(4)	K exp(−1)	K exp(5)	1	3	32	0.17	0.49	0.40
K exp(4)	K exp(0)	K exp(4)	1	4	16	0.14	0.24	0.33
K exp(4)	K exp(0)	K exp(4)	1	4	32	0.15	0.19	0.17
K exp(2)	K exp(−1)	K exp(5)	2	1	32	0.10	0.11	0.19
K exp(2)	K exp(−1)	K exp(5)	2	2	32	0.10	0.05	0.08

types. In a quarter, the two reference samples are of the first type, and in the other quarter, the two reference samples are of the second type. In the first and third examples, the reference samples have concentrations $K \exp(-1)$ and $K \exp(5)$. In the second example, the reference samples have concentrations $K \exp(1.5)$ and $K \exp(2.5)$. In the fourth example, the reference samples have concentrations $K \exp(0)$ and $K \exp(4)$.

The other aspects of the situation are common to all four examples. In each example, we let the second location of interest have concentration $\mu_1 + \sqrt{2}\delta g(\mu_1)/\sqrt{N}$, where N is the number of batches and $\delta = 8$. Equation 1 shows that the difference in concentrations is $\delta/\sqrt{N}$ times the standard deviation of $x_{1t} - x_{2t}$ under the condition that both locations have concentration μ_1. In this section, the case is considered in which the routine-duplicate pairs are drawn from a different lake each time a batch of samples is formed. For this reason, these concentrations are assumed to be drawn at random from a log-normal distribution. In particular, these concentrations are given by $K \exp(2 + \psi)$, where ψ is normally distributed and has mean 0 and variance 1. Three realizations of the routine-duplicate concentrations were generated, and all three were used in each example. The within-batch error is characterized by

$$g(\mu) = \sigma(\mu + \tau)^{1-\lambda} \tag{12}$$

where $\lambda = 0.5$, $\tau = 16$, and $\sigma = 0.1$. The value of $g/(\mu + \tau) = \sigma(\mu + \tau)^{-\lambda}$ is much less than 1, which is required by equation 5. The among-batch error is zero.

Our procedure for testing the null hypothesis, which is that the two locations of interest have the same concentration, consists of minimizing the F ratio appropriate for the case in which (λ, τ) is known over the confidence region for (λ, τ) obtained from the statistic in equation 9. The numerator of the F ratio is the mean square of the differences between corresponding transformed measurements for the two locations of interest. Under the null hypothesis, this mean square involves only the within-batch error and thus is an estimate of twice the within-batch variance. The denominator of the F ratio is the estimate of twice the within-batch variance obtained from the quality assurance samples.

In the search for the minimum, a candidate value of (λ, τ) is used to transform the measured concentrations, and the transformed values are used to form the statistic z given by equation 9. If $|z| \leq 1.96$, then the F ratio given by

$$\frac{\sum_{t=1}^{N}(y_{1t} - y_{2t})^2}{2s^2N} \tag{13}$$

where

$$2s^2 = \frac{\sum_{t=1}^{N}(y_{3t} - y_{4t})^2 + \sum_{q_1=q_2}(y_{5t} - y_{6t})^2}{1.5N} \tag{14}$$

is a possibility for the minimum. This F ratio has N and $1.5N$ degrees of freedom. A search is made over the rectangular grid that proceeds from $\lambda = 0$ to $\lambda = 1$ in steps of 0.1 and proceeds from $\tau = 0$ to $\tau = 40$ in steps of eight. Furthermore, wherever a grid line crosses the boundary $|z| = 1.96$, the boundary is located by quadratic interpolation, and the F ratio is computed at the boundary. Typically, the minimum will occur on the boundary.

This test can be thought of as a search for a reasonable value of (λ, τ) that makes the difference between the two locations of interest seem most like measurement error. By the Bonferroni inequality, the significance level of this test is less than the sum of the significance level for the critical value to which the F ratio is compared and the significance level of the test for (λ, τ). By using our choice of 1.96 to compare with $|z|$, the second of these levels is 0.05. A level of 0.05 is also chosen for the F ratio. For $N = 16$, the critical value for the F ratio is 2.088, and for $N = 32$, the critical value is 1.684. Thus, the level of our test is less than 0.10.

We obtained six sets of results for the first example, three with $N = 16$ and three with $N = 32$. Each set consists of 100 trials run with the same realization of the routine-duplicate concentrations. Table I shows the proportion of trials in which the critical value was exceeded, that is, the proportion of trials in which the difference $\sqrt{2}\delta g(\mu_1)/\sqrt{N}$ was detected. The critical value was exceeded in 80, 82, and 79 trials for $N = 16$ and in 62, 63, and 69 trials for $N = 32$. These results can be compared to the case in which (λ, τ) is known. In this case, the F ratio has a noncentral F distribution having a noncentrality parameter $\delta^2 = 64$, and the probability of exceeding the critical value is 0.994 for $N = 16$ and 0.981 for $N = 32$. These results indicate how performance is affected by a lack of knowledge of the proper transformation. The performance degradation seems neither insignificantly small nor unmanageably large.

In the design of the quality assurance samples, the choice of concentrations for the reference samples is as important as the choice of the number of quality assurance samples. Consider the second example of the first situation. The only change from the first example is a shift of the reference samples to concentrations much closer to the concentrations of the locations of interest. Let the reference sample concentrations be $K \exp(1.5)$ and $K \exp(2.5)$. As shown in Table I, the critical value was exceeded in 95, 92, and 96 trials for $N = 16$ and 89, 83, and 86 trials for $N = 32$. With the reference samples closer together, the estimate of (λ, τ) will be less precise. Nevertheless, the results seem better than in the first example. Apparently, the fact that the reference samples are close to the samples of interest compensates for any lack of precision in the estimate of (λ, τ).

As the third example, consider the case in which the reference samples have the same concentrations as in the first example, but the concentration of the first location of interest μ_1 is changed to $K \exp(4)$. As shown in Table I, the critical value was exceeded in 28, 34, and 43 trials for $N = 16$ and 17, 49, and 40 trials for $N = 32$. In this case, lack of precision in the estimation of (λ, τ) has a serious effect. The effect of the lack of precision is increased in this example because the concentrations of interest are no longer centered among the quality assurance concentrations. Comparison of the three realizations shows a greater sensitivity to the distribution of routine-duplicate concentrations than in the previous examples. For $N = 32$, the first realization, which gave the lowest number of exceedances, has fewer high-concentration routine-duplicate pairs than the other realizations. Greater sensitivity might be expected when the concentration of interest is further from the center of the routine-duplicate concentrations.

The fourth example provides more insight into the effect of the concentrations of the quality assurance samples. Consider the case in which $\mu_1 = K \exp(4)$, and the reference samples have concentrations $K \exp(0)$ and $K \exp(4)$. Compared to the third example, the reference samples span a smaller range, and Table I shows fewer exceedances. In this case, a tighter grouping of the concentrations of the quality assurance samples leads to poorer performance.

Situation 2

We now consider the situation in which equality of the concentrations at the two locations is not plausible. In this situation, the among-batch error is important. Under the null hypothesis, we let the concentration at the first location be $\mu_1 = K \exp(2)$, and the concentration at the second location be $\mu_{11} = \mu_1 \exp[2\sin(\pi t/2)]$. This model might be interpreted as representing two locations: one of which has constant concentration and the other of which has seasonally varying concentration. In reality, the parameters of this model would have to be estimated from the measurements. In this section, the concentrations under the null hypothesis are assumed to be known. Under the alternative hypothesis, the first location has concentration $\mu_1 + \sqrt{2}\delta\sigma(\mu_1 + \tau)^{1-\lambda}/\sqrt{N}$, where σ, λ, and τ are the same as in the first situation. Thus, the difference between the null and alternative hypotheses is the same as in the first situation. To specify the among-batch error for the first example of this situation, we let $\eta_0 = 0$ and $\eta_1 = \sigma/\mu_1$. For the second example, we let $\eta_0 = 0$ and $\eta_1 = 5\sigma/\mu_1$. Otherwise, situation 2 is the same as example 1 of situation 1. The concentrations of the reference samples and the routine-duplicate pairs are the same as are the properties of the within-batch error.

To develop a test statistic for this comparison, we start with the case in which (λ, τ) and the ratio η_1/σ are known. Let $T(\mu)$ be the result of transforming μ according to equation 4. In the case of (λ, τ) and η_1/σ known, an F ratio can be formed from the variance estimate

$$\frac{1}{N}\sum_{t=1}^{N}\frac{[y_{1t} - y_{2t} - T(\mu_1) + T(\mu_{11})]^2}{1 + [\eta_1(\mu_1 - \mu_{11})/\sigma]^2} \tag{15}$$

and the variance estimate $2s^2$ given in equation 14. As shown in equation 10d, the ratio $(\eta_1/\sigma)^2$ can be estimated from the reference samples for which $q_1 \neq q_2$. The estimate of $(\eta_1/\sigma)^2$ is

$$\max\left\{(U - 1)/[\mu(q_1) - \mu(q_2)]^2, 0\right\} \tag{16}$$

where $\mu(q_1) - \mu(q_2)$ is the difference between the concentrations of the two types of reference samples, and

$$U = \sum_{q_1 \neq q_2}\frac{\left\{y_{5t} - y_{6t} - T[\mu(q_1)] + T[\mu(q_2)]\right\}^2}{Ns^2} \tag{17}$$

Replacing U by U divided by the 0.05 percentile of the F distribution with $N/2$ and $3N/2$ degrees of freedom gives the upper bound on a one-sided confidence interval for $(\eta_1/\sigma)^2$. If the 0.05 percentile of the F distribution is denoted by $F_{0.05}(N/2, 3N/2)$, then the one-sided confidence interval for $(\eta_1/\sigma)^2$ is bounded by

$$\max\left\{\frac{\dfrac{U}{F_{0.05}}\left(\dfrac{N}{2}, \dfrac{3N}{2}\right) - 1}{[\mu(q_1) - \mu(q_2)]^2}, 0\right\} \tag{18}$$

Substitution of this upper bound into the F ratio and minimization over the (λ, τ) confidence region gives the test statistic. By the Bonferroni inequality, the level of the test is less than 0.15.

As shown in Table I for situation 2, example 1, the critical value was exceeded in 10, 11, and 19 trials for $N = 32$. The inclusion of the among-batch error and the spreading of the concentrations of interest over a range reduces our ability to distinguish real differences from measurement error.

As another example of this situation, the relationship between σ and η_1 is changed to $\eta_1 = 5\sigma/\mu_1$. The effect of this change is to increase the size of the among-batch error with respect to the difference to be detected because the difference between the null and alternative hypotheses is scaled by σ. As shown in Table I, the critical value was exceeded in 10, 5, and 8 trials for $N = 32$. The increase in among-batch error further degrades performance.

The foregoing examples show a strong relationship between the concentrations of the quality assurance samples and the adequacy of the error assessment. This relationship might not be as severe were another statistical test chosen. For example, the results in the third example of the first situation might be improved if the statistic z in equation 9 were replaced. One might guess that this statistic is not as powerful as it could be because this statistic uses only the ranks of the pair averages and not the other information available on the reference-sample concentrations. Recall that the test statistic for the first situation is formed by minimizing an F ratio over the (λ, τ) confidence region obtained from the statistic z. Consider the (λ, τ) value at which the minimum is achieved and the squared differences d_i^2 for this (λ, τ) value. To illustrate the potential for improvement in example 3, we compare the mean of squared differences for the upper reference-sample pairs with the mean of squared differences for the other pairs. In the 100 trials for the first realization with $N = 32$, the ratio of these two mean squares has a mean of 0.19. This value seems low because the power transformation is supposed to give constant standard deviation. If the true value of (λ, τ) were used to form the squared differences, then this ratio would be F-distributed with 8 and 40 degrees of freedom. The 0.05 point for this distribution is 0.33. In these 100 trials, the value 0.33 is exceeded only 10 times. This shows that the statistic in equation 9 permits transformation parameters λ and τ that seem to contradict the data. This statistic permits transformations that correspond to higher than reasonable error variances at high concentrations and thus the appearance that the differences between the samples of interest are due to measurement error.

Despite the possibility that equation 9 could be improved, the foregoing examples show that a moderate number of quality assurance samples with proper concentrations can provide adequate estimates of the error properties despite the complexities of these properties. Moreover, the foregoing test statistics can be used as a basis for study design with only the risk that the number of quality assurance samples specified will be too large. The use of the foregoing test statistics in the design of an environmental study requires that the investigator deal with the uncertainty in what concentrations will be encountered. The choice of quality assurance samples that will be adequate in any eventuality is not easy. A variety of possible relations between the concentrations of interest and the proposed set of quality assurance samples must be explored. The Monte Carlo experiments described suggest how this might be done.

Some Generalizations

The problem posed in this chapter, the assessment of error with two heteroscedastic components, has been solved with a combination of power transformation, variance component estimation, and variance function estimation. Each of these areas is addressed by a large amount of statistical literature. From this literature, one might obtain better methods for the problem posed and methods for problems defined by somewhat different sets of assumptions.

Statistical inference for power transformations has been set in many different contexts. The context considered in this chapter is the one discussed by Carroll and Ruppert (8). Power transformations have been offered as a method for achieving a simple linear model, homoscedastic error, and normally distributed error. In this chapter, power transformations were applied only to achieve homoscedastic error, not to simplify the environmental model. The environmental model, except for a few unknown parameters, is supposed to be given with the measurements and must itself be transformed when the measurements are transformed.

The method discussed is limited by our assumption that a power transformation will make the within-batch error homoscedastic. One can easily imagine the need to deal with a more general functional dependence of the variance on the concentration. Generalization might not be easy. Our method for fitting the transformation does not require any specific assumption about the true concentrations of the sample pairs. All that is required is that the ordering by pair averages bears some resemblance to ordering by increasing concentration. Alternative methods that involve the estimation of the dependence of the variance on the concentration from the apparent relationship between the observed differences and averages do not seem to have this advantage.

If the two error components were homoscedastic, then variance component estimation would be appropriate. Miller (9) listed five alternatives for separating the among-batch and within-batch contributions to the variance. These alternatives suggest various ways to use equation 10 to obtain estimates of η_0 and η_1. Anderson (10) discussed not only estimators but also designs for estimating variance components.

In addition to variance component estimation, variance function estimation is part of the assessment of the among-batch error. Davidian and Carroll (7) compared various approaches to variance-function estimation. Some of their conclusions are not appropriate to the problem considered in this chapter because these conclusions are based on the use of both the residuals from the model and the spread of the replicate observations to estimate the variance function. In the problem considered here, the residuals from the environmental model are to be compared with the spread of the replicate observations to see if a discrepancy exists. Nevertheless, some of their approaches, when extended to the case of two components, should be applicable.

This review of the statistical literature pertinent to the problem posed raises the question of whether the results discussed previously might overstate the need for quality assurance samples because these results are not based on the optimal statistical methods. Specific evidence of this possibility accompanies the discussion of our Monte Carlo experiments. Further research might produce a method that gives tighter confidence regions for (λ, τ) or might produce less conservative significance levels than the Bonferroni inequality.

In environmental studies, estimation, not testing, is the proper statistical approach. Although results on the needed quality assurance samples obtained through testing scenarios should be useful, such results might lead to inadequate quality assurance samples because the actual statistical approach is more complicated. Some examples of the complications in an environmental study were discussed by Stewart-Oaten (*11*).

In addition to alternative methods for analysis of the data, alternative designs can be suggested. In particular, the question of the benefits of including more quality assurance samples in each batch is important. How does increasing the quality assurance samples in each batch change the number of batches needed? How can more quality assurance samples in each batch be used to assess more complicated error models? These questions seem worthy of investigation.

The error model considered in this chapter with its two heteroscedastic components will often have to be generalized. The case of bulk sampling is mentioned in the introduction. In this case, a within-batch subsampling component must be added to the model to account for the difference between pairs of duplicate subsamples, which are subject to the subsampling error, and pairs of reference samples, which are not. The case of measurements made by several laboratories also requires a more complicated model. This case raises a variety of issues not considered in this chapter. Differences between laboratories often involve more than just another error component. The differences might involve fundamentally different measurement methods. The bias of one method with respect to another must sometimes be addressed in terms of the scientific theory behind the methods. Often, two laboratories will exhibit totally different error properties and thus will require separate error estimates. When comparisons between measurements are the only objective, the simplification that comes with the selection of a single laboratory might be worthwhile.

Abbreviations and Symbols

δ	scale factor in the true-difference specification
η_0, η_1	parameters of the among-batch error
λ	power parameter of the power transformation
μ	true concentration
μ_I, μ_{II}	concentrations for locations of interest
$\mu(q_1)$	concentration of reference sample of type q_1
$\mu(q_2)$	concentration of reference sample of type q_2
σ	parameter of the within-batch error
τ	shift parameter of the power transformation
a_I	average of transformed measurements
c^+, c^-	statistics for choice of transformation
d_i	difference between transformed measurements
$g(\mu)$	standard deviation of the within-batch component
$h(\mu)$	standard deviation of the among-batch component
K	scale factor in the concentration specification
m	integer part of $p/2$, $[p/2]$

p	number of nominally identical samples
q_1, q_2	functions of t that indicate reference sample type
S_j	normalized partial sum of squared differences
t, N	batch index, number of batches
$T(\mu)$	transformed concentration
U	estimate of $\{\eta_1[\mu(q_1) - \mu(q_2)]\}^2/\sigma^2+1$
var	variance
x	measurement of concentration
x_{1t}, y_{1t}	measurement for the first location of interest
x_{2t}, y_{2t}	measurement for the second location of interest
x_{3t}, y_{3t}	measurement on a member of a routine-duplicate pair
x_{4t}, y_{4t}	measurement on a member of a routine-duplicate pair
x_{5t}, y_{5t}	measurement on reference sample of type q_1
x_{6t}, y_{6t}	measurement on reference sample of type q_2
y	transformed measurement of concentration
z	statistic for choice of transformation

References

1. Linthurst, R. A.; Landers, D. H.; Eilers, J. M.; Brakke, D. F.; Overton, W. S.; Meier, E. P.; Crowe, R. E. *Characteristics of Lakes in the Eastern United States*; U.S. Environmental Protection Agency: Washington, DC, 1986; Vol. 1; EPA/600/4-86/007a.
2. Liggett, W. S. *Proceedings of the Thirteenth International Biometric Conference, Invited Papers*; Biometric Society: Washington, DC, 1986.
3. Box, G. E. P.; Cox, D. R. *J. R. Stat. Soc. Ser. B.* **1964**, *26*, 211–252.
4. Koch, W. F.; Liggett, W. S. In *Detection in Analytical Chemistry: Importance, Theory, and Practice*; Currie, Lloyd A., Ed.; ACS Symposium Series 361; American Chemical Society: Washington, DC, 1988.
5. Durbin, J. *Biometrika* **1969**, *56*, 1–15.
6. Cox, D. R.; Lewis, P. A. W. *The Statistical Analysis of Series of Events*; Menthuen: London, 1966; pp 45–51.
7. Davidian, M.; Carroll, R. J. *J. Am. Stat. Assoc.* **1987**, *82*, in press.
8. Carroll, R. J.; Ruppert, D. *J. Am. Stat. Assoc.* **1984**, *79*, 321–328.
9. Miller, R. G. *Beyond ANOVA, Basics of Applied Statistics*; Wiley: New York, 1986; pp 96–99.
10. Anderson, R. L. In *A Survey of Statistical Design and Linear Models*, Srivastava, J. N., Ed.; North Holland: Amsterdam, 1975; pp 1–29.
11. Stewart-Oaten, A. *Oceans '86 Conference Record*; Marine Technology Society: Washington, DC, 1986; Vol. 3, pp 964–973.

RECEIVED for review January 28, 1987. ACCEPTED June 4, 1987.

Chapter 13

Modern Sampling Equipment: Design and Application

Lorance H. Newburn

Automatic wastewater samplers are used extensively by water quality monitoring agencies and wastewater dischargers who must comply with permit requirements. The basic mode of operation of these samplers has changed very little since 1977. A sound understanding of the limitations of these designs and of the characteristics of the flow stream to be sampled is necessary in order to obtain a representative sample. Although studies have been conducted to evaluate samplers, few studies have been published on correct application. Some general guidelines for correct application are discussed.

B ROAD-BASED NEED FOR AUTOMATIC WASTEWATER SAMPLERS was created within the framework of the Federal Water Pollution Control Act of 1972 (*1*). In order to achieve the Act's objective of restoring and maintaining the chemical, physical, and biological integrity of the nation's waters, wastewater effluents had to be accurately characterized both in quality and quantity. Wastewater sampling became an intensive activity by those discharging wastewater and by those agencies charged with compliance enforcement. The emphasis of this activity originally centered on conventional pollutants. The parameters of greatest interest were biological oxygen demand (BOD), chemical oxygen demand (COD), solids, conductivity, pH, oil and grease, fecal coliform, heavy metals, and a few organic compounds. In 1977, the Clean Water Act was amended (*2*). The scope of wastewater sampling was broadened to include toxic pollutants and the pretreatment of effluents by industries. The Consent Decree identified 129 toxic materials discharged by industries grouped into 34 categories, which became the target of a massive campaign to eliminate these substances from the nation's waters.

1173–6/88/0209$06.00/0 © 1988 American Chemical Society

Water Sampling History

Traditionally, water was sampled by a simple dipping procedure that is very dependent on methodology. The sample collected may or may not be representative of the flow at the time the sample was taken. Also, water quality and flow rate in a given flow stream can vary considerably from one moment to the next. Collection of frequent aliquots is required to obtain an accurate representation of the flow over a given time period. As a result, manual sampling became a labor-intensive activity, and coupled with the increased concern for unbiased samples, those people interested in sampling found themselves increasingly dependent on automatic samplers.

Many manufacturers recognized the need for automated samplers, and a wide variety of designs was offered. With respect to sample representativeness, the basic difference in these designs was in the sample collection system. Samplers were offered that used low-, medium-, or high-speed peristaltic pumps, in which the pump operates from a timer, or pump rotations are counted to control volume. Also offered were samplers using vacuum pumps where a metering chamber is used to control volume; mechanical dippers filled by immersion; evacuated bottles, where each bottle has a tube connecting to the source; and pneumatic-ejection systems, where a submerged metering chamber is filled by the hydrostatic pressure of the liquid, and the resulting sample is forced to the sample container by gas pressure.

Automatic sampler use did not solve all the problems, however (3). The equipment varied considerably with respect to design, sample intake velocity, method of sample collection, versatility, and durability. Sampler users often lacked the skills necessary for proper application. It was not unusual for samples taken by one device to vary considerably from samples collected by another device in the same flow stream. Depending on the sampler selected, either overstating or understating the concentration of certain constituents was possible. An enforcement agency using one brand and model of sampler often obtained varied results from a discharger using the same brand and model. Apparently, samplers needed to be studied and methodologies needed to be standardized so accurate data could be obtained.

Sampler Evaluations

The first major sampler evaluation was published by Shelly and Kirkpatrick in 1973 (4). This study surveyed about 50 different samplers from 30 manufacturers for suitability in storm or combined sewer sampling applications. They concluded that no single unit could be universally applicable and suggested what improvements, if any, could be made that would make these units as ideal as possible. They also concluded that obtaining a representative sample requires more than the sampler itself. Their evaluation did not include side-by-side comparisons of commercial samplers but did suggest that such a comparison should be made. The U.S. Environmental Protection Agency (EPA) compared low-, medium-, and high-speed peristaltic pump units; a vacuum unit; an evacuated-bottle unit; a pneumatic-ejection unit; manual samples taken with a three-level Van Dorn-type sampler; and conventional dipping that followed U.S. Geological Survey field procedures. No statistical difference was found between vacuum samplers and medium- and high-speed peristaltic samplers.

A study comparing samplers was made by Harris and Keffer of the Region VII EPA office and published in June 1974 (5). They evaluated about 15 different samplers. The study compared vacuum-type samplers, which typically have a high initial intake velocity, and samplers that use peristaltic pumps having low to moderate intake velocities. The test data showed that vacuum samplers produced higher BOD and COD concentrations and disproportionately higher solids concentrations than samplers using peristaltic pumps. Vacuum samplers produced higher values than the manual flow-weighted grab samples, which were selected as the standard. They concluded that high-vacuum, high liquid intake velocity samplers were more effective in capturing solids than peristaltic pump samplers. They suggested that slower acting peristaltic pump samplers were either not capturing settleable materials or that particle settling velocities were higher than the liquid intake velocities after introduction to the intake line. The question was raised as to how vacuum samplers could produce higher results than their flow-weighted grab sample standards (Harris, D. and Keffer, W., personal communication). Some of the peristaltic pump samplers demonstrated 90%–95% sampling efficiency; the vacuum samplers demonstrated efficiency values in excess of 100%, and some samplers demonstrated efficiencies as high as 167%. One suggestion made to the authors was that if the strainer of a vacuum sampler was allowed to rest on the bottom of the flow stream where sediment had collected, the high intake velocity would scour those sediments from around the strainer; artificially enrich the sample; and produce higher than normal BOD, COD, and solids values. After some consideration, the authors agreed that this situation was probably true and that care should be exercised in positioning the strainer of a vacuum sampler or any other sampler having a high intake velocity so that no scouring effect would occur.

Another study was published in January 1975 by Barkley et al. (6). Whereas the Harris and Keffer report was based on comparisons in field use, this study was conducted in a laboratory under artificial conditions by using a tank with a mixer so that greater control could be exercised over the test media. Sixteen different samplers were put through an extensive series of tests and scored on 21 factors. Vacuum samplers were not studied in this series of tests, but some testing was done shortly afterward. The authors concluded that samplers with high intake velocities tend to scour sediments from around the strainer if care is not exercised in positioning the strainer above the sediments.

A separate study conducted by the U.S. Forest Service in Arcata, CA, dealt with contamination of successive samples in portable sampling systems (7). The results showed that a vacuum sampler using a metering chamber having greater wetted surface area produced more cross-contamination than a peristaltic pump sampler. This result contrasts a study conducted by the St. Anthony Falls Hydraulic Laboratory (8), which concluded that no evidence of cross-contamination was found in the same model of vacuum sampler used in the Forest Service study.

Another study (9) was conducted to determine the effect of certain variables on the recovery of volatile organic compounds. The authors concluded that high-speed suction-lift samplers cause volatile organic compounds to outgas. The authors recommended the use of moderate-speed peristaltic pumps to minimize this effect.

Although these studies may show the superiority of one sampler over another, the basic difference between these units is in liquid intake velocity and the wetted surface area of the liquid transport system. Vacuum sampler supporters argue that the higher intake velocity helps keep solids in suspension. Supporters of peristaltic pump samplers argue that intake velocities of 1.5–3 ft/s are more ideally suited for obtaining a representative sample because these samplers do not tend to scour sediments from around the strainer. As mentioned, data from the Harris and Keffer report (5) and research conducted by Barkley et al. (6) tend to support those favoring peristaltic pump samplers. Because of their greater wetted surface area, the metering chambers used in vacuum samplers may be a source of cross-contamination between samples (7, 8). The use of a metering chamber as a means of controlling sample volume is an advantage where the level of the stream being sampled changes considerably from one sample to the next. This situation would be a problem only when composite samples are collected because the sample could be biased by a variation in the volume of each aliquot placed in the sample bottle. Therefore, a number of factors need to be considered when selecting a sampler.

Ideal Sampler Features

Several attempts by those studying samplers have been made to define what would be considered the "ideal" sampler (4–6). As mentioned, some features are mutually exclusive, and a sampler incorporating all these features cannot be made. For those who must sample a number of locations, more than one type of sampler will probably be required. Features that the ideal sampler should have include the following:

1. alternating current (ac) or direct current (dc) operation with adequate dry battery energy for 120 h of operation at 1-h sampling intervals

2. suitability for suspension in a standard manhole and accessibility for inspection and sample removal

3. total weight including batteries under 18 kg (40 lb)

4. sample collection intervals from 10 min to 4 h

5. capability for flow-proportional and time-composite samples

6. capability for collecting a single 9.5-L (2.5-gal) sample or collecting 400-mL (0.11-gal) discrete samples in a minimum of 24 containers

7. capability for multiplexing repeated aliquots into discrete bottles

8. a single intake hose having a minimum ID of 0.64 cm (0.25 in)

9. intake-hose liquid velocity adjustable from 0.61 to 3 m/s (2.0–10 ft/s with dial setting)

10. minimum lift of 6.1 m (20 ft)

11. explosion-proof materials and electronics

12. watertight exterior case to protect components in the event of rain or submersion

13. exterior case capable of being locked and including lugs for attaching steel cable to prevent tampering and to provide security

14. no metal parts in contact with waste source or samples

15. an integral sample container compartment capable of maintaining samples from 4 to 6 °C for a period of 24 h in ambient temperatures ranging from −30 to 50 °C

16. with the exception of the intake hose, capability of operating in a temperature range from −30 to 50 °C

17. purge cycle before and after each collection interval with sensing mechanism to purge in the event of plugging during sample collection, followed by collection of complete sample

18. field repairability

19. interchangeability between glass and plastic bottles, particularly in discrete samplers

20. sampler exterior surface light in color to reflect sunlight

This list of features was developed before the amendment of the Clean Water Act, in which the emphasis in sampling shifted to toxic materials. One other feature, therefore, has to be included in the list: the ability to be readily adapted for toxic-pollutant sampling. For toxic pollutants, the sample should contact only tetrafluoroethylene (Teflon), glass, and medical-grade silicone rubber if a peristaltic pump is being used in the sample transport system.

Sampler Characteristics

According to the EPA's manual on complying with the sampling requirements published in 1979 (10), about 100 models of portable automatic sample collection devices existed. These devices varied widely in levels of sophistication, performance, mechanical reliability, and cost. No sampler on the market was considered ideal. Now, several samplers come close to being ideally suited. All of the major suppliers of portable samplers offer units that operate on ac or dc, can be suspended in a manhole, weigh less than 40 lb empty, offer sampling intervals from 1 min to at least 4 h, are capable of flow-proportional or timed-interval sampling, are convertible from sequential to composite sampling, are capable of multiplexing several samples into one discrete bottle, offer a 0.25-in ID suction line with a weighted strainer, have a minimum lift of 20 ft, are capable of being locked, have no metal parts in contact with the sample, operate over temperature ranges from −30 to 50 °C, and purge the sample line before and after sampling.

Samplers on the market differ, sometimes significantly, in sample intake velocity (none has an adjustable intake velocity), explosion-proof qualities, watertightness, insulation qualities, and ability to be repaired in the field. No manufacturer offers an ac- or dc-powered sampler that is certified to be explosion-proof. This accomplishment would require the addition of considerable weight to the sampler that would put it over the recommended 40-lb maximum. Pneumatically operated samplers currently being offered are explosion-proof. One sampler offers the advantage of exceptionally high lift (as high as 300 ft). Pneumatic-ejection units, however, may lead to cross-contamination of samples.

Watertightness is one definite requirement of a portable sampler. Manholes often surcharge, rivers rise, and samplers frequently flood. The waterproof qualities of commercially available samplers vary considerably, even though most manufacturers claim their samplers to be watertight. Moisture damage is the most frequent cause of sampler failure. If the purchase of a sampler is being considered, the purchaser should carefully examine the product beforehand and determine what kind of seal protects the electronic parts from moisture. One manufacturer rates their controller NEMA 6, which means the sampler can withstand submersion under 6 ft of water for at least 30 min.

Not all samplers are thermally insulated. Those that are insulated vary in insulation quality. The EPA research center in Cincinnati conducted insulation quality tests on several samplers and can provide data on their results.

Field repairability of samplers must necessarily be limited to replacement of expendable items or certain mechanical parts. As for the electronic parts, all samplers must be repaired by a skilled technician who has access to the proper test equipment. This requirement usually means shipping the unit back to the factory. Some manufacturers offer samplers that have removable control units; therefore, only the controller must be sent in for repair.

Two types of samplers are on the market: discrete and composite samplers. With a discrete sampler, an individual sample can be collected and retained in a separate container for future analysis. Most current manufacturers of discrete samplers offer units containing at least 24 bottles. With a true composite sampler, small aliquots can be taken at frequent intervals, usually over a 24-h period; collected in a single container; and held for future analysis. Of the two sampler types, the discrete sampler can determine what the constituents of the sample are and approximately when they were discharged. Discrete samplers are more widely used by water quality enforcement agencies for this reason. Discrete and composite samplers are available in various configurations depending on the particular application. Most samplers are dc-powered portable units, but some are ac-powered refrigerated units.

Most samplers are capable of gathering either timed-interval samples or samples collected proportional to flow. Timed-interval samplers have a fixed interval of time between each aliquot or sample. Flow-proportioned samples are based generally on equal increments of flow as measured by an associated flow meter. Most samplers require a simple contact closure signal from the flow meter, and some accept a 4–20-mA analog signal. A flow-proportioned composite sample, in which small aliquots are collected in a single container over small increments of flow, provides the most representative sample of the flow over a given time period. However, if the flow interval is too large, events such as the dumping of small tanks might pass by the sampler without being detected. For example, if the flow interval is 25,000 gal per aliquot and a 5000-gal tank is dumped, those 5000 gal could pass by the sampling point undetected.

Two other modes of operation of interest to individuals using discrete samplers include nonuniform time intervals and time override of flow-proportioned sample collection. Nonuniform time intervals give the user the option of programming a different time between each sample. This option is of particular interest to those studying hydrologic events, combined sewer overflows, and other events where extreme variation in flow or constituent concentration occurs. Each individual time interval can be programmed separately by the user. In the time-override mode, individual samples are composited into a single bottle at equal increments of flow volume as measured by the associated flow meter. However, at a programmed increment of time, the distributor is indexed to the next sample bottle, the next sample is placed in that bottle, and the compositing of samples is renewed. A time-override sample is used when the investigator desires to collect a sequential series of composite flow samples, each over a known, elapsed time interval while ensuring that at least one sample will be collected during the time interval, regardless of the flow volume.

Another important feature of a sampler is durability. A careful examination of the materials of construction is important to ensure that the sampler continues to perform as expected in highly humid, corrosive environments. All hardware should be stainless steel. Plated or painted hardware will not last very long in most sampling conditions. The case material should be of acrylonitrile butadiene styrene (ABS) plastic, glass fibers (Fiberglas), or some other material that is highly resistant to moisture and corrosive gases. As previously mentioned, the control unit must be able to withstand accidental submergence, and the exterior construction should be able to withstand the elements of nature without compromising the integrity of the sample. Sample bottles must be manufactured of a material that will not alter the composition of the sample. Glass bottles must be used in toxic-pollutant sampling applications. Polyethylene or polypropylene bottles are acceptable for most general-purpose applications.

Medical-grade silicone rubber tubing must be used in peristaltic pumps to avoid contamination of the sample by organic peroxides used in the manufacturing of conventional grades of silicone rubber. A test conducted by the EPA determined that short lengths of medical-grade silicone rubber do not alter samples in any way (9). When sampling for toxic pollutants, the suction line must be made of tetrafluoroethylene (Teflon). For general-purpose applications, the suction line can be poly(vinyl chloride), but should be food-grade to prevent the introduction of phenolic compounds.

User Knowledge

Probably the single greatest factor influencing the representativeness of the sample lies in the user and how the sampler is applied. To aid the user in complying with various federal and state regulations concerning water quality, a number of guides are available (10, 11, 12). But because of the many variations of conditions in which samplers can be used, coverage of them all in these guides is impossible. The collection of samples from a flowing waste stream requires a sound understanding of the problems inherent in sample collection if representative samples are to be obtained. Ideally, to get a true representative sample, the entire discharge should be collected, and then an aliquot of the thoroughly mixed discharge should be taken for analysis. This situation is impractical because millions of gallons of effluent are generally involved, and thorough mixing is impossible. Because of this incapability, the next best means of obtaining a representative sample is to collect a flow-proportioned composite sample. The key to representativeness of the sample is the method used to collect each aliquot.

Sample Intake Position

Before using an automatic wastewater sampler, the user should understand the characteristics of the flow stream and where in the stream the sample should be taken. The best discussion on this topic was contained in correspondence received in 1977 from Keffer, who authored one of the sampler studies (5) mentioned. Keffer pointed out that technical literature was lacking in standard guidance procedures for sampler installations or for sampling of any type. He stated that his opinion was not supported by carefully designed research activities but was a summary of his experience at annual sampling efforts at 400–500 facilities per year for over a 6-year period. He pointed out that the flow regime is a composite of a velocity distribution superimposed on a solids distribution, both of which vary with the physical condition of the site and vary in size, shape, and specific gravity of the solids particles. The shape and velocity of the flow stream determines the degree of turbulence. Those solids having a specific gravity much greater than water tend to settle in the bottom of the flow stream at low velocities or suspend near the bottom at higher velocities. Only where much turbulence occurs are these heavier solids uniformly distributed. Solids having densities much greater than water are generally not organic. Solids having specific gravities only slightly greater than water are usually organic and remain suspended in the flow but form layers or strata in smooth-flowing channels. Solids lighter than water and oils float and are almost always organic. Even where turbulence occurs in a flowing channel, oils tend to float. Thus, a flowing waste stream is generally highly stratified, nonhomogeneous, and presents a less than ideal medium from which to take a representative sample.

Keffer stated that from information currently available, a single-point intake is not likely to be satisfactory. Current assessment of state-of-the-art sampling methods suggested that a fixed intake located at 60% of the stream depth in an area of maximum turbulence and having an intake velocity equal to or greater than the average wastewater velocity at the sample point provides the most representative sample. This technique ignores contribution from bedload (i.e., sediment layers) or floatable solids. The selection of sampling at 60% of depth is based on the velocity and sedimentation concentration charts shown in Figure 1. Just below a hydraulic jump, the outlet of a flume or the nappe of a weir generally is the area of greatest turbulence in a flow stream. Attempting to sample at an intake velocity equal to or greater than the average wastewater velocity is difficult to achieve, and Keffer's routine practice was to maintain a transport velocity in excess of 2 ft/s. Placing the strainer of a high-velocity intake sampler on the bottom of a smooth-flowing stream where sediments have accumulated causes these sediments to be scoured from around the strainer. The sample is thus artificially enriched with solids.

Since 1980, two studies on positioning of sample intakes have been published. McGuire et al. (13) concluded that designers of water quality monitoring programs should realize the effects of sample intake positions on parameter concentration and should take into account the objectives of the study to determine the correct intake position. Reed (14), on the other hand, concluded that zones of natural mixing such as hydraulic jumps are unsuccessful in providing the necessary degree of mixture to obtain a truly representative sample. Therefore, some degree of calibration is necessary before accurate data can be obtained.

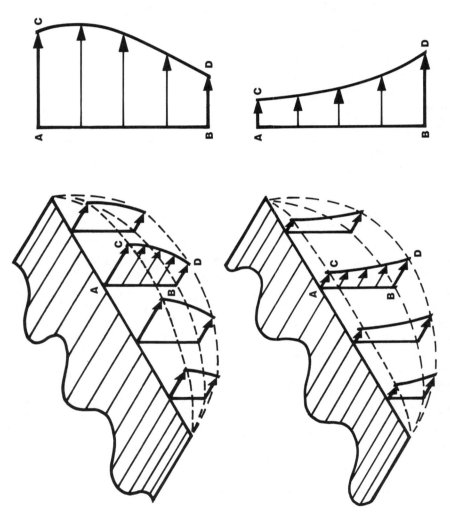

Figure 1. Velocity distribution (top) and sediment concentration (bottom) in a flowing stream or sewer.

References

1. "Federal Water Pollution Control Act Amendment of 1972"; Public Law 92–500, 1972.
2. "Clean Water Act of 1977"; Public Law 95–217, 1977.
3. Shelly, P. E. *Am. City & County* **1980**, *95*, 35–38.
4. Shelly, P. E.; Kirkpatrick, G. A. *An Assessment of Automatic Sewer Flow Samplers*; U.S. Environmental Protection Agency. U.S. Government Printing Office: Washington, DC, 1973; EPA 122–73–261. (Updated as *Sampling of Water and Wastewater*, 1976; EPA-600/4–77–039.)
5. Harris, D. J.; Keffer, W. J. *Wastewater Sampling Methodologies and Flow Measurement Techniques*; U.S. Environmental Protection Agency: Kansas City, MO, 1974; EPA 907-974-005.
6. Barkley, J. J.; Peil, K. M.; Highfill, J. W. *Water Pollution Sampler Evaluation*; Army Medical Bioengineering Research and Development Laboratory: Fort Detrich, MD, 1975; AD/A–009–079.
7. Thomas, R. B.; Eads, R. E. *Water Resour. Res.* **1983**, *19(2)*, 436–440.
8. Wood, A. "Test and Evaluation of a Portable Discrete Suspended Sediment Sampler for Manning Environmental Corporation"; External Memorandum No. 150; St. Anthony Falls Hydraulic Laboratory. University of Minnesota: Minneapolis, MN, 1977.
9. Ho, J. S. Y. *J. Am. Water Works Assoc.* **1983**, *75*, 583–586.
10. *NPDES Compliance Sampling Inspection Manual*; U.S. Environmental Protection Agency. U.S. Government Printing Office: Washington, DC, 1979; MCD–51.
11. *Handbook for Sampling and Sample Preservation of Water and Wastewater*; U.S. Environmental Protection Agency. U.S. Government Printing Office: Washington, DC, 1976; EPA-600/ 4–76–049. (Revised as EPA-600/4–82/029, 1982; addendum, 1983.)
12. Lauch, R. P. *Application and Procurement of Automatic Wastewater Samplers*; U.S. Environmental Protection Agency: Cincinnati, OH, 1975; EPA-670/4–75–003.
13. McGuire, P. E.; Daniel, T. C.; Stoffel, D.; Andraski, B. *Environ. Manage. (NY)* **1980**, *4*, 73–77.
14. Reed, G. D. *J. Water Pollut. Control. Fed.* **1981**, *53*, 1481–1491.

RECEIVED for review January 28, 1987. ACCEPTED April 6, 1987.

Chapter 14

Preservation Techniques for Organic and Inorganic Compounds in Water Samples

Jerry Parr, Mark Bollinger, Owen Callaway, and Kathy Carlberg

Preservation techniques are used to minimize changes between collection and analysis. Physical changes such as volatilization, adsorption, diffusion, and precipitation, and chemical changes such as photochemical and microbiological degradation are minimized by proper preservation. The preservation process encompasses both field and laboratory activities and includes techniques such as chemical addition, temperature control, and the choice of sampling containers. The lack of consistent, validated guidelines for preservation techniques and holding times for similar sample types has resulted in uncertainty among those who routinely collect and analyze samples. These numerous and often conflicting guidelines affect data quality, laboratory operations, and the cost of analytical determinations.

A NALYTICAL SAMPLES are in a chemically dynamic state at the time of collection. At the moment the sample is removed, the chemical processes that affect the sample may deviate from what occurs in situ for many reasons. Some of these reasons are obvious, others are more obscure. For example, a sample collected from a well source is exposed to conditions significantly different from the conditions underground. In the process of collection, the sample is often exposed to ambient light and its temperature most likely has changed. Consequently, photochemical reactions may take place, and the temperature-dependent kinetics of other types of reactions will be altered. Exposure to atmospheric conditions above ground will lead to changes in the dissolved gases in the sample. The presence of oxygen may initiate oxidation of

1173–6/88/0221$06.00/0 © 1988 American Chemical Society

some chemical species. As the sample is exposed to these changes in its new environment, the pH of the sample will likely change. Thus, the act of sampling may, for at least some finite period, alter the nature of the subset sample that is intended to be representative of the water source.

This dynamistic process does not end after the sample has been carefully transferred into some vessel for subsequent delivery to an analytical laboratory. In fact, the vessel may contribute to the process by adsorbing certain components of the sample. This adsorption permits additional loss of volatile compounds or introduces extraneous compounds into the sample. Complete stability of the sample through preservation cannot be totally achieved for every constituent in a sample, nor can all constituents be stabilized with the same degree of success. This situation is an increasing concern as the list of analytes determined by individual analytical methods grows longer and more complex, while detection limits reach lower levels.

The preservation process encompasses both field and laboratory activities and includes a variety of techniques. The practices and techniques recognized as resulting in the best stabilization of the sample are described in general terms in this chapter. Additional information concerning appropriate preservation techniques has been presented (1–4).

General Practices for Minimizing Changes

Preservation techniques are selected on the basis of their ability to minimize changes in order to best preserve the integrity of the sample after collection. Analytes present in a given sample can change for a variety of reasons. The most common changes that preservation techniques attempt to minimize are physical changes such as volatilization, adsorption, diffusion, and precipitation, and chemical changes including air oxidation, photochemical changes, and microbiological degradation. These changes are minimized with a variety of techniques, including sample container use, chemical additions, and temperature control.

Volatilization

Volatilization refers to the physical process in which volatile species can be lost to the atmosphere. The process is dependent on the vapor pressure of the analyte to be measured, the temperature of the sample, and the surface area.

For the purposes of this discussion, volatilization refers to the volatilization of organic molecules. Other species that undergo volatilization include gases such as hydrogen sulfide and hydrogen cyanide. However, volatilization of these species is controlled by preservation techniques described elsewhere.

Volatilization can be minimized by a simple technique. This technique involves containing the sample in a vessel with no headspace. Thus, any contact with air is prevented, and the development of an equilibrium of volatile compounds between the surface of the sample and the headspace based on vapor pressure is prohibited.

This technique is generally used for analyses of volatile organic compounds by gas chromatography or gas chromatography–mass spectrometry. However, this preservation process is sometimes overlooked in the measurement of other parameters (e.g., total organic carbon or total organic halides) that can include volatile components. Furthermore, as in most preservation techniques, the containerization process is of no use if the sample was improperly treated prior to preservation by agitation, vacuum filtration, or other procedures that volatilize the analytes of interest.

Adsorption and Absorption

Once a sample has been removed from its natural environment, the equilibrium between the sample and its environment is disturbed. Components in the sample can therefore undergo changes in response to the new environment. For example, components can adsorb, sometimes irreversibly, onto the walls of the sample containers. The two most common examples involve interactions of metal ions with glass surfaces and adsorption of oils onto container walls.

Metals can be irreversibly adsorbed onto glass surfaces. Therefore, for metals, the usual sampling approach is to collect the samples in plastic containers and thus eliminate the glass contact. Also, nitric acid is added to the sample to lower the pH to less than 2, which keeps the metal ions in solution. This situation is useful because the formation of metal hydroxides or hydrated oxides is insignificant under these conditions.

Oils present a more difficult problem. The adsorption of oils onto container walls cannot be easily prevented. Oils are likely to irreversibly adsorb onto the sides of plastic containers. Therefore, samples for organic parameters are collected in glass bottles. This *containerization* process allows the oils to be removed, typically with an organic solvent rinse. Another technique that is useful for samples that have low organic content is emulsification with a sonic probe. An aqueous solution of the oil in the sample is thus created.

Samples also have the propensity for absorbing gases from the atmosphere. Samples can absorb air components such as oxygen or carbon dioxide as well as vapor-phase species that might be present at the site (e.g., volatile organic compounds). Absorption of air components can have a significant impact on the sample by initiating air oxidation (e.g., sulfide to sulfate), chemical changes, or in the case of carbon dioxide, by changing the conductance and pH of the sample. (This process is a primary reason for field measurement of pH and conductance.) Absorption of other components can lead to false positive reporting of the absorbed component.

These changes can be minimized by expeditious preservation of the sample, and in the most extreme case, by eliminating exposure of the sample to the atmosphere. Field sample blanks can be used to determine if critical organic analytes are being absorbed into the samples from the air at the site.

Diffusion

Organic molecules such as phthalate esters and other plasticizers can diffuse through the walls of plastic sample containers or through bottle caps. This process is controlled by collecting samples in glass containers and by using bottle caps or liners that minimize this process. Tetrafluoroethylene (Teflon) liners are especially helpful in controlling contamination from diffusion processes.

Precipitation

Components in a sample may form salts that precipitate in the container. This process results from the interaction of components present in the sample due to a change in the sample environment (e.g., pH), or from the reaction of components in the sample with components in the air environment.

The most common occurrence is the precipitation of metal oxides and hydroxides resulting from reactions of metal ions in the sample with oxygen. Precipitation of these species is essentially eliminated by addition of nitric acid until a pH less than 2 results. The combination of a low pH and an excess of nitrate ion ensures that the metal ions stay in solution. Other acids (e.g., hydrochloric or sulfuric) tend to give anions that could enhance precipitation because of the low solubility of some chlorides and sulfates.

Chemical Changes

Components in samples can undergo a variety of chemical changes. Most of the highly specific preservation procedures were developed to control specific chemical changes. Some of these techniques are described in this chapter. Describing all of the techniques that could be required for every possible analyte is impossible. Most of these techniques are adequately described or referenced in the analytical method associated with that analyte.

Free chlorine in a sample can react with organic compounds to form chlorinated species (5). Obviously, this process is of most concern in samples containing free chlorine. Typically, samples of municipal drinking water and treated wastewaters are the most likely samples to contain free chlorine. For these sample types, the chlorine should be chemically removed by the addition of sodium thiosulfate.

Samples collected for determination of species such as cyanides or sulfides require preservation to ensure that the chemical equilibrium is strongly biased in one direction. Typically, these samples are preserved by pH control or by addition of an anion that will precipitate the component of interest. Thus, sodium hydroxide is added to samples collected for cyanide measurements.

This addition raises the pH to ensure that hydrogen cyanide gas cannot be evolved. Likewise, when sulfuric acid is added to samples collected for ammonia determinations to lower the pH, the stable ammonium ion is formed.

Photochemical Changes

Components in the samples can undergo changes associated with light-catalyzed reactions. The most common example is the photooxidation of polynuclear aromatic hydrocarbons. These changes are minimized by the collection of samples in amber glass containers.

Microbiological Degradation

Samples can contain organisms that may degrade organic components in the samples. These degradation changes are generally minimized by pH and temperature control or by chemical addition. Extreme pH conditions (low or high) and low temperatures are effective for minimizing degradation. Addition of toxic chemicals to the sample (e.g., mercuric chloride and pentachlorophenol) can kill the microorganisms and preserve the sample effectively, although these preservatives are not commonly used because of their inherent environmental hazard.

Preservation Process

Proper preservation of analytical samples requires planning of critically timed activities performed in both the field and in the laboratory. Both the field and laboratory personnel must work in concert to ensure that meaningful results can be obtained from the laboratory. These measures include not only field preservation, but additionally the proper documentation, packaging, shipping, and storage of the samples.

Planning

The first and most crucial step in ensuring that samples are properly preserved is the planning of the events that will occur between sampling and analysis. This process necessitates a discussion between the laboratory staff and the sample collection crew to address factors such as sample containers, on-site versus lab analyses, documentation, preservatives, contingency plans, cleaning of sample containers, and shipping logistics.

Sample Containers

In addition to preservation considerations, the decision as to the appropriate sample containers must be made by considering several factors:

- the cost of the containers and associated costs for shipping the samples to the laboratory,
- the cleanliness of the containers,
- the ability to factor in quality control activities (replicate samples),
- the ease of use for field applications.

For example, a 5-gal carboy might suffice on technical merits, but is obviously not practical for an analysis of fluoride. Generally, selecting bottles to meet the specific objectives of each project is more appropriate. Coding of these bottles by type, number assignment, color, or other means enhances the ability of the field crew to focus their efforts on sampling activities.

Communication between the laboratory and field crew should result in a clear understanding of who will supply the sample containers, how the sample containers will be cleaned and used, and what preservatives will be present in the bottles. For example, Table I illustrates the bottles required to collect a sufficient amount of properly preserved groundwater to measure the 260 chemicals listed in Appendix IX to the *Code of Federal Regulations*, Title 40, Part 264 (40 CFR 264).

Table I. Recommended Containers and Preservatives
for Appendix IX Groundwater Monitoring

Sample Container	Bottle Number	Preservation	Minimum Sample Size (mL)	Compound	Recommended Holding Time
Glass (3 × 40 mL)	11	4 °C	40 each	volatile organic compounds	14 days
Glass (5 × 1 L)	12	4 °C	1000 each	semivolatile organic compounds, pesticides, dioxins	7 days until extraction, 40 days after extraction
Polyethylene	4	2 mL of 50% HNO_3 to pH <2	500	metals	6 months
Plastic	6	2 mL of 50% NaOH to pH <12 at 4 °C	500	cyanide	14 days
Plastic	7	1 mL in zinc acetate, 1 mL of 50% NaOH to pH <9 at 4 °C	250	sulfide	7 days
Plastic	1	4 °C	100	fluoride	28 days

On-Site Analyses and Related Activities

On-site analyses and other field procedures should be discussed with the laboratory staff. For example, if field filtration of groundwater samples is performed, the laboratory should supply bottles with nitric acid preservative for trace metal analyses. On the other hand, if the filtration is performed by the laboratory, the bulk samples without preservative should be returned to the lab.

Documentation

Sample labels and chain-of-custody records should clearly identify both the sample and the treatment that has been applied to the sample. The documentation activities include recording the date and time of collection, the sample name, and the preservatives added as well as any field treatment of the sample such as filtration.

Contingency Plans

The laboratory and the field crew should both recognize that well-planned activities can go astray. An understanding of the appropriate activities to perform if problems arise should be made clear. Contingencies for occurrences such as lost sample containers, low-volume samples, and interferences should be defined during the planning stage.

Sample Shipments

The timely shipment of samples to the laboratory is an activity that requires planning. The shipment schedule and method of shipment should be specified. Arrangements for answering questions that may arise (e.g., contact, phone number, and location) should be defined in advance of the sampling effort. Contingency plans should be discussed. Samples should be shipped to the laboratory as soon as possible to minimize the time between collection and analysis. This step is especially critical for the analysis of chemical constituents such as volatile organic compounds that have holding times of 1 week or less.

Coordination of Activities

Proper preservation requires a coordination of activities between the field crew and the laboratory staff. The field sampling crew has the responsibility for ensuring that the laboratory receives properly preserved samples. A number of factors must be addressed to ensure that this process is performed properly.

The field activities start with a check of all sample bottles, preservatives, and labels on the site to ensure that everything agrees with the sampling plan.

The field crew must understand the sample preservation requirements. They should know what on-site measurements and sample pretreatment steps are to be performed and should have the equipment to perform these tests. Tests for certain species, such as residual chlorine and pH, should be performed in the field. If dissolved metals are to be determined, the field crew should filter samples in the field before adding the nitric acid preservative. The field crew should have checks to ensure that the preservation process was performed properly. For example, if the process calls for addition of 2 mL of sulfuric acid to obtain a pH of less than 2, the field crew should check the pH after addition to make sure that the pH is less than 2.

All sample bottle labels, chain-of-custody records, and other pieces of documentation must be clearly completed. All chemical additions to the samples must be recorded. This documentation is as important to the overall success of the sampling and analysis effort as proper preservation of the samples. Laboratory decisions that affect the handling of the samples are dependent on knowledge of field procedures used in collection of the samples.

The responsibility of the field samplers does not stop when the sample bottles are filled. The field crew must ensure that the samples arrive at the laboratory in an expeditious manner. The shipment process involves packaging of the samples to prevent breakage and to conform to federal regulations. If samples are to be kept cold, proper shipping containers and packaging (e.g., ice or cold packs) must be used to maintain these conditions until samples are received in the laboratory.

The laboratory should be notified of incoming sample shipments. This notification should include method of delivery, expected arrival time, and airbill number if known. The laboratory and field crew should agree on how to handle unusual conditions such as international shipments, delivery after working hours, and holiday shipments.

The sample shipment should include all appropriate documentation. The laboratory must be able to schedule all analyses that require immediate attention because of sample holding-time constraints or rush reporting requirements based on documentation contained within the shipment.

The laboratory staff have the responsibility for ensuring that all laboratory aspects of the preservation process are achieved. This responsibility includes proper storage of the samples within the laboratory, performance of analyses within prescribed holding times, and clear documentation of all activities.

The laboratory process starts with a clear understanding of the sampling events and analytical requirements. The laboratory staff should be prepared to perform the various activities that will be necessary. These activities start with receiving the samples. All sample shipments should receive immediate attention. Laboratory receipt of samples involves unpacking the samples; validation of accompanying chain-of-custody documentation, if any; noting the condition upon receipt; and arranging for proper storage pending analyses. Any discrepancies or uncertainties noted at this stage should be clearly communicated to the appropriate laboratory and field staff for resolution.

Once the samples are scheduled for analyses, the laboratory has a continued responsibility for tracking all laboratory activities on each sample so that the analytical results can be correlated to the original sample collected in the field.

Holding-Time Considerations

A discussion of preservation techniques would be incomplete without addressing sample holding times because preservation techniques are used to maximize holding times. The American Society for Testing and Materials (ASTM) defines the holding time as "the period of time during which a water sample can be stored after collection and preservation without significantly affecting the accuracy of analysis" (6).

Holding-time constraints affect both field and laboratory operations. Samples must be shipped to the laboratory as soon as possible after collection to facilitate the laboratory in meeting holding times. The laboratory must have sufficient equipment and staff to guarantee appropriate scheduling of analytical tests to meet holding times. Obviously, the cost of field and analytical work can be greatly affected by very restrictive holding times.

Various regulatory agencies and other groups have attempted to determine the holding times that should be applied to water samples for specific analyses. The most frequently cited holding times are those promulgated under the Clean Water Act in 40 CFR 136 for the analyses of wastewater relating to National Pollutant Discharge Elimination System (NPDES) permits (7). Holding times for many of the parameters listed in 40 CFR 136 have also been established under other programs. For example, holding times are cited in many SW–846 methods, which are the analytical methods used to comply with the Resource Conservation and Recovery Act (RCRA) regulations, and rigid holding-time requirements are specified for laboratories performing work under contract to the U.S. Environmental Protection Agency in Superfund site investigations. In addition, methodologies specified by the ASTM, the U.S. Geological Survey, the American Public Health Association, the American Water Works Association, and the Water Pollution Control Federation also have holding-time requirements.

For the most part, the holding times established in these various programs are consistent. However, some inconsistencies do exist that can create confusion as to the selection of appropriate holding times. For example, holding times for volatile organic compounds range from 5 days from sample receipt for Superfund work, to 7 days from sampling for NPDES permits, to 14 days from sampling for RCRA groundwater analyses.

The basic premise that holding times are used to help maintain the integrity of a sample appears to have been lost in the establishment of general regulatory guidelines. The major concern over this trend is the lack of clear evidence to show that holding times have been selected based upon meeting the ASTM definition. In fact, evidence suggests that holding times can be extended for some parameters (8).

Because data reviewers can easily determine if holding times were achieved, the evaluation of the overall technical quality of laboratory data is increasingly based solely on this single criteria, which is of questionable scientific validity.

If holding times are used in judging the validity of analytical data, the holding times that are established must have some scientific basis. The maximum time that a sample can be held before compromising the integrity of the analysis is dictated by the matrix (e.g., surface water, groundwater, or wastewater), the properties and concentration of the substance being determined, and the preservation technique employed.

Holding times that are established without considering all of these factors can impose unnecessary constraints and costs on both field and laboratory activities and ultimately lead to the rejection of scientifically valid data.

References

1. *Handbook for Sampling and Sample Preservation of Water and Wastewater*; U.S. Environmental Protection Agency. U.S. Government Printing Office: Washington, DC, 1982; EPA–600/4–82–029.
2. *Test Methods for Evaluating Solid Waste*, 3rd ed.; U.S. Environmental Protection Agency. U.S. Government Printing Office: Washington, DC, 1986; SW–846.
3. *Standard Methods for the Examination of Water and Wastewater*, 16th ed.; U.S. Environmental Protection Agency: Washington, DC, 1985.
4. *Guidelines for Collection and Field Analysis of Groundwater Samples for Selected Unstable Constituents*; U.S. Geological Survey: Reston, VA, 1976.
5. Rook, J. J. *Water Treat. Exam.* **1974**, 23, 234.
6. *Standard Practice for Estimation of Holding Time for Water Samples Containing Organic Constituents*; American Society for Testing and Materials: Philadelphia, PA, 1987; ASTM D4515–85.
7. *Code of Federal Regulations*, Title 40, Protection of the Environment, 1986; Part 136, Table II.
8. Friedman, L. C.; Shroeder, L. J.; Brooks, M. G. *Environ. Sci. Technol.* **1986**, 20, 826.

RECEIVED for review February 13, 1987. ACCEPTED May 6, 1987.

Chapter 15 ——————————————————————

Sampling Groundwater Monitoring Wells
Special Quality Assurance and Quality Control Considerations

Robert T. Kent and Katherine E. Payne

Quality assurance and quality control for sampling groundwater monitoring wells begins by defining the hydrologic and geochemical characteristics of the aquifer. Variations in lithology, permeability, and geochemistry of naturally occurring waters within the aquifer should be identified and taken into account when designing and sampling the groundwater monitoring wells, and when drawing conclusions based on the analytical data of the collected sample. Any conclusions should also take into account the limitations of the accuracy of the analytical test method used to quantify the monitored species. Other factors to take into account are sources of sample alteration that include the degree of well purging during sampling, contamination or alteration of the sample by the sampling device, and absorption of air contaminants by the sample during collection.

Q UALITY ASSURANCE (QA) AND QUALITY CONTROL (QC) PLANS for groundwater monitoring should provide for documentation of potential sources of sample alteration during collection. This documentation includes the extent of well purging prior to sampling, careful selection of the device used to purge or sample the well, and implementation of the appropriate preservation and handling techniques of the samples collected. An effective QA and QC plan begins with well design and is maintained through review of the analytical data. These and other considerations are discussed in order to present

1173–6/88/0231$06.00/0 © 1988 American Chemical Society

new data and fresh opinions based on the in-house and field testing of well sampling techniques, interpretation of analytical data obtained from monitoring systems, and a review of available pertinent literature.

Hydrogeologic Controls on Groundwater Monitoring

The initial and perhaps most important QA and QC consideration in sampling groundwater monitoring wells is developing a thorough knowledge of the physical and chemical characteristics of the aquifer system. This knowledge includes identifying variations in lithology of the aquifer, the directions and velocity of groundwater flow, and the spatial and temporal variations in groundwater quality. The following are examples of ways in which a knowledge of the aquifer system may influence the design of the groundwater monitoring system, the achievable representativeness of a groundwater sample, and the interpretation of the sample analytical results relative to the hazardous waste management unit.

Under regulations set forth by the Resource Conservation and Recovery Act (RCRA), owners or operators of hazardous land disposal facilities are required to monitor the uppermost permeable limit underlying the facility. The lithology of uppermost sediments along the southern Gulf Coast of the United States consists of interbedded, discontinuous strata of interdeltaic silts, clays, and sands. In the course of monitoring a hazardous waste facility, one well within the facility monitoring system may be screened in a coarse sand and another well in the same monitoring system screened in a clayey sand, where both wells monitor the uppermost permeable unit underlying the regulated facility. Water samples collected from such wells have been shown to exhibit a significant variation in concentration of inorganic monitoring parameters as a result of the naturally occurring variations in lithology of the permeable zone. Where the variations in sediments within a monitored zone include variations in carbon content, significant variations may occur in the amount of adhesion of any organic waste constituents present in the monitored zone to the aquifer sediment. Proper documentation of lithologic variations within a monitored zone is crucial to the correct interpretation of sample analytical data.

Variations in the permeability of the aquifer have been demonstrated to affect the quality of the groundwater sample collected from the monitoring well. Wells of similar construction and design screened in variable lithologies may exhibit variations in recovery rates subsequent to purging due to the innate variations in permeability of the monitored unit. As a result, the concentrations of monitored species, both organic and inorganic, may vary between wells because of variations in aeration or chemical reduction of the sample in the wellbore during the recovery period. For example, a monitor well completed in a well-sorted coarse sand may be sufficiently purged within 15 min; thus, prompt collection of a water sample is allowed. On the other hand, a monitor well that is completed in a sandy silt may take several hours to recover sufficient water to collect a sample. Where the recovery rate of a well is believed to affect water quality results, the suspected influence should be verified by sampling the slowly recovering well. The degree of influence may then be more or less quantified and considered in the overall interpretation of the water quality data.

The occurrence of vertical gradients of flow between permeable strata within an aquifer system, if not properly accounted for, may result in the monitoring of water quality in multiple zones within one well. Figure 1 shows a single-screen completion across a multilayered aquifer. The upper aquifer is contaminated, but the lower aquifer is not. Under static conditions, the lower aquifer has a higher hydraulic head than the upper aquifer, and vertical cross flow in the well occurs from the lower aquifer to the upper aquifer. Assuming that the transmissivities of the aquifers are 2000 gal/day/ft and the storage coefficient is 0.0003, over 40 days of pumping this well at a rate of 15 gal/min would be required before a sample of the contaminated water in the upper aquifer could be obtained. To avoid cross flow between permeable strata, the screen section of any one monitoring well should be set to monitor one discrete permeable stratum. Identification of vertical components of flow in the aquifer system can be made with clustered or nested wells, which consist of a group of wells installed in the same immediate vicinity and screened at variable depths.

In addition to defining the hydraulic characteristics, the chemistry of the monitored zone(s) should be thoroughly defined prior to interpreting sample analytical data obtained as part of a RCRA monitoring program. A naturally occurring spatial variation in the salinity of connate waters is commonly found along the Texas Gulf Coast. During the coarse of monitoring a regulated waste facility in the Gulf Coast area, total dissolved solids were measured at 30,000 mg/L in one well. Water quality in wells located at the other end of the facility showed total dissolved solids of approximately 500–800 mg/L. On the basis of a statistical comparison of the concentrations of chromium between wells, the well yielding saline-quality water was interpreted to have been affected by the waste management unit. Monitoring wells were later installed at points between the freshwater and saline wells. A measurement of specific cations and anions, total dissolved solids, and specific conductivity revealed a naturally occurring spatial variation in fresh and saline-quality water in the region, in contrast to any plume of contamination originating from the waste management unit.

The degree of alteration in the chemistry of the groundwater sample during sample collection (e.g., oxidation, precipitation, and adsorption) has been shown to be influenced by the initial chemistry of the groundwater (e.g., initial Eh, which is a measure of the available elections in solution; pH; redox buffering capacity; and pH buffering capacity). Through laboratory simulation of sampling from a very shallow water table (less than 18 ft below ground), the amount of iron precipitation due to aeration of the sample during collection has been demonstrated to be significantly reduced in waters with a lower initial pH (1). Mixing of well water with the atmosphere causes aeration and oxidation of

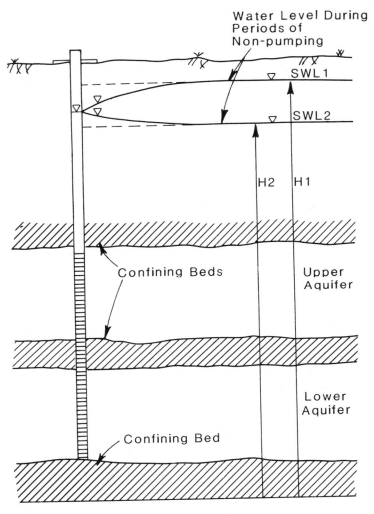

Figure 1. Two-layer aquifer system.

ferrous iron to ferric iron. Ferric iron rapidly precipitates as iron hydroxide and can adsorb other monitoring constituents including arsenic, cadmium, lead, and vanadium. Therefore, aside from the amount of aeration that may occur as a result of the sample collection procedure, the amount of iron precipitation resulting from aeration is dependent upon the innate quality of the groundwater.

In addition to considering the hydrologic and geochemical controls of the aquifer upon the groundwater sample, the flow behavior of the monitored species should be taken into consideration when designing the monitoring well. Figure 2 shows a cross section of a surface impoundment containing soluble

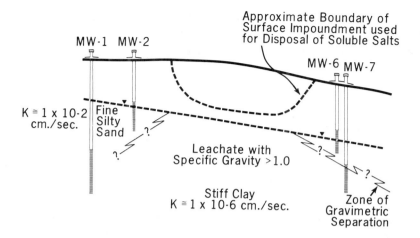

Total Dissolved Solids content
of samples from monitor wells

WELL	DEPTH	TDS
MW-1	45	830
MW-2	25	450
MW-7	48	95,000
MW-8	30	4,800
MW-9	30	3,900
MW-10	48	78,000

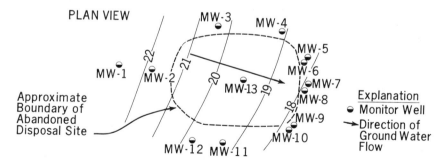

Figure 2. Flow behavior of leachate from a surface impoundment.

salts. As indicated by chemical analysis of groundwater from shallow and deep
cluster wells, the flow of leachate originating from the surface impoundment was
concentrated along the bottom of the aquifer because of the greater density of
the leachate relative to the density of the unaffected groundwaters. Without
consideration of the flow behavior of the leachate and without investigation of
the vertical extent of the aquifer, the contaminants may have gone undetected.

In conclusion, a thorough investigation of the hydrologic and geochemical conditions of the aquifer will result in a more representative monitoring system and will result in a more correct interpretation of groundwater quality data.

Sampling Strategy

The goal of a QA and QC program for sampling wells is to minimize alteration of the collected sample by specifying methods of sample collection and preservation and by documenting sampling procedures. These specified methods allow an interpretation of the data that takes into account variations or alterations occurring as a result of the sampling process. Areas where variations have been observed include purge time, sample holding time, sample collection methods, and sample preservation methods. In this section, specific examples are presented of potential sources of sample alteration that may occur during well sampling.

Well Purging

The purpose of well purging is to eliminate stagnant water in the wellbore and adjacent sand pack that may have undergone chemical alteration. This elimination allows the collection of a sample that is representative of the in situ quality of groundwater near a particular well.

Various methods for determining the necessary extent of well purging have been recommended. The U.S. Geological Survey (2) recommended pumping the well until temperature, pH, and specific conductance are constant. Schuller et al. (3) recommended calculation of the percent aquifer water pumped versus time based upon drawdown in the well. The U.S. Environmental Protection Agency (EPA) recommends removal of three well-casing volumes prior to sampling.

The extent of well purging will vary with the hydraulic properties of the water-bearing unit being monitored. Without proper consideration of the flow characteristics of the monitored unit, the integrity of the sample collected after purging could be compromised. Giddings (4) illustrated that emptying the wellbore of a well screened in a low-yield unconsolidated aquifer may result in a steep hydraulic gradient in the sand pack. This steep hydraulic gradient, in turn, can lead to the addition of clays and silts to the produced water; turbulent flow into the well; and a possible loss of volatile organic compounds in the produced water. Similarly, wells screened in very low-yield bedrock with fracture flow may also be bailed dry. If the water-bearing fractures or higher permeable zone is located near the static water table, the water will refill the well by cascading into the wellbore, and volatile compounds will be lost from the water. Purging of a high-yield aquifer that has major water-producing fractures or a highly permeable unit at the bottom of a screened section in an unconsolidated aquifer may result in limited purging of water higher up in the wellbore and subsequent sampling of water that has been stagnant in the well.

Removal of stagnant water from the wellbore before sampling is necessary to ensure that a representative sample is obtained. However, equally important are the hydraulic processes resulting from well purging. To minimize turbulent flow and sample alteration, each monitor well should be tested to determine the necessary extent and the appropriate rate of well purging prior to preparing a sampling program. Determination of the necessary extent of well purging can be based upon equilibration of groundwater indicator parameters during well evacuation. Figure 3 indicates changes in nine analytical parameters with pumping in a well that had been idle for 6 months prior to pumping. When pumping began, the partially reduced water surrounding the pump was discharged first. As pumping continued, formation waters were drawn into the

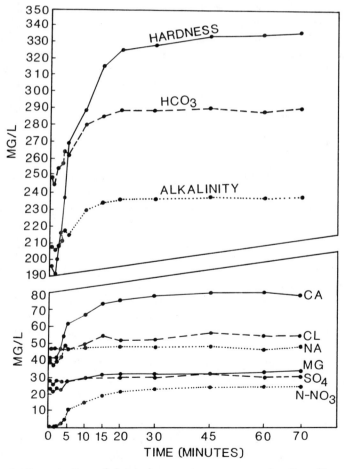

Figure 3. *Concentrations of chemical parameters versus pumping time. (Reproduced with permission from reference 3. Copyright 1981 R. I. Chapin.)*

wellbore, and the rate of change in the concentration of the chemical parameters decreased. Because of the low discharge rate of this well, the mixing of wellbore water and formation water continued for 45 min (5). To obtain a representative sample, the QA and QC sampling protocol for this well would specify a minimum period for well evacuation of 45 min prior to collection of the sample when the existing pump is run at the same flow rate used in the test.

Sampling Device

Commonly used sampling devices include electrical submersible pumps, positive-displacement bladder pumps, bailers, and suction-lift pumps. Choosing a sampling device is dependent upon site-specific criteria including compatibility of the rate of well purging with well yield, well diameter, limitations in the lift capability of the device, and the sensitivity of selected chemical species to the mechanism of sampling delivery.

Aeration or degassing of a sample can occur during withdrawal of a sample from a monitoring well. The introduction or loss of volatile organic compounds or gases (e.g., O_2, N_2, CO_2, and CH_4) in the groundwater sample can affect the solution chemistry of the sample and result in the change in speciation of both volatile organic compounds and other analytes of interest. The degree of aeration or degassing of the groundwater has been shown to vary with the type of sampling device employed.

A field evaluation of sampling devices was conducted in association with ongoing remedial action at the Savannah River plant in Aiken, SC (6). The electric submersible pump was chosen over various modified bailers, the bladder pump, and others because of its accuracy, precision, reliability, its ability to evacuate a well, and its moderate cost. However, levels of organic compounds at the Savannah River plant can be as high as 200,000 ppb. In this case, detection of organic compounds near analytical detection limits (i.e., 1–10 ppb) was not a criterion for choosing a sampling device. Where detection of organic compounds at low levels is desired, evaluation of sampling devices to determine the least potential alteration should be conducted.

Bailers are commonly used for both purging and sampling water from small diameter, shallow wells because of their relative low cost, portability, and ease of maintenance. Disadvantages of the bailer as a sampling device are mixing and the potential aeration or degassing of the sample during sample collection. The aeration is the result of repeated submergence and removal of the bailer during sampling, which may result in turbulent flow of water in the wellbore. Further aeration can occur as a result of pouring the collected sample from the top of the bailer into the sample bottles. Aeration of a sample when using a bailer can be minimized by gently lowering the bailer into the water when collecting the sample. Aeration or degassing can be further reduced with a bailer modified to include a bottom-draw valve. The device allows emptying of the bailer at a slow controlled rate; thus, aeration of the sample, which occurs during decanting, is avoided. Slight improvements in sample representativeness between conventional bailers and bottom-draw-type bailers have been documented by Barcelona et al. (7).

Field and laboratory testing of suction-lift and gas-displacement pumps indicate these pumps are consistently below average in terms of the accuracy of the sample delivered compared to other devices (8–10). The suction-lift pump employs application of a strong negative pressure that can cause degassing of the water sample. Gas-displacement pumps, typically air- or nitrogen-lift, can cause gas stripping of carbon dioxide and result in a change in initial pH of the carbon dioxide or cause gas stripping of volatile compounds.

Absorption of Air Emissions

In addition to the loss of volatile compounds due to degassing during sample collection, recent data indicate that organic compounds may be absorbed from the atmosphere into the water sample when decanting from the bailer to the sample bottle. In one case, 11 monitoring wells were sampled to determine the lateral and vertical extent of nitrotoluene and dinitrotoluene (DNT) isomers relative to surface impoundments containing DNT and DNT process byproducts. During collection of the samples, corresponding field blanks were collected at each monitor well to monitor potential absorption of organic compounds from the air by the collected water sample. The field blanks consisted of distilled water; the water was passed between two glass sample bottles approximately six times at the well site prior to collection of the well water sample. Water quality results for both the well sample and the field blank are included in Table I. The average percent variation in concentration of 2,4-DNT and 2,6-DNT in the groundwater, excluding outside values, as measured by the field blank, was 6% and 7%, respectively. The percent variation represents the potential DNT available for absorption, as indicated by concentrations measured in each corresponding field blank.

Sampling Device Construction Materials

Most of the wells presently used in RCRA monitoring programs are constructed of rigid poly(vinyl chloride) (PVC). Several studies have been completed to investigate the absorption and release of organic compounds by rigid PVC. Preliminary studies have led the EPA to recommend the use of well construction materials made exclusively of polytetrafluoroethylene (PTFE) or stainless steel as opposed to PVC (11). In fact, the quantities of the organic compounds absorbed by the PVC are low (<1 ng/cm^2) (12). A recent study by Reynolds and Gillham (13) documents absorption of organic compounds by PTFE tubing. In particular, the uptake of the compound tetrachloroethylene was noted within 5 min of exposure to the solution; quantification of the amount of absorption by the tubing was not reported (13). On the basis of these and other studies (14),

Table I. Organic Analyses of Groundwater and Corresponding Field Blanks
for 2,4-Dinitrotoluene and 2,6-Dinitrotoluene

| Well Number | Parameter | Concentration (mg/L) | | Percent Variation |
		Well Sample	Field Blank	
1	2,4-DNT	ND	ND	0
	2,6-DNT	ND	ND	0
2	2,4-DNT	1.88	0.014	0.7
	2,6-DNT	3.77	0.008	0.2
3	2,4-DNT	0.017	0.019	112.0
	2,6-DNT	0.024	0.013	54.0
4	2,4-DNT	0.098	0.002	2.0
	2,6-DNT	0.356	0.001	0.3
5	2,4-DNT	0.001	ND	0
	2,6-DNT	0.003	ND	0
6	2,4-DNT	0.306	0.001	0.3
	2,6-DNT	0.188	ND	0
7	2,4-DNT	0.011	ND	0
	2,6-DNT	0.083	ND	0
8	2,4-DNT	0.050	0.001	2.0
	2,6-DNT	0.356	0.001	0.3
9	2,4-DNT	0.004	0.002	50.0
	2,6-DNT	0.050	0.001	2.0
10	2,4-DNT	ND	ND	0
	2,6-DNT	ND	ND	0
11	2,4-DNT	ND	0.006	0.6
	2,6-DNT	ND	ND	0

NOTE: The abbreviation ND denotes not detected.

no real justification has been presented for the replacement of rigid PVC sampling devices with sampling devices utilizing PTFE tubing.

Distinguishing between the use of sampling materials constructed of rigid PVC versus materials constructed of flexible PVC is important. Plasticizers are added to PVC resin to yield a more flexible product. These plasticizers include phthalate esters, which have been shown to leach into water. Rigid PVC pipe with a National Sanitation Foundation (NSF) listing would not be expected to contain any more than 0.01 wt % of plasticizers (15). Flexible PVC can contain from 30 to 50 wt % of plasticizers. When monitoring for low levels of organic compounds, materials constructed of flexible PVC (e.g., tubing and sample bottles) can be appropriately substituted with PTFE, which does not require the addition of plasticizers for flexibility.

Sample Preservation

Water samples may undergo change with regard to their physical, chemical, and biological state during transport and storage. To preserve the integrity of a sample after collection, the samples are generally refrigerated or preserved by the addition of acid or alkaline solutions.

Despite these practices of stabilizing samples, a potential exists for alteration of a sample during transport and storage. Particular practices and areas of disparity that may contribute to the variance of apparent water quality during the sample holding period are as follows:

1. delaying filtering and preserving of samples until samples reach the laboratory,

2. aeration of the sample during filtration,

3. failure to filter samples prior to the addition of acid for preservation, and

4. the lack of necessary temperature reduction for successful stabilization of the sample during transport.

A field experiment has shown that the delay of preservation of samples can lead to variation in water quality analysis (*16*). In the experiment, multiple samples were collected from one monitoring well installed at an anaerobic lagoon and one monitoring well installed at an inactive sanitary landfill. Once collected, the samples were divided into four sets; the first set was preserved immediately, and the remaining sets were preserved 7, 24, and 48 h after collection by the addition of acid. Each of the collected samples was analyzed for calcium, iron, potassium, magnesium, manganese, sodium, and zinc within the EPA prescribed holding times specified for each of the parameters. Iron showed the most dramatic change in concentration. Seven hours after collection, the measured concentration of iron in the sample collected from the well located at the lagoon was 0.33 mg/L; the concentration of iron in the sample collected from the same well and preserved immediately was 11.6 mg/L. The change in iron concentration from the sample collected at the landfill showed a change from 5.74 to <0.08 mg/L between 0 and 7 h after collection and before preservation. Significant changes were also noted for magnesium, manganese, and zinc.

One possible explanation for the sample alteration is the aeration of the sample during transfer from the sampling device to the sample bottle or from the sampling device to a holding vessel prior to filtration, and prior to fixation of the metals by addition of acid. When groundwater is in a reduced state, the addition of oxygen via aeration can cause oxidation of ferrous iron to ferric iron and subsequent precipitation as ferric hydroxide. Once allowed to form, much of the ferric hydroxide will be removed by filtering prior to analysis.

A recent laboratory experiment measured the precipitation of iron from a collected sample with different filtration methods and different sampling devices. The filtration methods tested included on-line filtration, vacuum filtration following transfer from a holding vessel, and the same vacuum filtration procedure after a 10-min holding time. Sampling mechanisms used included a bailer, peristaltic pump, bladder pump, air- and nitrogen-lift pumps, and a submersible electrical pump. With each sample mechanism used, the samples handled by on-line filtration exhibited higher dissolved iron concentrations than samples transferred to holding containers prior to filtration. The 10-min holding period appeared to have no consistent effect on the concentration of measured iron as compared to immediate filtration from the holding vessel (17). The study indicates that the turbulence and associated aeration of the sample during filtering can significantly alter sample quality. In fact, the study indicates that aeration of the sample during filtration has at least as much impact on sample quality as the sampling device itself.

Many monitor wells are completed in low-yield, clay-rich sediments. Completing these wells in such a fashion that water samples can be collected free of sediment is impractical and in some cases impossible. EPA recommends field acidification of samples collected for metals analysis to a pH less than 2 (18). Acidification of unfiltered samples can lead to dissolution of minerals from clays in the suspended solids. Table II indicates that the measured concentrations of calcium and magnesium in samples acidified prior to filtration are directly related to the concentration of suspended solids. On the other hand, concentrations of calcium and magnesium in unacidified samples show no correlation to dissolved solids (19). This lack of correlation does not imply that samples should not be acidified, but rather that samples should be filtered prior to acidification. Otherwise, constituents of interest that may occur naturally in the formation matrix may be dissolved when acidified and result in a sample that is not representative of the water contained in the aquifer.

EPA states that preservation of samples by refrigeration requires that the temperature of collected samples be adjusted to a temperature of 4 °C immediately after collection and during shipment. The cooling rates of water samples chilled by ice and the temperature maintenance ability of frozen blue ice were recorded in order to observe the effectiveness of different types of ice in cooling samples and to determine the effort required to maintain sample bottles at 4 °C. In this study, 10 250-mL bottles and 12 500-mL bottles were filled with tap water. The initial temperatures of the samples were recorded. Thermisters (electronic thermometers) were inserted through small holes drilled

Table II. Addition of Acidic Preservative Prior to Filtering

Sample Number	Suspended Solids (mg/L)	Acidified Ca (mg/L)	Mg (mg/L)	Unacidified Ca (mg/L)	Mg (mg/L)
1	22,000	2442	55	44	18
2	18,500	1980	54	73	16
3	9700	1452	34	95	13
4	8600	1452	36	78	15
5	5200	915	33	134	20
6	3400	827	47	284	36
7	3100	704	27	101	19
8	2200	453	33	134	27
9	1900	286	18	78	13

in the center of each sample bottle lid. The bottles were placed in a 48-qt cooler and covered with two bags of ice, and their temperatures were monitored. Readings were taken every 10 min until the monitored bottles reached their desired temperature of 4 °C. These bottles were transferred to a precooled ice chest filled with blue ice. As indicated in Figure 4, the temperature of the samples dropped to 4 °C within 3 h. As shown in Figure 5, the blue ice was successful in maintaining the bottles less than 4 °C for 24 h. In contrast, when ambient temperature samples were placed in a 48-qt cooler and covered only with blue ice, the samples did not reach 4 °C (Figure 6).

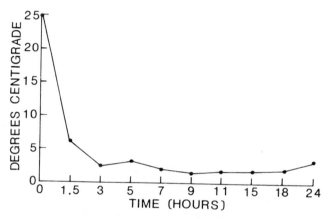

Figure 4. Field refrigeration of samples with water ice.

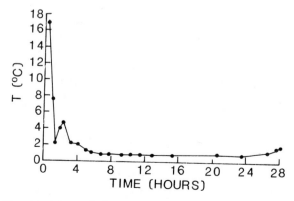

Figure 5. Bottles placed in crushed ice chilled to 4 °C and transferred to an ice chest prechilled with blue ice.

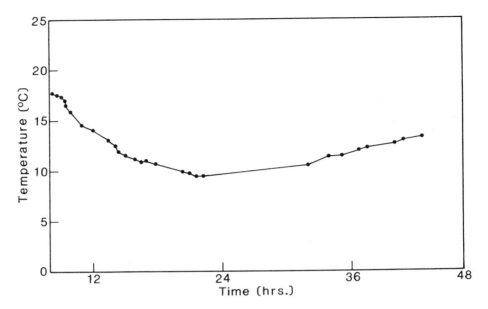

Figure 6. Field refrigeration of samples with blue ice.

This experiment suggests that when using blue ice for refrigeration, the samples must be initially chilled with wet ice. Samples can then be transported to the laboratory in either blue ice or wet ice. However, when using wet ice for an extended period, additional ice may need to be added to the ice chest to maintain the recommended temperature of 4 °C.

In conclusion, multiple avenues for sample alteration exist during collection, including the method of well purging, the method and device used to sample the wells, and the method of sample preservation. The alteration of a sample that occurs during sampling may be more or less quantified by collecting replicate samples a few days apart (20) (e.g., collecting samples from the monitoring system on Monday, collecting another set of samples on Thursday, and comparing the variability in analytical data between the two sets). When the variation is significant, the groundwater sampling protocol should be tested in the field (e.g., comparing variations in analyses resulting from different methods of filtration) to determine the source or sources of sample alteration.

Laboratory Test Methods

The limitations in accuracy of the test method used to quantify the analyte should be considered in any statistical or intuitive interpretation of the analytical data. Of particular concern is the accuracy of the gas chromatography–mass spectrometric (GC–MS) procedure. The method is commonly used to quantify specific organic waste constituents in groundwater samples collected from RCRA monitoring wells. The accuracy of the data obtained by the GC–MS method is dependent upon the experience of the analyst in identifying and quantifying the detected compounds. Furthermore, limitations occur in accuracy associated with the method of analysis, including efficiency of the solvent used in the extraction and the efficiency of the column itself. The degree of attainable accuracy is reflected in the surrogate-spike recovery limits established by the EPA for specific compounds. The surrogate spike consists of an aliquot of each sample that is spiked with a particular known compound of known concentration. The spike is analyzed to determine if the GC–MS equipment is operating at an acceptable level of accuracy. The EPA advisory limits of recovery for particular compounds range from 35% to as much as 140% of the known concentration of the aliquot. The variability resulting from the GC–MS analytical method may far outweigh the variability resulting from the sample collection method.

Summary

The development and implementation of a sampling and analysis plan for monitoring hazardous waste sites is a requirement of federal regulations. Although implementation of a plan does not guarantee that the analytical results accurately represent the in situ quality of the groundwater, such a plan may result in recognition of the potential changes that have occurred between collection and analysis. This information, when properly considered in reviewing the analytical data, may lead to a more accurate interpretation of the data once received.

The initial QA and QC consideration when sampling groundwater monitoring wells consists of developing a thorough knowledge of the hydrogeologic system. Documenting factors such as how the recovery rate of the well influences the sample or identifying lithologic variations in the monitored strata that may affect water quality will result in a more accurate interpretation of the sample analytical data. The inherent limitations of the accuracy of the analytical test method used to quantify the monitored species should also be considered when evaluating the sample analytical data.

Abbreviations

DNT	dinitrotoluene
GC–MS	gas chromatography–mass spectrometry
PTFE	polytetrafluoroethylene
PVC	poly(vinyl chloride)
QA	quality assurance
QC	quality control
RCRA	Resource Conservation and Recovery Act

References

1. Stolzenburg, T. R.; Nichols, D. G. *Proceedings, 6th National Symposium and Exposition on Aquifer Restoration and Ground Water Monitoring*; National Water Well Association: Dubin, OH, 1986; p 231.
2. *Guidelines for Collection and Field Analysis of Ground Water Samples for Selected Unstable Constituents*; U.S. Geological Survey: Washington, DC, 1976; Book 1, Chapter D-2.
3. Schuller, R. M.; Gibb, J. P.; Griffin, R. A. *Ground Water Monit. Rev.* **1981**, *1(1)*, 46.
4. Giddings, T. *Ground Water Monit. Rev.* **1984**, 253–256.
5. Chapin, R. I. M.S. Thesis, University of Texas at Austin, 1981.
6. Muska, C. F.; Colven, W. P.; Jones, V. D.; Scogin, J. T.; Looney, B. B.; Price, V., Jr. *Proceedings, 6th National Symposium and Exposition on Aquifer Restoration and Ground Water Monitoring*; National Water Well Association: Dubin, OH, 1986; pp 235–246.
7. Barcelona, M. J.; Helfrich, J. A.; Gibb, J. P. *Ground Water Monit. Rev.* **1984**, *4(2)*, 36.
8. Nielsen, D. M.; Yeates, G. L. *Ground Water Monit. Rev.* **1985**, *5(2)*, 83–99.
9. Schuller, R. M.; Gibb, J. P.; Griffin, R. A. *Ground Water Monit. Rev.* **1981**, *1(1)*, 44.
10. Barcelona, M. J.; Helfrich, J. A.; Gibb, J. P. *Ground Water Monit. Rev.* **1984**, *4(2)*, 38.
11. *RCRA Ground-Water Monitoring Technical Enforcement Guidance Document*; U.S. Environmental Protection Agency: Washington, DC, 1986.
12. Miller, G. D. *Proceedings, 2nd National Symposium on Aquifer Restoration and Ground Water Monitoring*; National Water Well Association: Worthington, OH, 1982; pp 236–245.
13. Reynolds, G. W.; Gillham, R. W. *2nd Annual Canadian/American Conference on Hydrogeology*; University of Waterloo, Waterloo, Ontario, Canada, 1985.
14. Current, G. M.; Thomson, M. D. *Ground Water Monit. Rev.* **1983**, *3(3)*, 68–71.
15. Barcelona, M. J.; Gibb, J. P.; Miller, R. A. *A Guide to the Selection of Materials for Monitoring Well Construction and Ground Water Sampling*; Illinois State Water Survey: Champaign, IL, 1983; p 40.
16. Schuller, R. M.; Gibb, J. P.; Griffin, R. A. *Ground Water Monit. Rev.* **1981**, *1(1)*, 46.
17. Stolzenburg, T. R.; Nichols, D. G. *Proceedings, 6th National Symposium and Exposition on Aquifer Restoration and Ground Water Monitoring*; National Water Well Association: Dubin, OH, 1986; p 216.
18. *Test Methods for Evaluating Solid Waste*, 2nd ed.; U.S. Environmental Protection Agency: Washington, DC, 1982.
19. Kent, R.; McMurtry, D.; Bentley, M. *Water Resources Symposium No. 12*; Center for Research in Water Resources: Austin, TX, 1985; 461–487.
20. Splitstone, D. E. *Proceedings of the National Conference on Hazardous Wastes and Hazardous Materials*; Hazardous Materials Control Research Institute: 1986; pp 8–12.

RECEIVED for review January 28, 1987. ACCEPTED July 1, 1987.

Chapter 16

Techniques for Sampling Surface and Industrial Waters
Special Considerations and Choices

James E. Norris

Special considerations and choices of sampling modes, sample locations, and sample treatment are described. The variations of these factors when applied to sampling industrial wastewaters, receiving streams, and their attendant biota and sediments are also addressed. Advantages and disadvantages of available options are discussed.

A NALYTICAL CHEMISTRY IS OFTEN the handmaiden of jurisprudence and political action in today's regulatory climate. Unfortunately, those who use the results of the analytical art most are often those who least understand its limitations. This chapter addresses such a fundamental limitation: the validity of a chemical analysis limited by the validity of the sample chosen for characterization. No amount of statistical transformation of data, application of sophisticated analytical methods, or imposition of onerous quality assurance and quality control measures in the laboratory can transform a bad (nonrepresentative) sample into a good one. *Sampling* is an attempt to choose and extract a representative portion of a physical system from its surroundings. The subsequent analytical characterization of the sample defines certain properties of the sampled system. This characterization involves extraction of the sample from the physical system of interest, which is the focus of this chapter; preservation and shipment of the field-collected sample; and in-laboratory subsampling of the field sample and subsequent analysis of the selected aliquot. The variables affecting the objectivity of this characterization are legion. This chapter addresses my experiences in the sampling of diverse water

1173–6/88/0247$06.00/0 © 1988 American Chemical Society

and wastewater streams over a period of 30 years. This chapter is not meant to be all-inclusive, but rather addresses certain situations where alternative sampling methods were evaluated and where some conclusions about these alternatives were drawn.

Sediment Sampling

Industrial wastewaters and waters in receiving streams are never pure solutions. Substantial suspended particulates are generally the rule, and rarely are these suspensions stable. When stream velocity and agitation decrease, particulates settle to the bottom. Native bottom particulates and fine clays are also exposed to the components in wastewaters via processes such as absorption, adsorption, and occlusion, which bind these components to bottom materials. Thus, such a stream bottom is characterized by variously intermixed layers of native material and waste sediments. This composite system is of interest in characterizing the impact of the wastewater on the local environment.

Sampling of such sedimentary materials is inevitably an exercise in creative grab sampling. The techniques most used fall into two broad categories: bottom grab (dredge) sampling and core sampling. Abundant variants and devices of each type are available. *Bottom grab samplers* have the advantage of obtaining a larger sample over a broader expanse of bottom. These samplers are generally easy to use. The greatest disadvantage of such samplers resides in the loss of finely divided particulates, which are carried away by outflowing water from the sampler. Indeed, for some streams, the bottom or near-bottom surface "fines" may be of most interest in stream quality characterization. In such circumstances, a *core sampler* represents a suitable alternative. When the closing mechanism or valve is engaged, waters in contact with the fines are entrapped and cannot carry the fines away. A major disadvantage of core samplers lies in the extremely small area of bottom encountered. As a general rule, more core samples of a bottom are required than bottom grab samples to provide sufficient and valid analytical data.

The aspect of maintaining the in situ vertical relationship of bottom sediment layers is generally academic. Bottom sediment layers are almost always so easily resuspended that sampling confounds and comingles such layers. This effect has been humorously compared to the Heisenberg uncertainty principle concerning the electronic level of matter. If, on the contrary, a compacted sedimentary bottom is sampled, or if an underlying clay layer is sampled, vertical integrity of layers can be maintained by using a core sampler having a split-spoon design. For the sorts of stream-bottom characterizations encountered in my experiences, core samplers have almost always been preferred; minimizing the loss of water-suspended fines has been important, and this technique meets that requirement.

Fish Sampling

No dissertation on techniques for sampling surface and industrial waters is complete without an excursion into fish collection and sampling. In the regulatory environment of the 1980s, toxicity-based effluent discharge limits are paramount, and the question of bioaccumulation arises logically. For example, the impact on indigenous fish species in receiving streams of both pesticide manufacturing discharges and pesticide agricultural runoff is often addressed in a fish collection and analysis campaign.

The collection of fish is the field of commercial fishermen. My experience has been that in all but the most limited of studies, the services of a commercial fisherman are an invaluable asset, especially where the study is limited to one or two varieties of fish in the indigenous population. However, where the sampling campaign is conducted by the investigator, some options can affect the success and ease of the collection effort. Among the various techniques for collection of fish, electric shockers and slat boxes present advantages over other alternatives such as hoop nets, gill nets, or trot lines. These alternative techniques are likely to obtain fewer samples per unit time (e.g., trot lines) or are likely to kill the specimen well before retrieval (e.g., gill nets). (As an alternative to repetitive stream collection of specimens, caged fish can be used for biouptake determinations where this use meets the objectives of the study.)

The preparation of the specimen is critical if reproducibility is required among multiple laboratories analyzing the fish tissue. If freezing the fish for preservation, shipping, and handling is necessary, then the whole fish should be frozen. A caveat should be noted here: If aluminum foil is used to wrap the fish for storage or transportation, the foil should first be washed with methylene chloride and thoroughly dried. The shiny side of the metal foil should not be in contact with the fish because this side is coated with a slip agent. Where the fillet of the fish is chosen to ascertain bioaccumulation of chemical species, the fillet should be manually removed, cubed, ground, quartered, mixed, and then split as samples for multilaboratory analysis.

Contrary to some popular protocols on the subject, use of a high-speed, high-shear blender to attempt homogenization prior to sample splitting is generally unwise. This technique almost always results in physical separation of fatty oils (lipids). If this step is undertaken, then splitting this nonhomogeneous multiphase sample into portions that truly represent the sampled mass will probably be impossible. Because so many organic compounds of regulatory concern are believed to accumulate preferentially in fatty tissue, lipid separation prior to sample splitting clearly should be avoided.

Sampling of Industrial Wastewater Discharges

The sampling of wastewater discharges is the subject of numerous scholarly studies, papers, monographs, and books. Information on techniques and available sampling instrumentation can be found in the U.S. Environmental Protection Agency's (EPA's) *Handbook for Sampling and Sample Preservation of Water and Wastewater (1)*. A more fundamental treatise has been published by the American Society for Testing and Materials (2). The three major sampling techniques available include grab sampling, composite sampling, and continual sampling.

In practice, *grab sampling* is almost always manual grab sampling. *Composite sampling* is usually accomplished by an automatic sampler taking periodic samples and compositing them in a jar or container. A *continual sampler* withdraws a sample constantly from a stream and accumulates the withdrawn volume for collection at a later time. In my experience, the most generally useful sampling technique has been composite sampling with an automatic sampler, which draws a constant sample volume at time intervals proportional to stream flow. In situations where organic compounds are those species of analytical interest, the automatic sampler commonly preferred is that which uses a peristaltic pump, tetrafluoroethylene (Teflon) tubing, and a cooled (4 °C) glass container for collection. Polyvinyl elastomer (Tygon) tubing should be avoided because of leachable phthalate plasticizers.

A continuous sampler, in concept, should give the most nearly representative sample of a water or wastewater flow. This statement is especially true for those automatic continuous samplers where the pumping rate is proportional to stream flow. Such samplers, however, generally require large collection containers and present a more challenging routine maintenance problem than the somewhat simpler automatic composite samplers.

The automatic composite sampler alluded to is recommended because of its ruggedness, ample but modest size, and ease of maintenance. Where sampling time intervals are proportional to stream flow, the compositing of constant-volume increments gives a sample nearly as representative as that obtained by the continuous sampler.

The grab sample is the least representative among the sampling choices available. If the grab sample is a manual grab sample, then it is also the simplest and easiest to obtain. In some situations, especially those dictated by regulatory protocols, the grab sample is the only choice available. For example, this situation occurs when sampling a wastewater for volatile organic compounds (VOCs) such as halocarbons and volatile aromatic compounds.

A single grab sample or a half-dozen grab samples over a 24-h period give only "snapshots" of the quality of the sampled stream. These samples can be representative at best of the stream condition during the several seconds of sampling. Where a stream is essentially constant-flow, spatially homogeneous in composition at any time and varying in composition only gradually over an extended period of time, a modest series of grab samples may give a fair approximation of stream characteristics. Real-world industrial wastewaters rarely meet the requirements of this idealized model.

Ironically, the officially sanctioned determination in wastewaters of chloroform, dichloroethanes, benzene, and dichlorobenzenes hinges significantly upon the least representative sampling technique chosen, the grab sample. In addition, effluent guideline limitations for such organic compounds consist of monthly averages and daily maxima. These limits imply that the measurements truly represent stream composition. The rationale for the regulatory requirement of grab samples is that VOCs are lost into the airspace of the 4 °C thermostated sample collection bottle if a continuous or automatic compositing sampler is used. Henry's law is usually considered, but one aspect rarely addressed is the dramatic decrease in partial vapor pressure of most VOCs in aqueous solution as the temperature drops from ambient levels to 4 °C.

The history of the sampled stream is of some importance in grasping the full significance of imputed VOC losses. Most wastewater discharges are warmer than their receiving streams. Indeed, many National Pollutant Discharge Elimination System (NPDES) compliance points measure a composite stream of warm, biologically treated effluent with warmed once-through noncontact cooling water. For example, consider a composite stream at 35 °C (95 °F) exiting a large header and dropping 6 ft into a concrete-lined ditch basking under a summer sun at 100 °F on its way to a Parshall flume (i.e., an artificial channel or chute for a stream of water calibrated to indicate flow rate from the water level in the chute) and an NPDES outfall sampling point. Are VOC losses at 4 °C in a darkened sample collection bottle really significant compared with the VOC losses occurring in the sampled stream?

This situation can be considered from another perspective: When three or six grab samples in 24 h are used to determine VOC concentrations in a wastewater, does the researcher gain more objectivity of results by minimizing assumed Henry's law losses, or does the researcher lose more objectivity of results by taking less representative samples? My experience has been that more objectivity is gained by using automatically composited flow-proportional samples thermostated to 4 °C than by using several grab samples.

This choice does not exist for regulatory compliance monitoring because grab samples are specified. However, the regulation does not address the situation of sampling upstream from a compliance point for purposes of control and characterization. Therefore, this situation is where the automatic compositing sampler can be profitably used.

Thus, the choice of sample type depends upon the intent of the sampling campaign, the regulatory proscriptions, the nature of the sampled stream, and the importance of the results of analyses of the collected sample.

Sampling of Surface Waters: Receiving Streams

The Clean Water Act addresses the premier position of water quality standards in the regulatory scheme. A given industrial or municipal discharge may meet all defined limitations of the NPDES permit. If, however, water quality of the receiving stream is being threatened, and established water quality standards are compromised by such discharges, then the discharges must be further abated, notwithstanding the best available technology.

As far as the mechanics of sampling are concerned, most of the options discussed in the preceding section are, a priori, applicable to stream or river sampling. When this decision-making process is encountered, however, the peculiarities of stream sampling become manifest.

Particularly for a river that is a navigable stream, the employment of fixed samplers (e.g., continual or composite) is generally out of the question. My experience and that of many of my colleagues is that the most practical river sampling program is accomplished from a boat at known sampling locations. The sort of sample collected is almost always a manual grab sample or a series of manual grab samples composited prior to analysis.

As far as choice of sampling locations, the EPA *Handbook for Sampling and Sample Preservation of Water and Wastewater* (3) addresses the various techniques for choosing appropriate locations. One technique especially appropriate for sampling rivers for chemical constituents is the *spatial gradient technique*. By applying this technique, the distance between points on a transect or grid in a grid pattern can be determined. This approach is as good as any, but caveats do exist. The use of this technique must be tempered with a knowledge of many factors such as effluent plume dissipation, mixing zones, segregation of wastewater discharges in the stream, and tidal effects.

For a valid determination of the impact of an outfall on a receiving stream, an upstream control point must be selected. This control point must be sufficiently upstream to be isolated from the effects of the discharge.

For some streams and rivers, even those having considerable flow and which empty into bays, oceans, or large inland lakes, a measurable tidal effect occurs. I have observed a river at low annual flow that was influenced so much by an incoming tide 80 river miles downstream that a wastewater discharge plume spread upstream for one-half of a mile. Indeed, for such coastal river zones, laminar flow in the river often has more dense brackish water incoming under an outgoing freshwater flow. For these reasons, the choice of an upstream control point must be carefully made. A control point located nearby has the advantage of convenience of sampling, but one several miles upstream is better insulated from any influence of the discharge.

Aerial photography, including infrared techniques, provides insight into the phenomena of in-stream segregation of an effluent plume, variations of mixing zones as a function of stream flow, and impacts of tidal effects. Together with surface observations, these data provide guidance in the choice of downstream sampling locations. Such data are most valuable when a modest library of aerial photographs and surface observations extending over a period of 1 or 2 years is available. For example, little useful information is gathered from a midriver sampling point 1 mi downstream from an outfall if the discharge is hugging the near bank of the receiving stream. Such anomalies are easily discerned by aerial photography and careful surface observation.

Against such a backdrop of knowledge of local conditions, good choices of sampling points can be made, and the water quality of the river or stream can be more objectively assessed.

Acknowledgments

I acknowledge the support provided by BCM engineers as well as the efforts and contributions made by J. English and K. Fitzgerald.

References

1. *Handbook for Sampling and Sample Preservation of Water and Wastewater*; U.S. Environmental Protection Agency. U.S. Government Printing Office: Washington, DC, 1982; EPA–600/4–82–029. (Addendum, 1983.)
2. *Annual Book of ASTM Standards*; American Society for Testing and Materials: Philadelphia, PA, 1986; Vol. 11.01, pp 130–139; Standard D3370–82.
3. *Handbook for Sampling and Sample Preservation of Water and Wastewater*; U.S. Environmental Protection Agency. U.S. Government Printing Office: Washington, DC, 1982; pp 195–200; EPA–600/4–82–029.

RECEIVED for review January 28, 1987. ACCEPTED May 5, 1987.

Chapter 17

Groundwater Sampling

James S. Smith, David P. Steele, Michael J. Malley, and Mark A. Bryant

Because more than 50% of this nation's water supply comes from groundwater, groundwater quality has become a priority. Accurate and precise analytical measurements cannot truly represent groundwater quality if the sampling alters the sample. The sampling of groundwater has numerous variables that affect the analytical results of the sample. These variables include the purging volume; the time interval between purging and sampling; the sample treatment, such as filtration; the sample storage (e.g., oxidation); the purging methodology; and placement of the purging device relative to placement of the sampling device in the well. Accurate groundwater measurements that can be interpreted over a time interval can be obtained if the sample regimen is consistent and developed according to the chemical information desired and the physical characteristics of the groundwater system.

G ROUNDWATER IS ONE OF OUR most important natural resources. As scientists, we are beginning to learn the characteristics of groundwater and how to protect this precious resource. This chapter is concerned with the reevaluations of groundwater sampling techniques based on many years of experience in sampling and interpretation of the analytical results.

Groundwater is sampled to determine its quality and how the quality changes with time. Researchers want to find out what chemical species are present in the groundwater and how much of each species is present. Later, the researchers will want to determine if these qualitative and quantitative aspects of groundwater have changed. If groundwater sampling and analysis are successful, then the results can be used to develop explanations or hypotheses concerning the effect of time on groundwater quality.

This situation sounds easy enough to accomplish. First, a well is drilled to the groundwater under investigation. Appropriate well construction techniques are used to avoid and prevent contamination of the aquifer to be studied. Then

1173–6/88/0255$06.00/0 © 1988 American Chemical Society

the well is developed to fulfill the U.S. Environmental Protection Agency's (EPA's) criteria for water turbidity (1). Three or more well water volumes are purged from the well. The groundwater is sampled, and the samples are sent to a laboratory for analysis. Finally, the scientists can relax and wait for the analytical results.

Nothing is wrong with this scenario as long as the analytical variability and the variability of sampling do not affect the results and therefore the scientists' ability to determine the actual changes in groundwater quality over a period of time. Experience has taught scientists to recognize the analytical variability, especially when the results are near the limit of quantitation. Despite following accepted practices, scientists have found that the variability in the sampling of groundwater is the major source of error in the measurement of groundwater quality. This broad generalization about groundwater sampling and analysis means that unless special care is taken to minimize the variability of the sampling procedure, the interpretation of groundwater quality measurements taken over a period of time are probably wrong. If this situation is true, then serious problems are associated with groundwater protection as it is presently practiced. This chapter adds additional information to the paper by Bryden et al. (2).

Groundwater Properties

Groundwater is a very complex matrix. Although each aquifer represents a different water quality, our experience has shown that all groundwater has two unique properties for natural water systems:

- The movement of groundwater through the aquifer precludes the transport of chemicals by particulates. Only substances soluble in the groundwater matrix are mobile within the aquifer.
- Groundwater is nearly oxygen-free. In other words, the dissolved ions tend to be in their most reduced state. High iron content of groundwater indicates that the water in the aquifer is carrying ferrous iron at neutral pH values. Ferrous iron groundwater cannot be aerobic groundwater.

In other words, a groundwater sample cannot contain any particulate matter, and must be protected from air at all times if the sample is to be truly representative. If groundwater samples that contain particulates and have been exposed to the air are analyzed, then the results obtained do not represent the water within the aquifer.

How critical are these properties to the analysis of groundwater quality? On a macroscopic scale, we have observed unit pH changes as ferrous iron is oxidized by the oxygen in the air to the insoluble ferric iron. Corresponding changes in specific conductance occur also. But more important is how particulates in the sample and our oxidation of the sample affect "trace" analysis of organic compounds and metals. What does trace analysis mean?

This question appears silly to the environmental scientist accustomed to parts-per-billion data. We have become anesthetized with parts-per-billion thinking. A trihalomethane concentration of 100 ppb is considered a large amount. For toxicity evaluations, this statement is true. But the measured amount is very small (i.e., 100 ng/mL). This amount (100 ng) can be translated to one ten-millionth of the amount of sugar that can be placed on a dime. The point is that the measurement of low turbidity [5 nephelometric turbidity units (NTU)] is not an appropriate indicator of particulate-free water for the detection of parts-per-billion levels for pollutants. Therefore, the turbidity criteria are appropriate for well development only and cannot be used as an indicator parameter for particulate-adsorbed pollutants that are suspended in the groundwater sample. This situation is like using a truck weighing station to indicate the presence or absence of a dime. We have observed too many results in which the quantitative amount exceeds the solubility of the substance. The analytical work is correct, but the conclusion that the substance is in the groundwater is dubious.

No matter how much care is taken in well construction to avoid contamination from the surface or near-surface area, trace amounts will get into the aquifer region of the well. Maximum care should be planned and executed for the removal of surface and subsurface contamination before well construction.

Techniques for Proper Groundwater Sampling

The best possible analysis of groundwater constituents results from the removal of particulates from all samples and in keeping all unpreserved samples oxygen-free until analysis. Even then, special instructions to the laboratory may be required to obtain accurate results. For instance, groundwater having a high ferrous iron content should be analyzed for the acid–neutral semivolatile fraction instead of the prescribed methodology of EPA's Method 625 (3) or contract laboratory program's method (4). This situation is due to the acidic compounds being entrapped into the iron precipitate when sodium hydroxide is added to the sample before the base–neutral extraction. Following the base–neutral extraction, the acidification of the sample does not release the acid compounds, and poor recoveries are observed.

The toughest question to answer is "How should groundwater be sampled to obtain a sample that accurately represents the aquifer?" Our recommendations include use of the following:

- purge-volume test,
- consistent sampling protocol,
- anaerobic sampling and sample handling conditions, and
- filtration of all samples.

Purge-Volume Test

We used EPA's recommended three well water volumes for purging a well before sampling. In connection with this protocol, we monitored pH and specific conductance as indicator parameters in the field. In almost every case, we obtained a constant pH and specific conductance reading before the three well water volumes were removed. Therefore, we have always thought that we have fulfilled EPA's and our own quality criterion that the stagnant well water be replaced by fresh groundwater from the aquifer. We have assumed that the chemical parameters or the quality of that fresh groundwater is a constant. However, working with several groundwater systems that are contaminated with volatile organic solvents demonstrated that this assumption is grossly inaccurate.

By performing a purge-volume test to analyze for the parameter of most concern, we found some very interesting phenomena involving volatile organic solvents. For example, near the Delaware River, where the saturated zone is connected to the river and the tides associated with the river, more than six well volumes must be purged before the concentration of trichloroethylene becomes constant (see Table I, sample numbers 1–5).

In a more complex mixture of volatile organic solvents, the purge-volume test of a well showed three trends. Two solvents started at relatively high concentrations and decreased with purging to a "not detected" value. Two other solvents that were "not detected" at the three well water volumes purged were detected at the four well volumes purged and increased until constant values were obtained near the 10 well water volumes purged. The fifth volatile organic solvent detected in this well remained at a constant concentration value

Table I. Purge-Volume Test Results for Trichloroethylene

Sample Number	Volume Purged (Well Water Volumes)	Concentration (ppb)	Time after Purging	Water Column Location
1	0	not detected		
2	3	75		
3	6	150		
4	12	170		
5	24	170		
6		170		bottom
7		100	4 min	top
8		100	4 h	
9		10	24 h	

throughout the test. The information gained from the purge-volume test was very important in the site groundwater evaluation. Any inconsistency in well purging would not only influence the quantitative result, but also would change the qualitative suite of chemicals observed. If the well purge-volume test was omitted, then the interpretation of this well's data would probably draw the familiar conclusion of laboratory error(s).

The purge-volume test is valuable because it allows choice of the best purging conditions to obtain the most consistent analytical results. This test is especially important in the sampling and analysis of groundwater for volatile organic solvents.

Consistent Sampling Protocol

The word that keeps making its way into our recommendations on groundwater sampling is *consistency*. Consistency is the key to obtaining data that can be compared over a time interval. Consistency includes purge volume, purge rate, time between purge and sampling, level of water sampled, and sample handling.

The purge volume is selected after the purge-volume test. This purging of the well will be the same every time the well is sampled. The purge rate will also be the same for each sampling. The maximum purge rate is advantageous. Ideally, the well should be emptied first, and then after the groundwater achieves equilibrium, the second well volume should be purged.

One of the least consistent elements of groundwater sampling is the time between the purging of the well and the sample removal from the well (*see* Table I, sample numbers 7–9). Within 24 h after purging, we observed no trichloroethylene in samples from a well containing groundwater that contained 200 μg/L of trichloroethylene immediately after completion of purging. This observation strongly suggests that sampling should occur as soon after purging as possible. In any sampling protocol, the time between the completion of purging and the sampling of a well must be constant each time the well is sampled.

One of the most perplexing problems that we have encountered concerns the depth within the well that water is purged and sampled. This problem occurs in wells where the groundwater cannot be removed completely by the purging pump. The water column in the well decreases, but the well is not pumped to dryness. With the purging pump in the bottom of the well, we observed a case in which the concentration of trichloroethylene was 50% greater at the bottom of the well as compared to the concentration near the top of the water column in the well (*see* Table I, sample numbers 6 and 7). Even though several possible explanations can be offered for this observation, we feel that the lack of mixing within the well water volume contributed the major effect. The result of this experience is that the position of well water sampling should be as close as possible to the point of well purging. From sampling to sampling, the sampling point within the well water column must be the same.

Anaerobic Sampling, Sample Handling Conditions, and Sample Filtration

We believe sample handling must be drastically changed in order to obtain the most representative sample of the groundwater in the aquifer. The sample must be free of particulates and must not be exposed to air. Any sample that is not preserved should be placed into a sample container without any headspace similar to the present volatile organic analysis (VOA) sample handling protocol. Every sample except the VOA sample can be pressure-filtered under a nitrogen atmosphere through a 0.45-μm polyvinylidene fluoride or polytetrafluoroethylene filter medium. Organic samples are included in those parameters to be filtered.

The best situation would be the use of the purging pump as a sampling pump, too. To sample, insert a 0.45-μm filter into the polytetrafluoroethylene (Teflon) tubing outlet line. The filter would not aerate the sample if the outlet tube is always full of groundwater sample. This methodology delivers a filtered sample that is exposed to a minimum amount of air.

Groundwater monitoring is at the stage where the variations in the chemical data due to sampling can be reduced. The benefit of this work is a more standardized sampling methodology for groundwater. This standardization will produce a consistency in sampling that will allow time variations of groundwater parameters to be seen above sampling variations. This procedure will produce more accurate analytical data and a truer representation of the water within an aquifer. Application of this procedure leads to a better understanding of groundwater and a more cost-effective program to protect this precious resource. Groundwater quality data can thus be interpreted accurately; pollutant migration, seasonal fluctuations of parameters, and groundwater cleanup technologies can then be understood. Research on groundwater sampling should be the top priority in groundwater environmental protection.

References

1. "RCRA Ground-Water Monitoring Technical Enforcement Guidance Document"; U.S. Environmental Protection Agency. U.S. Government Printing Office: Washington, DC, 1986.
2. Bryden, G.; Mobey, W.; Robine, K. *Ground Water Monit. Rev.* **1986**, *6(2)*, 67–72.
3. *Fed. Regist.* **1984**, *49(209)*, 153–174.
4. "U.S. EPA Contract Laboratory Program Statement of Work for Organics Analysis Multi-Media–Multi-Concentrations"; Exhibit D, October 1986; U.S. Environmental Protection Agency: Washington, DC, 1986.

RECEIVED February 13, 1987. ACCEPTED April 29, 1987.

SAMPLING AIR AND STACKS

Chapter 18

Effects of Environmental Measurement Variability on Air Quality Decisions

John G. Watson

The variability, or uncertainty, of environmental measurements can have a major effect on the decisions made about an environmental issue. Alternatives are evaluated prior to making a decision by using combinations of measurements and models to quantify the ramifications of those alternatives. Both measurement and model uncertainty affect the decision-making process, but these uncertainties are seldom quantified and are rarely part of the decision. Uncertainty estimates are used in the decision-making process by ignoring uncertainty, defining margins of safety, or balancing costs against benefits.

A IR POLLUTION FROM ANTHROPOGENIC SOURCES is a fact of life today and has been for thousands of years. Air pollution probably came into being when the first fire was lit. As humans discovered new ways to apply the conversion of fuel into energy, energy into sustenance and leisure, and sustenance and leisure into more people, air pollution emissions increased. The first recorded decision to control air pollution emissions was issued in 13th century London as a royal proclamation prohibiting the use of coal (1). Though the "measurements" on which this decision were based (e.g., observations of black chimney plumes, reduced visibility, soot deposits on clothing, and sickness and death due to respiratory disease) were imprecise and variable, the regulatory decision was undoubtedly correct.

Most of the environmental decisions before the latter half of the 20th century were made without the benefit of extremely precise measurements. These decisions were fairly obvious, however, and did not require such

1173–6/88/0263$06.00/0 © 1988 American Chemical Society

precision. Today, this situation is no longer the case. The inherent variability of environmental measurements can make the difference between alternative outcomes to a decision-making process. This variability must therefore become an integral part of that decision making. The objectives of this chapter are to categorize the types of environmental decisions made with respect to air quality, specify the uses of air quality measurements in decision making, and summarize the uses of measurement variability in decision making. Examples specific to suspended particulate matter are supplied to clarify certain concepts. These examples are intended to be illustrative rather than comprehensive; full examination of past and present air quality decisions is beyond the scope of this chapter.

Types of Air Quality Decisions

Many approaches can be applied to the study of the environment, and these approaches often yield differing conclusions and alternatives for environmental control. The decision maker must select the best of these alternatives, and the "best decision" normally is defined as that which is the least disruptive and most cost-effective. The decision maker is often confronted with alternatives in four aspects of air quality: effects determination, identification of pollutants causing effects, source attribution, and emissions controls.

Effects Determination

The first air quality decision involves the perception that a problem exists and is primarily caused by an atmospheric constituent. An effect can be one of human respiratory distress, damage to materials or plants and wildlife, visibility impairment, or a perceived nuisance. These effects only become problems when they are defined as such by the general population in which they occur, as evidenced by the differences in environmental concern between developed and developing countries (2). Environmental observations are used to determine the existence of a problem (e.g., the number of sicknesses and deaths in a given period). These observations are highly variable. Exposure to atmospheric constituents is one of many possible causes in most cases.

Identification of Pollutants Causing Effects

In order to be classified as an air pollutant, an atmospheric constituent must have the chemical and physical properties and be present in sufficient quantities to cause an unacceptable effect. The decision to make such a classification is based on measurements that establish the cause and effect relationship, and these measurements exhibit large uncertainties. For example, the U.S. primary air quality standard for suspended particulate matter has been revised to consider size classification as opposed to total suspended particulate matter in response to more precise measurements of particle entry into the human lung (3). The original measurements on which the earlier standard was based (4) have been improved over the intervening time period (5, 6), and the decision regarding the tolerable physical properties and concentrations of suspended particulate matter is different.

Source Attribution

Excessive exposures to air pollutants can only be reduced by curtailing their emissions from anthropogenic sources. Because many such sources usually exist, decisions must be made with respect to the contributions from each one to ambient concentrations. These decisions are based on emissions, meteorological, and ambient air measurements, all of which have substantial measurement uncertainties. The relationships between emissions and receptors are embodied in *air quality models*, which introduce additional uncertainty because they cannot be perfect representations of reality.

Emissions Controls

Many alternatives exist with respect to reducing emissions from major contributors, and the appropriate decision, from a purely cost-effective standpoint, is that which achieves the greatest reduction in ambient concentrations per dollar invested in emissions controls. These decisions rely on measurements of control efficiency and the assumption that a reduction in emissions will be accompanied by a proportional reduction in ambient concentrations. Large uncertainties are associated with these measurements and assumptions. For example, even though road dust has been identified as a major contributor to suspended particulate matter in many urban areas, and street-sweeping programs have been implemented to reduce road dust, this control has not been shown to be very effective (7, 8). Nonlinear photochemical processes may cause oxidant and secondary particle levels to increase at certain locations in response to decreases in nitrogen oxide emissions (9).

Air Quality Models

Measurements are always used in models to arrive at a decision. The word *model* is used in its broadest sense to include all of the methods used to interpret environmental measurements. The most widely used air quality models for each of the decision categories described in the previous section are descriptive, covariational, classificatory, mechanistic, and inferential. The effects of environmental measurement uncertainty on decisions cannot be separated from the models in which those measurements are used.

Descriptive Models

Descriptive models summarize the spatial, temporal, and statistical distributions of individual observations. The model results show which air pollutants are reaching levels of concern and are the basis for defining the existence of a problem. Descriptive models require a large number of measurements in space and time. For example, an annual geometric average of total suspended particle concentration in excess of 75 $\mu g/m^3$ has been deemed unacceptable by national ambient air quality standards, and a model of one 24-h measurement taken every 6th day has been deemed adequate to determine whether or not this standard is violated in a given year (*10*).

Covariational Models

Covariational models calculate measures of association between two or more variables and are often used to assign effects to a pollutant species or a pollutant to an emissions source. The output data of these models, such as correlation coefficients, are high absolute values when variables change in the same manner over a period of time or over a geographical area; the output data of these models are low absolute values when covariation is lacking. By themselves, these models establish only whether the values for a set of variables change in the same way. Ware et al. (*11*), for example, pointed to many confounding variables not related to air pollution that could cause similar variabilities among morbidity, sulfur dioxide concentrations, and suspended particulate matter found to be statistically significant in several epidemiological studies.

Classificatory Models

Classificatory models define categories based on the achievement of certain measured values for a set of attributes. Each category has a specific set of decisions associated with the members of that category. The prevention of significant deterioration regulations provided by the Clean Air Act Amendments of 1977 (*12*), for example, automatically impose certain air quality modeling, monitoring, and emissions requirements on 28 stationary source types that have the potential to emit more than 100 tons/year of any criteria pollutant. Other source types may emit up to 250 tons/year without invoking this decision-making process. All proposed sources are classed based on emissions calculations derived from environmental measurements.

Mechanistic Models

Mechanistic models, also known as *source-oriented models*, contain mathematical descriptions of the interactions among variables derived from fundamental physical and chemical laws. They directly relate a cause to an effect and can be used to hypothesize the effects of alternative decisions. Air quality dispersion models that use meteorological and emissions data to calculate ambient concentrations (*13*) are the most common example of these models.

Inferential Models

Inferential models, also known as *receptor-oriented models*, can be derived either from fundamental physical and chemical laws or from empirical observations. Inferential models differ from mechanistic models in that their primary set of input data is that which is measured at the receptor, not at the source. The chemical mass balance receptor model (*14*), for example, is a U.S. Environmental Protection Agency (EPA) model that calculates contributions to ambient concentrations of suspended particulate matter from a large number of chemical species measured on the receptor sample and in the source emissions. These models are most often used to decide which source types are the major contributors to unacceptable pollutant levels.

Uses of Measurements in Models

All models are simplified descriptions of complex systems, and these simplifications are effected by making certain assumptions. A model is only valid to the extent that these assumptions are complied with in a specific application (*15*). This definition leads to two sources of uncertainty in a model result: *measurement uncertainty*, which results from the variability of the environmental measurements used as model input data, and *model uncertainty*, which is caused by deviations from the model assumptions (*16*). Either one of these uncertainties may be the dominant cause of uncertainty in the model result used for decision making. Environmental measurements are used not just as model inputs, but also to derive fixed constants, to test compliance with model assumptions, and to estimate measurement uncertainty.

Model Input Measurements

Every model requires some data on which to operate for the period of time being examined. For example, mechanistic air quality models may require boundary and initial conditions of the precursor, intermediate, and end-product species as well as three-dimensional wind fields and atmospheric stability estimates. The chemical mass balance receptor model requires ambient concentrations and source compositions for a number of chemical species. These variables are considered input data when measured for the time simulated.

Parameter Measurements

Parameters are constants that are not supplied to the model on a case-by-case basis, as are input measurements. Parameters are obtained by a theoretical calculation, by measurements made elsewhere and assumed to be appropriate for the place and time being studied, or by tacit assumption that the value of a variable is constant or negligible. Reaction rates, emissions rates, transformation rates, dispersion parameters, and source compositions are common parameters in mechanistic and inferential models. Values for these variables are rarely measured over the period of time being modeled because of lack of feasibility or excessive measurement costs. Parameters normally carry higher levels of variability than input data because parameters are not specific to each case being studied.

Model Validation Measurements

Additional measurements that may not appear in the model as input data or as parameters can be used to determine the extent of deviations from model assumptions. The use of measurements in this way is the first step in quantifying model uncertainty. For example, Gordon et al. (17) applied the chemical mass balance model to a subset of chemically speciated suspended particulate matter concentrations. They then calculated the concentrations of the remaining species and compared these calculations with their measured values to evaluate the validity of their source contribution estimates. In this way, they found that several key species must be included in the model to comply with the chemical mass balance model assumption of linearly independent source compositions.

Measurement Uncertainty Measurements

The magnitude and distribution of variability in the model input data, parameters, and validation measurements can only be determined by replicate measurements of the same variables. Most models assume that a single point measurement represents a spatial or temporal distribution, and an estimate of the variability over space or time is needed if an adequate value for measurement uncertainty associated with a model result is to be derived. Measurements "collocated" with a resolution finer than the temporal and spatial scales of the model can be used to estimate the uncertainty of input data and parameters. These uncertainties can also be estimated from periodic performance tests of each measurement method if analytical variability is larger than spatial or temporal variability (18).

Quantifying Uncertainty

Incorporating the uncertainty of environmental measurements into the decision-making process presupposes (1) that the uncertainty associated with an individual measurement can be quantified, (2) that these measurement uncertainties can be propagated through the models to provide a reasonable confidence level around the model result that is used to justify the decision, and (3) that these measurement uncertainties can be compared with model uncertainties to determine which is of greater concern. Several methods have been applied to the quantification of uncertainty, though these methods are largely unproven.

Uncertainty of an Individual Measurement

These uncertainties are ones of precision, accuracy, and validity (*19*). *Precision* is a measure of the variability of measurements of the same quantity by the same method. Precision can be determined from the standard deviation of repeated measurements and is usually assumed to follow a normal distribution. Precision is the easiest uncertainty to quantify because some form of performance testing or replicate analysis comprises the quality control of most measurement networks. Very few data bases for ambient or source measurements report these precisions, however. The sulfate regional experiment (*20*) data base, for example, has been widely used to decide when and where high levels of particulate sulfate occur and to test air quality models precisely because this data base is one of the few measurement programs with extensive quantification of measurement precision.

Accuracy is the difference between measured and referenced values, and this difference is expected to be within the precision interval for the measurement to be deemed accurate. Primary standards, reference materials, and equivalency protocols have been established by the National Bureau of Standards and the EPA for several pollutants measured at sources and in ambient air in order to quantify the accuracy of these measurements (*21*). Standardization methods are still inadequate to establish the equivalency of many measurements, however. Decisions regarding the attainment of the proposed national air quality standard for particulate matter having diameters ≤ 10 μm, for example, are still clouded by demonstrated differences among simultaneous measurements by different sampling systems (*22, 23*), even though each of these systems has passed prescribed equivalency tests.

Measurement validity consists of two components, method validity and sample validity. Method validity requires the identification of measurement method assumptions, the quantification of effects of deviations from those assumptions, the ascertainment that deviations are within reasonable tolerances for a specific application, and the creation of procedures to quantify and minimize those deviations in a specific application. Sample validity consists of procedures that identify deviations from measurement assumptions for each measurement. Each individual value needs to be designated as valid, valid but suspect, or invalid based on predefined criteria. The validity of a measurement is the most difficult uncertainty to determine and is often the cause of very high or low measurement values. For example, a recent examination of the highest inhalable particulate concentrations in EPA's inhalable particulate network (*24*) demonstrated that 10 out of the 50 samples selected for intensive study yielded mass concentrations that were inconsistent with the sample chemical composition and size distribution. Without these nonroutine measurements that allowed the measurement validity to be questioned, the highest concentrations might have been attributed to pollution sources instead of to anomalies in the measurement process.

Measurement Uncertainty in Models

As difficult as estimating the uncertainty of an individual measurement is, combining these estimates from a number of measurements that are acted upon by a decision-making model is even more challenging. Analytical and computationally intensive methods have been applied to this task.

If a model output can be expressed as some function of a set of environmental measurements, and if the errors in those measurements are random and normally distributed about the true value, then an analytical expression for the model output variable can be derived from the theory of maximum likelihood (25). For example, such expressions have been derived for the Gaussian plume (16) and the chemical mass balance (26) air quality models, which are commonly used to decide which particulate emitters are the major contributors to ambient concentrations. These analytical error propagation methods make numerous assumptions, however, and the range of practical applications for which they are valid remains to be determined.

Computationally intensive methods rely on repeated applications of the model to input data that have been altered in a way that mimics the uncertainty caused by the measurement process. The distribution of model results derived from these repeated applications defines the portion of their uncertainty caused by the variability of the environmental measurements. *Monte Carlo methods* (27) perturb all input data with random numbers drawn from an assumed error distribution. Though this method is very effective, a large number of trials are required to achieve statistical stability, and the error distributions of input data are often unknown. The *bootstrap method* (28) recalculates statistical parameters, such as those yielded by description, classification, and correlation models, from many randomly selected subsets of the model input data set. The fundamental assumption of this method is that measured data sets define their own distributions of random error. This method does not require an assumed distribution, though extension of this model to mechanistic and inferential models is not apparent, and many trials are still required for statistical stability. The Fourier amplitude sensitivity test (29) associates each uncertain model input variable with a specific frequency in a Fourier transform space of the model variables. The model equations are solved for discrete values of the Fourier transform variables. The uncertainty of the model result is a function of these Fourier coefficients. This method requires fewer computations than the Monte Carlo and bootstrap methods. Each of these computationally intensive methods is based on a number of assumptions that may not be complied with in actual practice, and the methods require substantial additional testing to define the limits of their application.

Model Uncertainty

Model uncertainty is nearly impossible to quantify with current technology. No air quality model is ever applied in a situation without some deviations from its basic assumptions. A testing protocol for identifying those deviations and determining their significance involves (1) explicitly listing those assumptions as part of the model derivation from basic principles, (2) analytically introducing deviations from those assumptions and quantifying the effects on model results, (3) performing controlled experiments to verify model components, (4) comparing model results with those of a more complete or more widely accepted model, (5) comparing model results with ambient measurements, and (6) introducing unique tracers into the modeled system (30). Each of these methods has been applied to each of the decision-making models described, though none of these applications has adequately quantified model uncertainty over the range of typical applications in which decisions are made.

Using Uncertainty in Making Decisions

As already noted, the variability of environmental measurements on which decisions are based is rarely quantified and reported. This measurement uncertainty is even more rarely propagated through decision-making models and weighed against a quantitative estimate of model uncertainty. Even if these indicators of measurement variability were available, how could they be used in the decision-making process? Quantitative uncertainty estimates are ignored, used to describe worst-case situations, or used to weigh costs against benefits.

Ignoring Uncertainty

When measurements and model outputs are calculated to two or more significant figures without the specification of uncertainty intervals, the implicit assumption is that these intervals do not exist. The decision is based only on the value at hand with no consideration of whether that value overestimates, underestimates, or is totally unrelated to the true value. Many day-to-day decisions are made by ignoring the effects of variability in environmental measurements.

Worst Case

When uncertainty is quantified, postulating the value for a measurement or model output that will result in the most deleterious outcome is possible. If this outcome is still acceptable, then a decision about a proposed action can be made with confidence. Conversely, if the best possible outcomes are unacceptable, then a negative decision about the proposed action can be easily justified. For example, national ambient air quality standards are set well below the level at which any deleterious effects to the most sensitive segments of the population have been observed. Most air quality modeling conducted prior to permitting a new source is performed under worst-case emissions and meteorological conditions by using models that are expected to overestimate rather than underestimate ambient concentrations. This use of uncertainty in decision making is adequate for the majority of day-to-day environmental decisions such as those involving emissions source locations and emission controls. The worst case is an improbable event, however, and decisions based on this analyses are often more restrictive than is necessary for adequate environmental protection.

Cost–Benefit Analysis

Analytical frameworks have been developed to integrate uncertainty estimates from measurements and models into the decision-making process. These frameworks also integrate the four categories of air quality decisions cited at the beginning of this chapter and attempt to balance the uncertainties associated with one category against those of the other categories. The outcomes of these frameworks are positive (i.e., benefits) and negative (i.e., costs) measures associated with each alternative decision. These frameworks explicitly separate costs and benefits because the relative weight attached to each is really a subjective judgement provided by the decision maker. These frameworks are crude at this time and are rarely used as a fundamental basis for decisions. They do provide a conceptual structure for integrating decision categories, models, and uncertainty estimates.

A *decision tree approach* (31) identifies all of the possible outcomes of a series of decisions and assigns a cost and a probability to each of these outcomes. Estimates of measurement and model uncertainty are part of the decision tree and are used to compute the potential environmental costs of delaying a decision as well as the potential benefits of uncertainty reduction. This approach is appealing because the value added to the decision-making process is determined by more precise environmental measurements. This model could be used to optimize monitoring networks by selecting only those observations, locations, and measurement methods that would make a difference in the selection of outcomes in the decision tree. These decision trees must be custom-designed for each environmental concern, and the cost of such a design may exceed the cost of an incorrect decision arrived at by a worst-case analysis.

A *risk assessment* truly integrates models used in each decision-making category. For example, a computer code for air quality risk assessment has been implemented with quantitative modules for emissions, transport, exposures, and effects (32). The output of these risk assessments is in terms of a probable number of deaths, sicknesses, or number of dollars spent. Alternative decisions regarding emissions controls, locations of emissions sources with respect to exposed populations, and the physical and chemical nature of emissions can be evaluated by comparing the outputs for each alternative. Uncertainty is incorporated both implicitly, by expressing results in terms of statistical probabilities rather than as direct outcomes, and explicitly, by propagating uncertainties from one model to another by using the methods described in this section.

Conclusions and Future Research

The variability of environmental measurements has an important effect on the decisions based on those measurements. The uncertainty that results from that variability should be a formal part of the decision-making process. This uncertainty will not be incorporated until the measurement uncertainties are quantified as part of the measurement process and then propagated through the models that are normally used as the justification for decisions. Methods have been proposed for estimating and propagating these uncertainties, but these methods are largely untested.

Further research is needed to determine the range of applicability of uncertainty estimation methods in practical situations. Additional research is needed to derive new methods for incorporation into measurement and modeling practice. Finally, frameworks need to be developed and tested that will allow the uncertainties associated with different decision categories to be balanced against each other for the purposes of specifying additional environmental measurements that will reduce the uncertainty of an environmental decision.

References

1. Halliday, E. C. WHO Monograph Series No. 46; World Health Organization: Geneva, Switzerland, 1961.
2. Watson, J. G.; Broten, A. R.; Smith, S. M.; Chow, J. C.; Shah, J. J. *Proc. 76th Annu. Meet. Air Pollut. Control. Assoc.* Atlanta, GA, 1983.
3. *Fed. Regist.* **1987**, *52(126)*, 24634–24749.
4. *Air Quality Criteria for Particulate Matter*; U.S. Department of Health, Education, and Welfare: Washington, DC, 1969.
5. *Air Quality Criteria for Particulate Matter and Sulfur Oxides*; U.S. Environmental Protection Agency: Research Triangle Park, NC, 1980.
6. Swift, D. L.; Proctor, D. F. *Atmos. Environ.* **1982**, *16*, 2279.
7. Portland Road Dust Demonstration Project; Oregon Department of Environmental Quality: Portland, OR, 1983.
8. Gatz, D. F.; Wiley, S. T.; Chu, L. C. *Characterization of Urban and Rural Inhalable Particulates*; Illinois Department of Energy and Natural Resources: Springfield, IL, 1983.

9. Roth, P. M.; Reynolds, S. D.; Tesche, T. W.; Gutfreund, P. D.; Seigneur, C. *Environ. Int.* **1983**, *9*, 549.

10. Akland, G. G. *J. Air Pollut. Control Assoc.* **1972**, *22*, 264.

11. Ware, J. H.; Thibodeau, L. A.; Speizer, F. E.; Colome, S.; Ferris, B. G. *Environ. Health Perspect.* **1981**, *41*, 255.

12. *Prevention of Significant Deterioration Workshop Manual*; U.S. Environmental Protection Agency: Research Triangle Park, NC, 1980.

13. Turner, D. B. *J. Air Pollut. Control. Assoc.* **1979**, *29*, 940.

14. Williamson, H. J.; DuBose, D. A. *User's Manual for Chemical Mass Balance Model*; U.S. Environmental Protection Agency: Research Triangle Park, NC, 1983.

15. Goodall, D. W. In *Mathematical Models in Ecology, The Twelfth Symposium of the British Ecological Society*; Blackwell Scientific: Blackwell, England, 1972; p 173.

16. Freeman, D. L.; Egami, R. T.; Robinson, N. F.; Watson, J. G. *J. Air Pollut. Control Assoc.* **1986**, *36*, 246.

17. Gordon, G. E.; Zoller, W. H.; Kowalczyk, G. S.; Rheingrover, S. H. In *Atmospheric Aerosol: Source/Air Quality Relationships*; Macias, E. S.; Hopke, P. K., Eds.; ACS Symposium Series 167; American Chemical Society: Washington, DC, 1981; pp 51–74.

18. Watson, J. G.; Lioy, P. J.; Mueller, P. K. In *Air Sampling Instruments for Evaluation of Atmospheric Contaminants*, 6th ed.; Lioy, P. J., Ed.; American Conference of Governmental Industrial Hygienists: Cincinnati, OH.

19. Hidy, G. M. *Environ. Sci. Technol.* **1985**, *19*, 1032.

20. Mueller, P. K.; Hidy, G. M. *The Sulfate Regional Experiment: Report of Findings*; Electric Power Research Institute: Palo Alto, CA, 1983.

21. Greenberg, R. R. *Anal. Chem.* **1979**, *51*, 2004.

22. Rodes, C. E.; Holland, D. M.; Purdue, L. J.; Rehme, K. A. *J. Air Pollut. Control Assoc.* **1985**, *35*, 345.

23. Wedding, J. B.; Lodge, J. P.; Kim, Y. J. *J. Air Pollut. Control Assoc.* **1985**, *35*, 649.

24. Rogers, C. F.; Watson, J. G. *Proc. 77th Annu. Meet. Air Pollut. Control Assoc.* San Francisco, CA, 1984.

25. Bevington, P. R. *Data Reduction and Error Analysis for the Physical Sciences*; McGraw–Hill: New York, 1969.

26. Watson, J. G.; Cooper, J. A., Huntzicker, J. J. *Atmos. Environ.* **1984**, *18*, 1347.

27. Tiwari, J. L.; Hobbie, J. E. *Math. Biosci.* **1976**, *28*, 25.

28. Efron, B. *SIAM Rev.* **1979**, *21*, 460.

29. McRae, G. J.; Tilden, J. W.; Seinfeld, J. H. *Comput. Chem. Eng.* **1982**, *6*, 15.

30. Javitz, H. S.; Watson, J. G. In *Receptor Methods for Source Apportionment: Real World Issues and Applications*; Pace, T. G., Ed.; Air Pollution Control Association: Pittsburgh, PA, 1986; pp 161–174.

31. Balson, W. E.; Boyd, D. W.; North, D. W. *Acid Deposition: Decision Framework*; Electric Power Research Institute: Palo Alto, CA, 1982; Vol. 1.

32. Eschenroeder, A. Q.; Magil, G. C.; Woodruff, C. R. *Assessing the Health Risks of Airborne Carcinogens*; Electric Power Research Institute: Palo Alto, CA, 1985.

RECEIVED for review January 28, 1987. ACCEPTED May 5, 1987.

Chapter 19

Airborne Sampling and In Situ Measurement of Atmospheric Chemical Species

Roger L. Tanner

Common aspects of airborne sampling and analysis procedures are discussed whereby the composition of gaseous, aerosol, and condensed water droplet phases may be characterized in three dimensions in the boundary-layer atmosphere. The scope includes a review of techniques for isolating a particular phase during airborne sampling, techniques for continuous in situ measurements of trace gases and aerosols, and means for collection of condensed-phase samples. Time resolution considerations for airborne measurements are discussed, and some examples comparing real-time and integrative measurements are included.

S AMPLING AND IN SITU MEASUREMENT of chemical species in ambient air are important components in studies of atmospheric chemistry. Most major pollutant emission sources, and dry deposition sinks as well, are located at the Earth's surface. Therefore, the boundary layer of the atmosphere must be well-mixed vertically, or lacking that, airborne (or tower) measurements must be made to characterize the composition of gaseous, aerosol, and condensed water phases in three dimensions in the boundary-layer atmosphere. Special considerations apply in the sampling and in situ measurement of atmospheric constituents aloft. However, airborne sampling and analysis procedures have aspects in common, whether the focus of the measurement program is global tropospheric studies in remote locations, heterogeneous and homogeneous photochemical smog studies, or precipitation scavenging and acidic deposition studies. These common aspects are the central theme of this chapter, which is not designed as a comprehensive review of continuous

1173–6/88/0275$06.00/0 © 1988 American Chemical Society

methods for the species of interest, but as a guide to the principles and application of these methods in the atmospheric environment.

Throughout this chapter, an in situ chemical measurement in the atmosphere is defined as follows: the collection of a representative sample of a definable atmospheric phase and the determination of a specific chemical moiety in that phase, in situ (in its original location), with definable precision and accuracy (1). This definition is required because most constituents of interest in the atmosphere are not inert ingredients sequestered in a single phase but reactive trace constituents that may be distributed between two or more phases depending on the physical and chemical properties of the atmosphere being sampled. The scope of this chapter includes discussion of the following: (a) techniques for isolating a particular phase during sampling from an airborne platform; (b) techniques for in situ, continuous measurement of trace gases including ozone, nitrogen oxides [e.g., NO_y, NO, NO_2, HNO_3, HONO, peroxy-acetyl nitrates (PANs), and organic nitrates], sulfur dioxide, ammonia, nitric acid vapor, carbon monoxide, and hydrogen peroxide; (c) techniques for continuous measurement of aerosol species and size distributions; (d) means for collection of condensed aqueous-phase samples (e.g., cloud and rain liquid water, supercooled droplets, ice clouds, and snow); and (e) measurement of related physical parameters. Measurement time resolution questions will also be discussed in the context of what constitutes a "real-time" measurement, and some examples of comparisons of real-time and integrative measurements of the same atmospheric species will be given.

Separation of Phases

Air pollutants exist, usually at trace levels, in the atmosphere in one or more physical states depending on the presence or absence of condensed phases. That is, air pollutants may be present as gases or vapors, as surface or bulk constituents of aerosol particles and droplets, or as constituents of cloud liquid water (including supercooled water) droplets, cloud ice particles, and various forms of precipitation. Proper sampling requires that a single phase be sampled if inferences are to be made from the chemical composition of that sample. This requirement is particularly important when, under the atmospheric conditions sampled, a pollutant species may be distributed between more than one phase (e.g., nitrate and nitric acid between gas and aerosol phases depending on the temperature and the pH of the aerosol droplet) (2). A few representative methods for separation of phases during atmospheric sampling from an airborne platform are discussed in this section.

Cloud droplets, aerosols, and gaseous phases are usually separated on the basis of differences in aerodynamic properties. In the absence of a condensed aqueous phase, sampled air is simply filtered to remove aerosol particles by impaction, diffusion, or interception onto the filter surface, and the air stream is then analyzed for the gaseous constituent of interest. Several assumptions must be valid for this sampling procedure to yield accurate results. Gaseous constituents to be analyzed must not react irreversibly with the filter surface or with collected aerosol particles; otherwise negative errors will result. Collected aerosols must also not release or retain gaseous constituents upon changes in sampling conditions. Both of these problems have been observed in continuous and integrated filter analyses of nitric acid (3, 4) as a result of the phase equilibrium that exists between nitric acid vapor and particulate ammonium nitrate in ambient air (1, 5). The occurrence of phase equilibration during sampling may also affect the accuracy of chemical determinations done on the collected aerosol phase, but these determinations are integrative measurements and thus not the subject of this chapter.

In the presence of a cloud liquid water phase, most water-soluble aerosol particles are incorporated (scavenged) into the cloud water phase, which is a major sink process for removal of sulfate from the atmosphere whenever the scavenging clouds actually produce precipitation. This process can be observed directly in airborne sampling (6) by simultaneously recording aerosol light scattering with a nephelometer and cloud liquid water content with a heated wire probe (Johnson–Williams sensor or King probe). The cloud droplet size distribution (approximately Gaussian and centered about 10–20-μm diameter) overlaps the coarse-particle regime of ambient aerosols but may be separated from the fine, accumulation-mode aerosols by impaction techniques. Airborne sampling of gases and aerosols may be conducted continuously in the presence of cloud liquid water (i.e., in cloud interstitial air) by removing the cloud droplets with a centrifugal rotor (7), cyclone, or other impaction device and then directing the air to the continuous gaseous or aerosol instruments. This cloud-free air may also be filtered of any nonscavenged fine-mode aerosol at sampling rates as high as 1 m^3/min.

Sampling of Condensed Phases

Direct sampling for liquid water in clouds may be done from an aircraft with a multiple slotted-rod collector (8) inserted through the skin of the aircraft; the water is collected by gravity inside the aircraft and subsequently analyzed. Chemical measurements that can be performed continuously on collected clouds will be discussed briefly in the next section. Rainwater can also be collected aloft in a similar way, but because rain droplet size distributions are considerably larger (typically 0.1–2-mm geometric mean diameter), a single, large slotted rod is used for collection of rain droplets.

In practice, several problems are apparent when using these collection devices (9). The collection efficiency, even for a single rain or cloud droplet size distribution, depends on the physical location of the collector on the aircraft fuselage. Collection efficiency may also vary with aircraft speed and angle of attack because of physical distortion of the collector, which occurs because collectors are usually made of somewhat flexible plastic. Finally, although the rain collector is observed to collect very little cloud water, the converse is not true and cloud water samples collected from precipitating clouds provide data that are more difficult to interpret (10).

Collection of supercooled cloud water or precipitation presents few sampling difficulties because any available surface outside the slip stream of the aircraft will serve as a freezing site for the supercooled water. Sampling ambiguities result from melting the collected ice from the collection surface for subsequent chemical analysis. This result is caused by the fact that two phase transitions have already been performed on the sample (liquid-to-solid on the collection surface and solid-to-liquid during the melting), and such phase transitions may result in phase exclusion or chemical transformations of certain trace constituents. For example, the solubility of SO_2 in ice is a subject of active investigation (Lee, Y.-N., Brookhaven National Laboratory, unpublished data).

Collection of snow from airborne platforms is, on the other hand, very problematical at this time. Several types of cyclone-based devices have been investigated, but the performance of these devices is generally much poorer than predicted; thus, further investigation in this area is warranted.

Continuous Gas-Phase Techniques

Techniques for the continuous, in situ measurement of trace gases from an airborne platform are well-developed in some cases and still in the preliminary development stage for others. The commonly used technique for each gas considered (O_3, CO, NO_y, SO_2, NH_3, NO_2, PAN, and H_2O_2) is described in this section, and the discussion moves from well-established techniques to more recent, state-of-the-art innovations.

Ozone

Measurement of ozone may be made in real-time from airborne platforms at ambient air concentrations by using the ethylene chemiluminescence technique (11). Observed precision and accuracy of ±10% or better in the 20–200-ppb range are usually more than adequate for most applications, and the observed time resolution of 1–3 s is exceptionally good. In fact, a similar design, the nitric oxide chemiluminescence instrument for ozone, having a time resolution of about 0.1 s, has been used for aerial eddy correlation flux measurements (12). These types of measurements are of great interest in dry deposition studies.

Carbon Monoxide

CO is nearly always measured by using a nondispersive IR spectroscopic technique. IR light attenuation is compared with that observed in a reference cell containing a known amount of CO in excess of expected measured values. The limit of detection (LOD) for conventional instruments mounted on an aircraft is about 50 ppb, time resolution is 10 s, and precision and accuracy at >10 times the LOD are about $\pm 10\%$. An instrument with improved performance characteristics for airborne use has been reported (*13*).

Nitrogen Oxides

Air free of aerosols and condensed-phase atmospheric water (clouds and precipitation) may be analyzed for a variety of nitrogen oxide and oxyacid species. Measurements of nitric oxide (NO) and other nitrogen oxides and oxyacids are most frequently made by using techniques based on the chemiluminescent reaction of ozone with NO (*14*). Chemiluminescence is produced when a portion of the excited-state NO_2 molecules formed in the O_3–NO reaction luminesce in the spectral region beyond ~ 600 nm. Instruments for NO_y based on ozone chemiluminescence consist of a chamber for mixing ambient air with excess ozone and a window for viewing the filtered chemiluminescence >600 nm with a red-sensitive photomultiplier tube. Nitrogen species other than NO are converted to NO by passage over a heated catalyst, which is usually Mo at 375 °C.

Commercial instruments having two parallel sampling paths are available for real-time analysis of NO and NO_y, but their LOD is only about 2 ppb. Airborne applications in nonurban regions in which frequently $[NO_y] < 2$ ppb require the modification of commercial instruments to improve the signal-to-noise ratio. Optimizing the chemiluminescence detector response has been achieved (*15, 16*) in several ways: improving light collection efficiency with polished, gold-coated reaction chambers; using low-noise, red-sensitive photomultiplier tubes; and increasing the intensity of chemiluminescent light by operating at high-volume flow rates under reduced chamber pressure in an enlarged chamber at higher ozone concentrations. Background ("zero-air" response) stability has been improved through the use of a prereactor, in which sampled ambient air is contacted with ozone prior to admission to the reaction chamber. This zero-air response compensates for the effects of quenching gases and spectrally interfering compounds in ambient air.

Determination of gaseous nitric acid concentrations may be made from airborne platforms by filter (*17*) or at the surface by diffusion-denuder (*18*) techniques with time resolution of 15–30 min. However, the only available real-time technique uses a two-channel chemiluminescence analyzer in which one sample stream passes through a nylon filter prior to the Mo catalyst. Nylon removes HNO_3 and other acidic gases, but transmits other nitrogen oxides and organonitrogen compounds with high efficiency, with the probable exception of HONO (*19*). Nitric acid is determined continuously from the difference in the two NO_y channels with and without the nylon filter. Serial sampling of HNO_3 by this method with a single-channel instrument is not appropriate because of serious adsorption effects exhibited by nitric acid vapor under ambient air sampling conditions.

Significant difficulties arise in the determination of NO_2 using the ozone chemiluminescence approach because the Mo catalyst nonspecifically converts several nitrogen oxides and oxyacids to NO in addition to NO_2. Two separate approaches to specifically convert NO_2 to NO have been taken. Ferrous sulfate has been used for NO_2 conversion to NO (20), but humidity-dependent adsorption effects have been reported for this catalyst when PAN is present (21). Photolytic conversion of NO_2 to NO has been reported for surface measurements of NO_2 (22), and this method is promising for future airborne applications.

An entirely different approach to real-time NO_2 determination based on observation of the surface chemiluminescence of NO_2 in the presence of an aqueous luminol solution has been reported (23). A commercial instrument based on this principle has also been introduced, and investigations now in progress will help to establish its suitability for airborne, real-time NO_2 measurements in the presence of other NO_y species.

Sulfur Dioxide

Most real-time measurements of SO_2 made from an aircraft have used a modification of a commercial flame photometric detector (FPD) (24), although a recent commercial version of a pulsed-fluorescent instrument for SO_2 is promising. The sensitivity of the commercial FPD must be enhanced by the addition of a known background of a sulfur compound, usually SF_6 (60 ppb) in the hydrogen fuel gas supply (25). The resulting LOD then approaches 0.2 ppb for SO_2 (10-s time constant), which is adequate for most ambient applications. Sulfur gases (mostly SO_2 in ambient air) are determined with the FPD after removal of aerosols by filtration onto materials that are SO_2-inert (polytetrafluoroethylene (Teflon) or acid-treated quartz).

In the airborne use of the real-time FPD for SO_2, controlling the mass flow of sampled air is essential. This controlling is done with a constant-pressure sampling manifold, or by inserting mass-flow controllers in the hydrogen and exhaust-gas lines (the exhaust-gas lines are placed downstream of a trap for removal of condensable water). The sensitivity and especially the baseline (zero-air) signal are quite sensitive to the H_2/O_2 ratio in the burner. Calibrations with an instrument modified as described are linear with <10-ppb full-scale sensitivity, and this modified instrument has been used successfully on many clear- and cloud interstitial-air sampling missions (10).

Ammonia

Methods for determination of ammonia (NH_3) in real-time from an aircraft platform are few in number and have been successfully used in only a few studies. A continuous method for NH_3 with a time resolution of a few minutes has been reported (26). This method is based on Venturi collection, derivatization to an isoindole, and determination by fluorescence in a flow-through fluorimeter. The LOD reported (0.3 ppb) is adequate for surface sampling (27) but was seldom exceeded in airborne missions with this apparatus. Use of an IR heterodyne radiometer for vertical profile measurements of NH_3 from a surface location has also been reported (28); this technique is complex and the instrumentation expensive, but it does have the advantage of being able to determine ammonia levels in the vertical direction for some cases. Ammonia levels, determined by a tungstic acid denuder with ozone chemiluminescence detection of thermally evolved NO_y (29), have been measured aloft (30) with <10-min sampling times, but real-time use is probably not possible.

Gaseous Peroxides

Measurement capabilities for gas-phase peroxides (e.g., H_2O_2, and methyl hydroperoxide and peracetic acid when present) have lagged behind developments in methods for peroxides in atmospheric liquid water samples, despite the ease with which H_2O_2, at least, can be scrubbed from air samples. This ease is due to the difficulties in collecting $H_2O_2(g)$ without generating "artifact" peroxide by aqueous-phase chemistry involving, for example, ozone and radical species. Recently, three methods have been demonstrated for artifact-free collection of H_2O_2 based on prompt derivatization and analysis (31), or ozone-removal techniques (32, 33), and H_2O_2 has been directly observed by a diode-laser absorbance technique (34) as well. Collected peroxide is determined by peroxidase-catalyzed p-hydroxy phenylacetic acid fluorescence dimer (31, 32) or hemin-catalyzed luminol (33) techniques. Airborne measurements of gaseous H_2O_2 have been reported for only one of these gas-phase techniques (35) by using continuous, in situ determination of collected peroxides with a time resolution of <2 min. An intercomparison of all gaseous peroxide methods currently under development was conducted near Los Angeles, CA, in August 1986, and the results of this study (to be reported at the 3rd International Conference on Carbonaceous Particles in the Atmosphere, Berkeley, CA, Oct 1987) should help establish the best method for use in future airborne studies.

Continuous Aerosol Techniques

Several techniques exist for the continuous, in situ determination of aerosol number, size distribution, and mass. (Techniques for mass determination are derived from light-scattering measurements with a nephelometer.) These techniques include the electrical aerosol analyzer; optical particle counters; optical particle probes (PMS probes) for fine aerosols, clouds, and rain droplet distributions; and even near-real-time mass distributions using impactor separation with a piezoelectric balance. Most of these techniques have received substantial use on aircraft platforms and have effective time resolutions for size and mass distribution determinations of a few minutes. The exceptions are the nephelometer from which fine mass concentrations may be determined with a time resolution of about 1 s, and the PMS probes, which measure number distributions each second, and provide statistically valid size distribution data for fine aerosols and cloud droplets when averaged for about 5 min. Time resolution of the PMS probes varies with number concentration and phase of particles present (and aircraft speed) and is in the range from 0.1 s in liquid clouds to as long as 30 min for dry particles in the free troposphere.

Aerosol sampling techniques must take into consideration the aerodynamic properties of the sampled particles. That is, the aerosol must be sampled isokinetically if the original aerosol size distribution is to be maintained. This requirement need not be rigorously adhered to if only fine aerosol (particles <2 μm in diameter) is being sampled, but is critical if coarse-particle or cloud droplet analyses are to be performed.

Chemically specific measurement methods that can achieve continuous, near-real-time determinations of aerosol species are, in general, not available. The only significant exception to this statement is a FPD for the determination of aerosol sulfur (i.e., mostly sulfate in ambient air). (The FPD application to the measurement of ambient SO_2 was discussed previously.) Aerosol sulfur concentrations are determined by the FPD after removal of gaseous sulfur compounds by passage through a diffusion-denuder tube coated with lead(II) oxide or another equivalent sulfur gas sink. By using sensitivity enhancement by SF_6 addition to the H_2 fuel and a constant-pressure inlet or mass-flow controls to reduce the sensitivity of the baseline to pressure changes, a real-time FPD detector for aerosol sulfur can be used in airborne measurements with the same sensitivity and calibration factors as the analogous FPD-based system described previously for SO_2.

Time Resolution Considerations

The *time resolution* of an in situ measurement is the time required for the real-time sensor to reach 90% of the final response to a step change in concentration of the measured species. This time response consists of the residence time of the sampled air in the instrument, the time required for the trace constituent to reequilibrate with the inlet and intrainstrumental flow lines, and the time response of the instrumental sensor itself. For trace reactive gas species, the reequilibration time is usually the determining factor in establishing the overall response time in airborne sampling. Typical response times for airborne instruments vary from about 1 s for ozone (ethylene chemiluminescence) and light scattering (nephelometer) to about 3 s for nitrogen oxides (ozone chemiluminescence), and as much as 10–20 s for SO_2 and aerosol sulfur (FPD technique). As noted, the chemiluminescent technique cannot be used in serial fashion with a nylon filter to determine nitric acid because the reequilibration time is excessively long for this highly reactive gas; a two-channel system is used instead. Step changes in nitric acid concentration require substantially longer time periods for reequilibration than observed with other nitrogen oxides (e.g., NO and NO_2).

Time resolution considerations are somewhat altered for integrated sampling techniques such as the tungstic acid technique for ammonia or the impinger–fluorescence analysis technique for peroxides. Here, a fixed amount of analyte is required for analysis; thus, lower air concentrations translate directly to longer time resolution. Other techniques such as the continuous concurrent scrubber technique for peroxides require an in situ derivatization procedure with mixing of reagents prior to determination; hence, the time resolution is determined by the relevant chemical and mixing kinetics. Thus, it is important in evaluating any method's time resolution for airborne application to establish the limiting factors: sample volume, reequilibration volume, or postcollection factors. Only then can the optimum choice of method for a particular application be made.

Comparison of Real-Time and Integrative Measurements

Comparison of data from real-time and integrative techniques employed on airborne platforms to measure trace species in the atmosphere are not numerous in the refereed literature. A few examples will be given in this section, although this list is not intended to be comprehensive.

Nitric acid data obtained by most published methods have been compared extensively with surface measurements (3, 4, 36), and new data from an intercomparison conducted in 1985 and reported at the 192nd meeting of the American Chemical Society (ACS) will be available in the literature. Airborne intercomparison data are much more rare, although one comparison of filter-pack (17) and real-time chemiluminescent HNO_3 techniques taken from the work of Kelly (37) has been reported (1). The agreement is good ($r^2 > 0.5$, the slope is about 0.75, and the intercept is not significantly different from zero), considering that half-hour averages of the real-time data are taken, the sampling lines are not identical for the two methods, and the filter-pack data are subject to some systematic artifacts (3). Precision and accuracy are estimated at $\pm 25\%$ on the basis on this study. Another report (38) of airborne use of this technique is less optimistic; further work in this area is strongly recommended.

Sulfur dioxide data have been obtained simultaneously by the FPD real-time method and by the impregnated filter approach as early as the mid-1970s, as part of the studies of SO_2-to-SO_4 oxidation in power-plant plumes. Filter-pack measurements of SO_2 and sulfate were used to deduce conversion rates and to deduce real-time data to define plume shapes and homogeneity of background

SO_2 levels. Systematic attempts to correlate integrated averages of real-time data across the plume with filter data were few and generally not successful because of sampling and instrumental difficulties. Correlation of surface measurements of SO_2 by the same techniques has also been problematical because most large data sets have been acquired with the commercial FPD, which is insufficiently sensitive for background ambient measurements, even in northeastern North America. One exception is the comparison of filter and real-time SO_2 data obtained over a full year of measurements at Whiteface Mountain, NY, by Kelly (39). In that comparison, real-time SO_2 = 0.75(filter SO_2) + 0.60 ppb. This regression equation value had an r^2 value of 0.67 for 44 sampling periods. This agreement is considered good given the range of ambient SO_2 concentrations and the large temporal extent of the measurements. Care must also be taken in SO_2 intercomparisons to accurately zero the FPD and to account for drifts in the zero signal due to changing sampling conditions (e.g., barometric pressure, CO_2 concentration, and absolute humidity).

FPD-determined surface aerosol sulfate data have also been compared with filter measurements (40, 41). When artifact-free quartz filters were used and the sulfate levels were well above the FPD limit of detection, the agreement was quite acceptable: The slope was $1.0 \pm 20\%$, $r^2 > 0.9$, and the intercept was variable but $<2\text{-}\mu g/m^3$ of sulfate.

In summary, several techniques are available for real-time, in situ measurement of trace atmospheric species from airborne platforms. However, much work remains to be done to improve the techniques available, to introduce new techniques for use in aircraft (in particular, in situ or near-real-time methods for reactive hydrocarbons and aldehydes), and to extend the range of species for which adequate intercomparisons have been performed.

Acknowledgments

I acknowledge the many helpful discussions with Peter Daum and Thomas Kelly. This work was performed as part of the Processing of Emissions by Clouds and Precipitation (PRECP) program under the auspices of the National Acid Precipitation Assessment Program (NAPAP) and of the U.S. Department of Energy under contract no. DE–AC02–76CH00016.

References

1. Tanner, R. L. In *Chemistry of Acid Rain*; Johnson, R. W., Ed.; ACS Symposium Series 349; American Chemical Society: Washington, DC, 1987.
2. Stelson, A. W.; Friedlander, S. K.; Seinfeld, J. H. *Atmos. Environ.* **1979**, *13*, 369–371.
3. Spicer, C. W.; Howes, J. E., Jr.; Bishop, T. A.; Arnold, L. H.; Stevens, R. K. *Atmos. Environ.* **1982**, *16*, 1487–1500.

4. Forrest, J.; Spandau, D. J.; Tanner, R. L.; Newman, L. *Atmos. Environ.* **1982**, *16*, 1473–1485.
5. Tanner, R. L. *Atmos. Environ.* **1982**, *16*, 2935–2942.
6. ten Brink, H. M.; Schwartz, S. E.; Daum, P. H. *Atmos. Environ.* **1987**, *21*, in press.
7. Walters, P. T.; Moore, M. J.; Webb, A. H. *Atmos. Environ.* **1983**, *17*, 1083–1091.
8. Winters, W.; Hogan, A.; Mohnen, V.; Barnard, S. ASRC Publication No. 728; Atmospheric Sciences Research Center, State University of New York at Albany: Albany, NY, 1979.
9. Huebert, B. J.; Baumgardner, D. *Atmos. Environ.* **1985**, *19*, 843–846.
10. Daum, P. H.; Kelly, T. J.; Schwartz, S. E.; Newman, L. *Atmos. Environ.* **1984**, *18*, 2671–2684.
11. Nederbragt, G. W.; Van der Horst, A.; Van Duijn, J. *Nature (London)* **1965**, *206*, 87.
12. Lenschow, D. H.; Pearson, R.; Stankow, B. B. *J. Geophys. Res.* **1982**, *87*, 8833–8837.
13. Dickerson, R. R.; Delaney, A. C. *Anal. Chem.* **1987**, *59*, in press.
14. Fontijn, A.; Sabadell, A. J.; Ronco, R. J. *Anal. Chem.* **1970**, *42*, 575–579.
15. Delaney, A. C.; Dickerson, R. R.; Melchoir, F. L., Jr.; Wartburg, A. F. *Rev. Sci. Instr.* **1982**, *53*, 1899–1902.
16. Tanner, R. L.; Daum, P. H.; Kelly, T. J. *Int. J. Environ. Anal. Chem.* **1983**, *13*, 323–335.
17. Daum, P. H.; Leahy, D. F. Report No. BNL-31381R2; Brookhaven National Laboratory: Upton, NY, 1985.
18. Shaw, R. W., Jr.; Stevens, R. K.; Bowermaster, J.; Tesch, J. W.; Tew, E. *Atmos. Environ.* **1982**, *16*, 845–853.
19. Sanhueza, E.; Plum, C. N.; Pitts, J. N., Jr. *Atmos. Environ.* **1984**, *18*, 1029–1031.
20. Kelly, T. J.; Stedman, D. H.; Ritter, J. A.; Harvey, R. B. *J. Geophys. Res.* **1980**, *85*, 7417–7425.
21. Tanner, R. L.; Lee, Y.-N.; Kelly, T. J.; Gaffney, J. S. Presented at the 25th Rocky Mountain Conference, Denver, CO, August 1983; Paper No. 102.
22. Kley, D.; Drummond, J. W.; McFarland, M.; Liu, S. C. *J. Geophys. Res.* **1981**, *86*, 3153–3161.
23. Wendell, G. J.; Stedman, D. H.; Cantrell, C. A.; Damrauer, L. D. *Anal. Chem.* **1983**, *55*, 937–940.
24. Garber, R. W.; Daum, P. H.; Doering, R. F.; D'Ottavio, T.; Tanner, R. L. *Atmos. Environ.* **1983**, *17*, 1381–1385.
25. D'Ottavio, T.; Garber, R.; Tanner, R.; Newman, L. *Atmos. Environ.* **1981**, *15*, 197–203.
26. Abbas, R.; Tanner, R. L. *Atmos. Environ.* **1981**, *15*, 277–281.
27. Kelly, T. J.; Tanner, R. L.; Newman, L.; Galvin, P. J.; Kadlacek, J. A. *Atmos. Environ.* **1984**, *18*, 2565–2576.
28. Hoell, J. M.; Harward, C. N.; Williams, B. S. *Geophys. Res. Lett.* **1980**, *7*, 313–316.
29. Braman, R. S.; Shelley, T. J.; McClenny, W. A. *Anal. Chem.* **1982**, *54*, 358–364.
30. LeBel, P. J.; Hoell, J. M.; Levine, J. S.; Vay, S. A. *Geophys. Res. Lett.* **1985**, *12*, 401–404.
31. Lazrus, A. L.; Kok, G. L.; Lind, J. A.; Gitlin, S. N.; Heikes, B. G.; Shetter, R. E. *Anal. Chem.* **1986**, *58*, 594–597.
32. Tanner, R. L.; Markovits, G. Y.; Ferreri, E. M.; Kelly, T. J. *Anal. Chem.* **1986**, *58*, 1857–1865.
33. Groblicki, P. J.; Ang, C. C. *Proceedings of a Symposium on Heterogeneous Processes in Source-Dominated Atmospheres*; Report LBL-20261; Lawrence Berkeley Laboratory: Berkeley, CA, 1985; pp 86–88.
34. Slemr, F.; Harris, G. W.; Hastie, D. R.; Mackay, G. I.; Schiff, H. I. *J. Geophys. Res.* **1986**, *91*, 5371–5378.
35. Heikes, B. G.; Kok, G. L.; Walega, J. G.; Lazrus, A. L. *J. Geophys. Res.* **1987**, *92*, 915–931.
36. Anlauf, K. G.; Fellin, P.; Wiebe, H. A.; Schiff, H. I.; Mackay, G. I.; Braman, R. S.; Gilbert, R. *Atmos. Environ.* **1985**, *19*, 325–333.

37. Kelly, T. J. Report No. BNL–38000; Brookhaven National Laboratory: Upton, NY, 1986.
38. Walega, J. G.; Stedman, D. H.; Shetter, R. E.; Mackay, G. I.; Iguchi, T.; Schiff, H. I. *Environ. Sci. Technol.* **1984**, *18*, 823– 826.
39. Kelly, T. J. Report No. BNL–37110; Brookhaven National Laboratory: Upton, NY, 1985.
40. Camp, D. C.; Stevens, R. K.; Cobourn, W. G.; Husar, R. B. *Atmos. Environ.* **1982**, *16*, 911–916.
41. Morandi, M.; Kneip, T.; Cobourn, J.; Husar, R.; Lioy, P. J. *Atmos. Environ.* **1983**, *17*, 843–848.

Received for review January 28, 1987. Accepted June 11, 1987.

Chapter 20 ————————————————————————

Sampling for Organic Compounds

John B. Clements and Robert G. Lewis

Important methods of sampling ambient air for organic compounds for subsequent analysis are discussed. The following methods of sample collection are presented along with the advantages and disadvantages of each: concentration of the component of interest onto solid sorbents or onto filter–sorbent combinations, cryogenic concentration, and integrated sampling into containers such as canisters or bags. Sampling problems involving breakthrough of the component of interest, formation of artifacts during sampling, and interferences from water collected from air during sampling are also discussed. The relative advantages of integrated sampling into containers versus concentration onto sorbents or filters are presented. The problem of determining the distribution of organic compounds in the atmosphere between the condensed and gaseous phases is also discussed.

A MBIENT AIR CAN CONTAIN a large number of organic compounds at concentrations that make them of concern as air pollutants. As a general rule, organic compounds are present in ambient air in very low concentrations. These low concentrations complicate the process of obtaining a sample for analysis. Sampling the low but extremely important levels of organic compounds in ambient air generally requires concentration from a rather large volume of air. This requirement intensifies the problems of separation of components of interest, detection limit, and sample contamination. A few techniques allow the direct analysis of organic compounds and do not require concentration, but these techniques have not been applied widely and will not be dealt with in this chapter. This chapter, which is not a comprehensive review, presents the current status of a number of important methods for sampling ambient air for organic compounds.

Vapor pressure and polarity are the two most important properties governing the success of sampling organic compounds from ambient air. For organic compounds having medium to high vapor pressures, components of interest are usually concentrated from large volumes of air with a solid sorbent material of some sort. Low vapor pressure components are normally associated with particles, and filtration of large volumes of air to collect the particles is the usual means of sample collection. In both cases, analysis requires the separation of the sample from the matrix and further separation into individual components for analysis. Highly polar organic compounds are very difficult to sample properly. These materials have a strong tendency to adhere to the sampling media and sampling media containers. Separation of the sample from its matrix and separation into individual components are difficult, and no satisfactory method for sampling polar organic compounds is available.

Although sample analysis will not be discussed in this chapter, the method of analysis needs to be kept in mind in the selection of sampling methodology. All methods of analysis require a preliminary separation in order to isolate the organic compound or compounds of interest. The limit of detection of the analytical method frequently governs how much sample is required, and the ever-present problem of interfering substances would be best dealt with, if possible, at the sampling stage by excluding the interfering substance.

Volatility

The target organic compounds can be characterized by their volatility, and the designations *volatile*, *semivolatile*, and *nonvolatile* have been suggested as three convenient categories for organic compounds in ambient air (1). The vapor pressure ranges suggested are $>10^{-2}$, $10^{-2}-10^{-8}$, and $<10^{-8}$ kPa for the volatile, semivolatile, and nonvolatile categories, respectively.

Volatile Organic Compounds

General Description of Methodology

Volatile organic chemicals have probably received more attention than any other group of organic chemicals present in ambient air. Important aromatic hydrocarbons, halogenated hydrocarbons, aldehydes, and ketones are members of this category. The most widely used procedure for sampling ambient air for these compounds is to pass large volumes of air, typically 2–100 L, through a solid sorbent material onto which the component of interest is adsorbed. Unfortunately, solid sorbents are not compound-specific, and unwanted compounds as well as the target compounds are collected and must be separated. Another concentration procedure undergoing a rebirth and rapidly gaining favor is the cryogenic condensation of volatile organic compounds from air. Modern automation techniques, sensitive gas chromatographic (GC) detectors, and the availability of rugged inert sampling containers have made this long-used method of collecting air samples quite attractive, and several recent monitoring studies successfully employed cryogenic condensation for collecting volatile organic compounds. For analysis, the condensate from cryogenic sampling is separated by GC and a determination is made by common detectors such as flame ionization, electron capture, and mass spectrometric detectors.

Sorbent Sampling

Sampling with solid sorbents is commonly employed to obtain volatile organic compounds from the atmosphere. Solid sorbent media can be divided into three categories: organic polymeric sorbents, inorganic sorbents, and carbon sorbents. Organic polymeric sorbents include materials such as a porous polymeric resin of 2,4-diphenyl-*p*-phenylene oxide (Tenax–GC) and styrene–divinylbenzene copolymer (XAD) resins. These materials have the important feature that minimal amounts of water are collected in the sampling process; thus, large volumes of air can be sampled. A major disadvantage is the inability of the polymeric sorbents to capture highly volatile organic compounds (e.g., vinyl chloride) as well as certain polar materials (e.g., low molecular weight alcohols and ketones).

Inorganic sorbents include silica gel, alumina, a magnesium aluminum silicate (Florisil), and molecular sieves. These sorbents are considerably more polar than the organic polymeric sorbents; thus, the efficient collection of polar organic compounds is permitted. Unfortunately, water is also efficiently captured, and this condition causes a rapid deactivation of the sorbent. For this reason, inorganic sorbents are not often used for sampling volatile organic compounds.

Carbon sorbents are relatively nonpolar compared to the inorganic sorbents, and water is less of a problem, although the problem may still prevent analysis in some cases. Carbon sorbents exhibit much stronger adsorption properties than organic polymeric sorbents. This superior adsorption allows the efficient collection of volatile organic compounds such as vinyl chloride. The strong adsorption of organic compounds on carbon sorbents is a disadvantage because the removal of most of the collected chemicals requires displacement with a solvent, and as discussed later, this procedure is not the isolation method of choice. Both the capture process and the compound recovery process must be considered when choosing the sorbent material for sampling of volatile organic compounds. One of two processes, thermal desorption or solvent extraction, is used. Thermal desorption is most useful for compounds having boiling points less than 300 °C; solvent extraction is most useful for compounds boiling above 150 °C and is used extensively for semivolatile organic compounds.

Tenax–GC and subsequent thermal desorption have been most widely used for sampling volatile organic compounds. This sorbent is discussed in more detail in the next section. Graphitized molecular sieves (e.g., Spherocarb and Ambersorb XE–340) can also be thermally desorbed. However, the high desorption temperatures required for these sieves make recovery of higher boiling volatile organic compounds difficult and increase the likelihood of thermal decomposition problems. Thermal desorption is an attractive approach in many cases because the entire sample is introduced into the analytical system. In solvent extraction, only a small portion (e.g., <1%–10%) can be introduced. The use of the entire sample in thermal desorption permits collection of smaller air volumes in order to obtain sufficient sample for analysis. In addition, the thermal desorption process is more easily automated and does not require disassembly of the sampling cartridge prior to analysis. Solvent extraction offers the advantage that the concentration of analyte introduced into the analytical system can be optimized, and replicate analyses of a sample can be allowed. These features are not available in the thermal desorption process. However, the thermal desorption procedure is the overall preferred means for isolating a sample from its matrix for analysis.

Unfortunately, sorbent sampling is fraught with problems. Contamination and interferences may be particularly troublesome. As mentioned previously, some popular solid sorbents have low capacities for volatile compounds, and the resulting breakthrough of target compounds before the completion of sampling leads to erroneous determinations of atmospheric concentrations (2). Finally, unknown reactions can occur during sampling and perhaps during analysis (3–6). These reactions can lead to the incorrect determinations of compounds not present and prevent the accurate determination of compounds actually present in the air sampled. These potential problems require that sorbent sampling be planned and executed with great care.

Tenax-GC Sampling

Tenax has been put to widespread use in sampling for ambient organic compounds. Tenax–GC, a polymeric solid sorbent originally developed as a support for GC, is the most widely used solid sorbent for ambient air sampling (7-9). Tenax–GC has the advantage of high thermal stability, which permits the thermal desorption of collected volatile materials. Tenax–GC also has the advantage of a low affinity for water, which means that little water from the atmosphere is collected to complicate sample isolation and analysis. Sampling with Tenax–GC, thermal desorption, GC separation, and sample analysis with mass spectrometry has been the most widely used system for obtaining the concentrations of volatile organic compounds in ambient air. Krost et al. (10) have developed and used this system for the determination of volatile organic compounds in the ambient air in a number of important studies.

Recent research has shown that quantitative interpretation of data from the collection of ambient organic compounds using solid sorbents is complicated to a greater extent than previously believed. Walling (*11, 12*) discussed the problems associated with exceeding the capacity of adsorbent, inadequate blank correction, and unwanted reactions occurring during the sampling and analysis process by using Tenax–GC sorbent sampling as a model. He proposed a distributed air volume approach as a powerful quality control technique to help determine if complications that would invalidate the sample occurred during sampling. In the distributed air volume approach to air sampling, several (generally four) samples are obtained simultaneously from the same parcel of air. An important stipulation is that a wide range of air volumes should be sampled by using different air sampling flow rates for the same starting and stopping times. On completion of the sampling and analysis, the concentration of each target compound in each sample is compared across all samples taken during the sampling period. The results are sorted into two groups: one in which the concentrations of a target compound do not vary across all samples beyond expected laboratory variation, and one in which variation in concentrations of a target compound does exceed preestablished laboratory analytical variation. Constant concentrations across a set of samples for a target compound is presumptive evidence that the average of the values is a valid estimate of the concentration of the target compound in the ambient air at the time of the sampling. Sets of concentrations that exceed normal variation are not accepted as providing a valid estimate of air concentration and are investigated to determine if a cause for the problem can be found.

Canister Collection

The condensation of volatile organic compounds from air into a cryogenic trap, particularly when combined with whole air sampling using appropriate containers, is an attractive alternative to sorbent sampling. Advantages of this technique are that a wide range of organic compounds can be collected and determined, contamination problems are greatly reduced, and consistent recoveries are generally obtained. An important limitation is the condensation of large amounts of moisture with the sample unless steps are taken to avoid the problem. This problem has been partially overcome.

A system consisting of collection of whole air samples in stainless steel canisters whose interior walls have been specially electropolished to prevent decomposition of the collected organic compounds, cryogenic concentration of an aliquot, and GC with detection by electron capture or flame ionization has been developed by McClenny et al. (*13*) and applied to monitoring studies. The cryogenic condensation of the sample and its separation and analysis have been automated (*14*). The problem of water corrupting the analysis has been dealt with by incorporating a drying step prior to chromatography. This solution is accomplished by passing the air sample through perfluorinated ionomer membrane (Nafion) tubing, which is a copolymer of tetrafluoroethylene and a fluorosulfonyl monomer, before the cryogenic condensation. Water vapor in a mixture of gases passing through the Nafion tubing permeates through the walls while most of the compounds of interest as air pollutants are retained. This retention phenomenon appears dependent upon polarity because other small polar molecules, such as low molecular weight alcohols, aldehydes, and ketones, also permeate to some extent through the Nafion tubing walls. For this reason, the system is restricted to the determination of relatively nonpolar organic compounds. Notwithstanding this restriction, the system can be used effectively to monitor important air pollutants such as halogenated hydrocarbons and volatile aromatic hydrocarbons.

Sampling using canisters can be accomplished either by evacuating the container in advance and then allowing sample to enter the container or by pressurizing a canister with sample air by using pumps with inert interior surfaces. The pump approach is preferred when integrated samples over a relatively long time period (e.g., 12 h) are desired.

Tenax–GC and Canister Comparison

Side-by-side comparisons of samplers based on canisters and on Tenax–GC have become a major area in the continuing methods development for volatile organic compounds. A comparison study in an unoccupied residence (15) proved to be a good example of the essential equivalency of the two sampling approaches in the absence of factors sometimes associated with ambient air sample collection (e.g., variations in temperature, water vapor content, and the presence of high levels of ozone). Side-by-side comparisons in ambient air have been carried out, but the analytical results have not yet been consolidated and interpreted to provide definitive conclusions. Ambient comparisons in the U.S. Environmental Protection Agency's toxic air monitoring system are a continuing effort to show the degree of equivalency between the two sampling techniques.

Semivolatile Organic Compounds

General Description of Methodology

Although semivolatile organic air pollutants have not received the broad interest as have volatile organic compounds, some very important members of the class have received considerable attention. Important class members include polynuclear aromatic hydrocarbons with four or fewer fused rings and their nitro derivatives, chlorobenzenes, chlorotoluenes, polychlorobiphenyls, organochlorine and organophosphate pesticides, and the various polychlorodibenzo-*p*-dioxins. As with volatile organic compounds, the concentrations of semivolatile organic compounds in the ambient air are so low that concentrating the target compounds from a large air volume is required. A large volume of ambient air is passed through an appropriate solid sorbent, or filter–sorbent combination, to collect the compounds of interest. Solvent extraction is the universal method for isolating the organic compounds from the sampling matrix, and the analytical determination is made by using GC with an appropriate detector.

The most popular sorbent for capturing semivolatile organic compounds is polyurethane foam. This material efficiently collects many semivolatile organic compounds from air and has the advantage of low resistance to air flow; thus, large volumes of air can be collected to obtain sufficient sample for analysis. Granular solid sorbents such as a styrene–divinylbenzene copolymer resin (Amberlite XAD-2); a porous, cross-linked styrene–divinylbenzene polymeric resin (Chromosorb 102); a porous styrene–divinylbenzene resin (Poropak R); Florisil; and silica gel have also been used, but these materials have a much higher resistance to air flow and for this reason are not as widely used as polyurethane foam.

A practical, field-usable sampler incorporating polyurethane foam or granular sorbents has been developed by modifying the well-known high-volume sampler used for collecting ambient particulate matter (16–18). In this modification, a cylinder of polyurethane foam or granular sorbent is inserted behind a glass or quartz fiber filter and the airflow rate is then reduced somewhat. Generally about 300 m^3 of air is sampled as opposed to approximately 2000 m^3 when the high-volume sampler is used in its normal configuration. Sampling periods as long as 24 h can be used, and the semivolatile compounds of interest are isolated by extracting both the filter and the vapor trap with an appropriate solvent. A number of monitoring studies have used this device to determine a variety of semivolatile organic compounds in ambient air (19–29). A battery-powered, low-volume version of this air sampling system has been developed and evaluated (18). This version is lightweight, portable, operates very quietly, and is suited for domiciliary air sampling or use as a personal air monitor. Both types of monitoring studies using the device have been reported (30, 31). This system is also convenient for outdoor use at remote sites where electrical power is not readily available (19).

Phase Distribution

Many semivolatile organic compounds have vapor pressures that allow them to exist in the atmosphere as vapors, in the condensed phase, or in both phases, depending on the ambient conditions at the time of sampling. For many purposes, determining the distribution of semivolatile organic compounds between the gaseous and condensed phase in the atmosphere is desirable. Attempts to accomplish this feat have been made by choice of the method of sampling. Unfortunately, no satisfactory method has yet been devised to unequivocally make this determination. Incorrect assumptions would be that the material collected on a filter in front of a vapor-collecting sorbent represents the amount of semivolatile organic compound in the condensed phase in the atmosphere, and the material collected on the sorbent represents the vapor phase of the semivolatile organic compound. These assumptions are faulty because as air passes through the filter during the sampling process, unknown amounts of the semivolatile organic material transfer from the filter to the sorbent. This transfer corrupts the use of the analysis of individual sampling components as a means to determine the distribution of phases, but does not invalidate the total determination. The phase distribution problem has been discussed by Lewis (1), and field experiments to evaluate the problem have been reported (32, 33).

Nonvolatile Organic Compounds

Nonvolatile organic compounds (i.e., compounds having vapor pressures less than 10^{-8} kPa), for all practical purposes, have negligible concentrations in the vapor phase in the atmosphere. These compounds are bound to solid particles, and the usual methods of sampling for atmospheric particles (e.g., filtration) are commonly used to obtain samples of nonvolatile organic compounds. By far the most used sampling system is the previously mentioned high-volume sampler. This sampling device, which has been described in detail (35), is widely used to sample large volumes of air, typically 1600 to 2400 m^3, by using a glass fiber filter as a collection medium. Extraction of the collected particulate matter with a solvent is used to isolate the organic compounds of interest, and thin layer chromatography or high-performance chromatography is used for separation and analysis.

Polynuclear hydrocarbons with more than four fused rings and their nitrogenous and oxygenated derivatives are by far the predominant members of the nonvolatile category. Benzo[a]pyrene, a potent carcinogen resulting from inefficient combustion, is the most studied nonvolatile organic compound, and its ambient concentrations have been widely determined throughout the world for many years. Benzo[a]pyrene has been a significant air pollutant in the past, but levels of this pollutant have been significantly reduced in recent years.

References

1. Lewis, R. G. *Proceedings of the EPA/APCA Symposium on Measurement of Toxic Air Pollutants*; Air Pollution Control Association Special Publication VIP-7; Air Pollution Control Association: Pittsburgh, PA, 1986; p 134.
2. Pankow, J. F.; Isabelle, L. M.; Kristensen, T. J. *J. Chromatogr.* **1982**, *245*, 31.
3. Pellizzari, E. D.; Demian, B.; Krost, K. *J. Anal. Chem.* **1984**, *56*, 793–798.
4. Walling, J. F.; Bumgarner, J. E.; Driscoll, D. J.; Morris, C. M.; Riley, A. E.; Wright, L. H. *Atmos. Environ.* **1986**, *20*, 51–57.
5. Bursey, J. T.; Wagoner, D. E.; McGaughey, J. F.; Homolya, J. B. Presented at the 192nd National Meeting of the American Chemical Society, Anaheim, CA, September 1986; paper ENVIRO 133.
6. Sievers, R. E. Comments at the National Symposium on Monitoring Hazardous Pollutants in Air, Raleigh, NC, April 1981.

7. von Wijk, R. In *Advances in Chromatography 1970*; Zlatkis, A., Ed.; University of Houston: Houston, TX, 1970; pp 122–124.
8. Pellizzari, E. D.; Bunch, J. E.; Berkley, R. E.; McRae, J. *Anal. Lett.* **1976**, *9*, 45.
9. Pankow, J. F.; Isabelle, L. M. *J. Chromatogr.* **1982**, *237*, 25–39.
10. Krost, K. J.; Pellizzari, E. D.; Walburn, S. G.; Hubbard, S. A. *Anal. Chem.* **1982**, *54*, 810–817.
11. Walling, J. F. *Atmos. Environ.* **1984**, *18*, 855.
12. Walling, J. F. *Experience from the Use of Tenax in Distributed Air Volume Sets*; APCA/ASQC Specialty Conference on Quality Assurance in Air Pollution Measurements, Boulder, CO, October, 1984; Air Pollution Control Association: Pittsburgh, PA, 1985.
13. McClenny, W. A.; Lumpkin, T. A.; Pleil, J. D.; Oliver, K. D.; Bubacz, D. K.; Faircloth, J. W.; Daniels, W. H. *Proceedings of the EPA/APCA Symposium on Measurement of Toxic Air Pollutants*; Air Pollution Control Association Special Publication VIP-7; Air Pollution Control Association: Pittsburgh, PA, 1986; p 402.
14. McClenny, W. A.; Pleil, J. D.; Holdren, M. W.; Smith, R. N. *Anal. Chem.* **1984**, *56*, 2947.
15. Spicer, C. W.; Holdren, M. W.; Slivon, L. E.; Coutant, R. W.; Shadwick, D. S. *Intercomparison of Sampling Techniques for Toxic Organic Compounds in Indoor Air*; U.S. Environmental Protection Agency: Research Triangle Park, NC, 1987; EPA-600/4-87-008.
16. Lewis, R. G.; Jackson, M. D. *Anal. Chem.* **1982**, *54*, 592.
17. Lewis, R. G.; Brown, R. A.; Jackson, M. D. *Anal. Chem.* **1977**, *49*, 1668.
18. Lewis, R. G.; MacLeod, K. E. *Anal. Chem.* **1982**, *54*, 310.
19. Lewis, R. G.; Martin, B. E.; Sgontz, D. L.; Howes, J. E. *Environ. Sci. Technol.* **1985**, *19*, 986–991.
20. Yamasaki, H.; Kuwata, K.; Miyamoto, H. *Environ. Sci. Technol.* **1982**, *16*, 189–194.
21. Thrane, K. E.; Mikalsen, A. *Atmos. Environ.* **1981**, *15*, 909–918.
22. Bidleman, T. F. In *Trace Analysis*; Lawrence, J. F., Ed.; Academic: New York, 1985; Vol. 4, pp 51–100.
23. Billings, W. N.; Bidleman, T. F. *Environ. Sci. Technol.* **1980**, *14*, 679–683.
24. Keller, C. D.; Bidleman, T. F. *Atmos. Environ.* **1984**, *18*, 837–845.
25. Erickson, M. D.; Michael, L. C.; Zweidinger, R. A.; Pellizzari, E. D. *Environ. Sci. Technol.* **1978**, *12*, 927–931.
26. Brenner, K. S.; Mader, H.; Stevele, H.; Heinrich, G.; Womann, H. *Bull. Environ. Contam. Toxicol.* **1984**, *33*, 153–162.
27. Atlas, E. L.; Giam, C. S. *Science (Washington, D.C.)* **1981**, *211*, 163–165.
28. DeRoos, F. L.; Tabor, J. E.; Miller, S. E.; Watson, S. C.; Hatchel, J. A. *Evaluation of an EPA High-Volume Sampler for Polychlorinated Dibenzo-p-dioxins and Polychlorinated Dibenzofurans*; U.S. Environmental Protection Agency: Research Triangle Park, NC, 1987; EPA-600/4-86-037.
29. Wilson, N. K.; Lewis, R. G.; Chuang, J. C.; Petersen, B. A.; Mack, G. A. *Analytical and Sampling Methodology for Characterization of Polynuclear Aromatic Compounds in Indoor Air*; presented at the 78th Annual Meeting of the Air Pollution Control Association, Detroit, MI, June 1985; paper 85–30A,2.
30. MacLeod, K. E. *Environ. Sci. Technol.* **1981**, *15*, 926–928.
31. Lewis, R. G.; Bond, A. E.; Fitz-simons, T. R.; Johnson, D. E.; Hsu, J. P. *Monitoring the Non-Occupational Exposure to Pesticides in Indoor and Personal Respiratory Air*; presented at the 79th Annual Meeting of the Air Pollution Control Association, Minneapolis, MN, June 1986; paper 86–37.4.
32. Coutant, R. W.; Brown, L.; Chuang, J. C.; Lewis, R. G. *Proceedings of the EPA/APCA Symposium on Measurement of Toxic Air Pollutants*; Air Pollution Control Association Special Publication VIP-7; Air Pollution Control Association: Pittsburgh, PA, 1986; p 146.

33. Coutant, R. W.; Brown, L.; Chuang, J. C.; Riggin, R. M.; Lewis, R. G., accepted for publication in *Atmos. Environ.*, 1987.
34. "Standard Practice for Application of the HI-Vol (High-Volume) Sampler Method for Collection and Mass Determination of Airborne Particulate Matter"; *Annual Book of ASTM Standards*; American Society for Testing and Materials: Philadelphia, PA, 1986; Designation: D4096–82; Vol. 11.03, Section II.

RECEIVED for review January 28, 1987. ACCEPTED April 30, 1987.

Chapter 21

Aerometric Measurement Requirements for Quantifying Dry Deposition

B. B. Hicks, T. P. Meyers, and D. D. Baldocchi

Dry deposition can sometimes be measured by intensive micrometeorological techniques, but these methods require either demanding chemical precision (e.g., for the gradient method) or rapid frequency response (e.g., for eddy correlation). For tower application, eddy correlation requires a frequency response of about 0.5 Hz in daytime, and about 5 Hz at night. For aircraft application, the usual specification is for 5–10-Hz response. For routine measurement programs, use is frequently made of less direct approaches, in which dry deposition is computed from atmospheric concentration and site-specific deposition velocity data. Available information limits such simplified approaches to only a few chemical species (e.g., O_3, SO_2, and HNO_3) and particulate sulfate, nitrate, and ammonium. In general, average concentration data can be used.

D RY DEPOSITION HAS DIFFERENT MEANINGS in different disciplines. To some workers, it is the particle fraction of the total exchange of airborne trace substances to the surface. To others, it is only the large-particle component of this transfer (i.e., the deposition of particles due to gravity). However, in most contemporary considerations, dry deposition involves both gas and particle exchange: *Dry deposition* is the aerodynamic exchange of trace gases and aerosols from the air to the surface as well as the gravitational settling of particles. In this context, dry deposition is viewed as a parallel to wet deposition, namely, as a mechanism by which airborne pollutants are transferred to the surface unaccompanied by hydrometeors.

In some locations and for some chemical species, dry deposition can greatly exceed wet deposition, and the opposite situation can also occur. In general, the uncertainty regarding dry deposition is a major impediment to the resolution of questions concerning the fate of emissions into the atmosphere, such as are of principal concern to the National Acid Precipitation Assessment Program (NAPAP).

Except in some special circumstances, existing technology cannot measure dry deposition rates of any trace atmospheric constituent in a routine way. Nor is there any generally recognized simple method by which dry deposition fluxes can be modeled. The difficulties involved have been reviewed extensively elsewhere (1-3). None of the special methods that have been used (including micrometeorology, snowpack accumulation, mass balance calculations, leaf washing, and throughfall studies) is appropriate for all species, nor is any method yet developed to the extent permitting application in other than intensive experiments. The main application of these methods is to develop relationships describing dry deposition, which can then be used to evaluate deposition either from other information generated by models or from measurements of concentrations and other variables at monitoring sites. No experimental methods are yet available to address the matter of dry deposition in highly complex terrain, although some techniques (e.g., throughfall approaches) can be used in circumstances more complex than can other methods. Micrometeorological methods are especially constrained to simple circumstances.

A key issue is the credibility of dry deposition measurements. In general, experimental methods fall into three convenient categories: measurements at the surface, measurements of the flux through the air to the surface, and estimates derived from knowledge of concentration and deposition velocity.

Measurements at the surface itself (e.g., throughfall and snowpack accumulation methods) provide the most direct quantifications of dry deposition, but are possible only for a few chemical species (mainly particulate) and for some surfaces.

For some chemical species, measurements of fluxes through the air to the surface (e.g., micrometeorological gradient and covariance methods) can be made accurately, but these methods are representative of surface values only if the surface is horizontally uniform and if conditions are not changing rapidly with time.

Estimates derived from knowledge of, or assumptions concerning, concentrations and deposition velocities are susceptible to errors arising from specification of both the appropriate deposition velocity (V_d) and the atmospheric concentration at some specified height ($[C]$), and also from errors associated with the failure of the assumption that $F = V_d[C]$, where F is the flux from the surface. Therefore, in this context, consideration should be given to several independent sources of error:

- errors arising from the fact that the deposition velocity concept is not always appropriate (e.g., in complex terrain, over patchy surfaces, or for pollutants potentially having a surface source such as NH_3 or NO_x);

- errors arising from the specification of [C], either by measurement or modeling;

- errors arising from the specification of V_d, either by estimation from field data or by computation in transport models; and

- errors arising from the incorrect assumption that the surface is a perfect sink (i.e., [C] $\neq$ 0 in the final receptor).

The present purpose of this chapter is not to examine the problems associated with dry deposition measurement, nor to review the possible techniques that offer promise at this time. These matters have been addressed in detail elsewhere (3). Instead, the purpose is to summarize the techniques presently used to estimate dry deposition from aerometric measurements. The intent is to examine the philosophies underlying the different networks presently being operated in North America, to explore the measurements being made in these networks, and to review the techniques by which the dry deposition data derived from these programs will be verified. The focus is on the in-air measurement and interpretation of concentrations, both on a routine monitoring basis and as a component of intensive research programs. The question of sampling requirements for liquid sampling (e.g., for throughfall studies) and for sampling of solids (e.g., for surface accumulation methods) will not be addressed. In all cases, however, a need arises for better chemical sensors suitable for use in intensive research applications, and for better understanding of the processes governing deposition and surface emission.

At present, application of the deposition velocity concept in measurement applications is limited to O_3, SO_2, HNO_3, and submicrometer sulfate and nitrate particles. For these species, there is need to consider the requirements of a routine measurement program based on network measurement of air concentrations. The following discussion concentrates on the methods by which air chemistry and critical supporting variables can be measured in order to estimate dry deposition of those chemical species for which current knowledge is best. Measurement of air concentrations of trace species for which deposition velocities are presently not well-known will not be considered in this chapter.

Needs for Intensive (Research) Measurement

The needs of a wide variety of measurement techniques have recently been summarized in the proceedings of a NAPAP workshop on dry deposition (3). In general, surface sampling methods provide detailed information on specific surfaces by overaveraging times that vary (according to the particular method) from hours to months. Methods based on micrometeorology (e.g., gradients and covariances) provide averages over larger spatial scales (e.g., of the order of 1 km for tower-based sensors, and 10–100 km for aircraft systems), but are typically limited to averaging times of about 1 h. In essence, these methods are designed to develop and "calibrate" simple inferential methods, by application of which larger areas and longer averaging times can be addressed.

All chemical species have well-recognized needs for sensors having rapid response. The application is in conjunction with "eddy correlation" measurements, in which vertical fluxes (F) are determined as the average covariance between concentration mixing ratios (X) and the vertical wind component (w).

$$F = D < w'X' >\tag{1}$$

where the primes denote deviations from mean values, and the angle brackets indicate time averaging. The symbol D denotes the density of air. The quantity F is necessarily statistical, and related statistical uncertainties due to sampling can be readily computed. In general, the turbulent structure of the atmosphere imposes a limit on the accuracy with which F can be determined. As a guiding rule, the standard deviation associated with any single evaluation of F is typically not less than 10% for a half-hour averaging period.

In practice, micrometeorological requirements for measuring eddy fluxes are somewhat flexible. Although eddy correlation is sometimes claimed to require exceedingly rapid response of the sensor system (often stated as many Hertz), in most field applications, 0.5-Hz response of the chemical sensor will permit adequate field measurement of the turbulent flux. The key consideration is that sufficient information is known about the high-frequency component of vertical turbulent exchange that corrections can be applied with confidence to account for the effects of some sensor response limitations (4). At night, however, considerably higher frequencies contribute to the turbulent exchange phenomenon and impose greater demands for high-frequency sensor performance. For aircraft operation, the central consideration is the speed of the aircraft; in most situations, 5–10-Hz response is adequate. Table I presents a summary of fast-response chemical sensors presently used in dry deposition studies.

All micrometeorological methods involve an assumption that the fluxes measured at the height of the turbulence sensors are the same as the fluxes to the underlying surface. To assure the validity of this assumption, measurements must be conducted only in conditions that are not changing with time, and at locations where the surface is horizontally uniform. These stationarity and "good-fetch" requirements are fundamental to all micrometeorological research and impose severe limitations on the general applicability of all micrometeorological techniques. In practice, the sensors must be mounted sufficiently high so

Table I. Summary of Some Chemical Sensors Used in Recent Studies of the Turbulent Exchange of Trace Gases

Chemical	Sensor	Use	Reference
SO_2	Flame photometry	EC	5
	UV absorption	EC	6
	Bubblers	G	7
NO	O_3 luminescence	EC	8
NO_2	O_3 luminescence	EC	9, 10
	Luminol	EC	11
O_3	NO luminescence	EC	12[a]
	NO luminescence	EC	13[b]
	Ethylene luminescence	EC	9, 14
	Bubblers	G	15
HNO_3	Nylon filters	G	16

NOTE: References are selected to identify typical applications of the devices, not the origin of the development. Abbreviations are as follows: EC indicates use in an eddy correlation study, and G indicates that the application involved the measurement of vertical gradients.
[a]Information is contained on towers only.
[b]Information is contained on aircraft only.

that no single surface element will unduly influence the flux being measured, and sufficiently low so that flux divergence terms can be ignored.

Special considerations arise in the case of particle fluxes. Some workers employ sensors capable of detailed resolution by particle size (17) but then confront problems concerning the counting statistics in each size range (18). Other workers (19, 20) sample by using a broad particle size range, thus reducing the adverse effects of unfavorable counting statistics but forfeiting the capacity to study particle deposition as a function of particle size. At this time, an outstanding need exists for a sensing system that would provide chemical-species-specific outputs as a function of particle size and have about 1-Hz response.

In some situations, gradient methods may be used. In such cases, vertical differences in air concentration of the chemical species in question are measured and interpreted by using an atmospheric diffusivity (K) derived from other sources:

$$F = DK[d<X>/d(z-d)]$$ (2)

where z is the height above the ground, and d is the height of the relevant zero plane and corresponds to the effective source or sink height of the material being considered. In practice, the requirement for accurate diffusivity information is a severe limitation because of difficulties that arise over forests or any similar surface for which the precise distribution of sources or sinks is not well-known (21). These difficulties are sufficient that gradient methods are often viewed with substantially less favor than the more direct eddy correlation methods. However, for some chemical species for which fast-response sensors are not available, no alternative remains but to employ gradient methods in studies carefully designed to avoid such problems. Table I includes references to some representative gradient studies of dry deposition, both for chemical species that can also be measured by eddy correlation (e.g., O_3) and for species for which gradient techniques are at present the only viable option (e.g., especially HNO_3).

The ability to measure a difference in concentrations between two levels in the air does not necessarily mean that deposition fluxes can be determined. In general, the differences that must be measured are of the order of 0.2%–5% of the concentration, depending on the choice of heights above the surface and on the deposition velocity of the material in question. In order to address questions concerning the processes involved in the air–surface exchange phenomenon, these vertical differences must be resolved with a precision typically better than 10%, so that the precision required of the chemical sensor is likely to be in the range 0.02%–0.5%.

Both eddy correlation and gradient studies are sufficiently difficult that the methods are normally tested by comparison against the requirement for heat energy balance at the surface. Thus, measurements are normally made of the sensible and latent heat fluxes (H and L_wE, respectively) by applying equations 1 and 2 to measurements of temperature and humidity. In this way, an objective test of the validity of the micrometeorological technique is possible by verifying that $H + L_wE = R_n - S - G$, where R_n is the net radiation flux, S is the storage of heat associated primarily with changes of canopy biomass temperature with time, and G is the conduction of heat into the ground. The steps involved in such a verification have been discussed elsewhere (22).

"Eddy accumulation" is frequently advocated as a method for avoiding the complexity of a fast-response chemical sensor, while retaining the benefits of direct flux measurement associated with eddy correlation. In concept, eddy accumulation profits from the use of a fast-response sampling system to replace the usual fast-response air chemistry sensor: air is sampled and stored or measured at a rate proportional to w' with separate samplers for updraft and downdraft systems. The difference between values of $<X>$ determined independently for the updraft and downdraft samples can be easily translated into a value of the desired eddy flux F. (In simple concept, an eddy accumulation system requires some system to correct for the impossibility of determining $<w>$ with precision.)

In practice, the promise of the eddy accumulation concept has yet to be fulfilled. Hicks and McMillen (23) simulated the technique and demonstrated the great difficulties associated with several critical components of any eddy accumulation system. Speer et al. (24) presented results of field tests of a prototype eddy accumulation system that appears to have operated with insufficient precision to measure trace gas-exchange rates.

Businger (25) reviewed the use of micrometeorological methods as tools for investigating dry deposition, and especially for measuring deposition velocities. The reader is referred to reference 25 for details of the techniques summarized previously and for discussion of a variety of additional methods that are derived from the basic principles presented here. Wesely and Hart (26) presented a discussion of the influence of random noise on the interpretation of micro-meteorological data. In essence, pollutant sensors are inherently noisy; ensemble averages are typically required to reduce statistical uncertainty associated with eddy fluxes derived from their use.

Needs for Routine Measurement (Monitoring)

Routine monitoring of dry deposition is presently in its infancy, and little assurance has been given yet that the exploratory efforts presently underway will eventually prove to be successful. Most monitoring activities call for the measurement of the atmospheric concentrations of appropriate chemical species plus supporting information on meteorological and surface properties from which values of the relevant deposition velocities can be estimated. In this context, the surface flux can be inferred from four possibilities:

a. from direct measurements of both [C] and V_d,
b. from measured V_d and values of [C] derived from other sources,
c. from measured [C] and values of V_d estimated from other information, or
d. from computed values of both V_d and [C].

In practice, options a and b are impractical because direct measurements of V_d are not possible except in the unusual circumstances that permit application of one of the more direct methods of flux measurement. Option c is selected for application in the various monitoring programs presently underway or being initiated. These various programs differ mainly in the way in which they attempt to specify V_d. The "concentration monitoring" approach focuses attention on the need for accurate concentration data from which dry deposition can later be estimated by applying some value of V_d based on external information such as land-use categories. A recent modification calls for V_d to be derived by extrapolation from reference data obtained by using research-grade methods at a subset of locations. The research network operation of NAPAP is based on this "inferential" method, whereby a central array of core research establishment (CORE) stations provides reference data for a larger network of satellite stations. As will be discussed later in this chapter, an intermediate approach is being adopted for the dry deposition monitoring program presently being set up as part of the National Trends Network (NTN) of NAPAP. Option d reflects the approach adopted in numerical modeling schemes.

The need to specify time trends at specific locations may be satisfied by recording air concentrations alone, although some insecurity is related to the tacit assumption that V_d remains constant at each site. When spatial differences are an issue, the concentration monitoring approach requires augmentation to provide information on the spatial variability of the deposition velocity.

Specification of Atmospheric Concentration at Some Specified Height

For some chemical species, existing air pollution monitors offer a potential solution to the problem of routine measurement. This situation is especially the case for O_3, for which UV absorption methods are well-developed and for which alternative methods are not competitive. In this case, the ready availability of commercial sensors capable of yielding accurate hourly averages of ozone concentrations is of considerable benefit; ozone concentration records reveal consistent diurnal variability that imposes a need to consider hourly data more than in the case of pollutants such as SO_2 (27).

No monitor exists for either SO_2 or NO_x that has the sensitivity or freedom from drift necessary to permit continuous reliable operation at remote or rural background sites. Furthermore, no capability is available to monitor hourly fluctuations in HNO_3 concentration with assurity. In all of these cases, alternative methods for field measurement are presently being developed and tested. The following summary is intended to provide some guidance concerning several of the methods presently under development.

Filter Packs. Several different configurations of filter packs are now in routine use. Many share similar characteristics: a polytetrofluoroethylene (Teflon) or similar prefilter intended to remove particles from the airstream followed by a nylon filter to remove HNO_3, and a cellulose final filter previously doped with potassium or sodium carbonate or bicarbonate, and sometimes treated with glycerol, to ensure a moist surface. After an extended study of filter packs, a special subcommittee of the National Atmospheric Deposition Program (NADP) recommended in 1981 that several changes should be made to filter packs before widespread deployment in a monitoring program in the United States. To permit first-order calculation of deposition of sulfate and nitrate aerosol, the use of an open-face filter was severely criticized because the presence of large particles could cause great uncertainty in the computation of an appropriate deposition velocity. One recommendation was that some technique be used to prohibit large (i.e., >2-μm diameter) particles from entering the sampling system. To protect against liquifaction of the final doped filter in high humidities, either the sampled airstream should be heated slightly or the sampler should be turned off when the humidity exceeds some predetermined value. The construction of the filter pack was recommended to be molded tetrafluoroethylene (Teflon) rather than machined from a cast block. Finally, the flow rate and pressure drop data obtained at critical points of the sampler should be recorded for later scrutiny and for accurate computation of the volume sampled. Filter-pack units developed as a consequence of these recommendations are now available from commercial sources.

Bubblers. A variety of bubblers was also examined by the NADP committee. Although many bubblers were constructed to provide assured long-term reliability, they shared the common feature of a liquid collection medium in which chemical speciation can be changed substantially. Bubblers were not recommended for long-term routine monitoring in the United States because of the anticipation that evaporation would be rapid and that the matter of speciation would present problems.

Denuders. Recent attention has been focused on denuders of a variety of styles. Denuder difference methods have been successful in many situations, and the recent development of annular denuders has caused considerable excitement. Denuders are perceived by some workers to suffer from some potential problems regarding the efficiency with which they scavenge species such as HNO_3, especially in humid conditions. At this time, denuders of many varieties are being developed, and it can probably be predicted with confidence that the current developmental programs will eventually be successful. A special category of devices operates on the same principles as the denuders but is not reliant on the assumption of complete scavenging of the material being measured from the sampled airstream. Such devices also show considerable promise and appear to be reliable at their present stage of development.

Passive Monitors. For purposes of estimating dry deposition, passive monitoring devices have probably not been explored adequately. Such devices are an extension of the "lead candle" approach: a sensitized medium is contained in a protected environment, and a slow leak connects its sampling volume to the ambient atmosphere. This slow leak may be as simple as a hypodermic needle or a set of very small holes in an inert cover plate. In principle, the transfer of the material being measured to the measuring medium is controlled by molecular diffusivity and by the geometry of the exposure. In sufficiently controlled conditions, an accurate average concentration number should be obtained.

Guidelines adopted for chemical measurements made in the networks of the routine monitoring and research programs of NAPAP, with emphasis on the nested-network operation of the National Oceanic and Atmospheric Administration's (NOAA), Department of Energy's (DOE), and Environmental Protection Agency's (EPA) CORE research program (28), are presented in the appendix at the end of this chapter. Concentration monitoring measurement philosophies, containing various stages of augmentation to assist in specifying appropriate values of V_d, are being used in the programs of other agencies [e.g., the EPA, the

Electric Power Research Institute (EPRI), and the Atmospheric Environment Service of Canada (AESC)], and are also discussed in the appendix.

Derivation of Deposition Velocity

Rather complicated mathematical formalism lies behind the relatively simple concept of a deposition velocity. The overall behavior can be described in terms of a resistance model, in which the overall resistance to transfer (i.e., the reciprocal of the deposition velocity) is made up of individual resistances associated with specific transport processes combined in series and in parallel as in an electrical circuit.

Deposition velocities are known to vary with the surface area index, the texture of the surface, its biological properties, and the prevailing meteorological conditions. Surface wetness due to condensation (i.e., dew) is known to be an important property (29) that promotes the transfer of water-soluble trace gases. For the case of particles, the role of surface wetness might well be to increase the efficiency of retention of impinging particles through the generation of liquid bridges between the particles and the deposition substrate.

Most often, values of the deposition velocity are derived by using a multiple resistance transfer model, which is somewhat limited in its generality but can be applied with gradually improving confidence if circumstances are well-selected. At present, deposition velocity evaluation techniques are being developed both to interact with long-range transport models and to be driven by field data for application in measurement programs. All such methods for calculating deposition velocities rely on separate evaluation of the important contributing resistances, followed by a calculation of their combined effects. For both modeling and measurement applications, canopy resistances are assumed to be made up of individual subresistances, visualized in both series and parallel arrays to simulate the roles of different in-canopy biological and physical processes (30, 31). The techniques used to evaluate deposition velocities are not yet precise, and in some aspects they are grossly deficient. A recognized goal of ongoing research is to identify important processes and to formulate them in terms of variables suitable for modeling and routine measurement.

The resistance model provides a framework for coupling individual processes, some depending on the nature of the pollutant (e.g., chemical species and particle size) and some being surface-dependent. The limitations of the framework need to be remembered. Effects associated with particle sedimentation need to be considered very carefully, and perhaps most importantly, the method has difficulty separating emission from deposition. Suppose a diluted sulfur plume moves across a sulfur-rich swamp. As the air moves over the swamp, its concentrations of sulfur will increase as a result of the emissions from the surface. At some stage downwind, a measurement of sulfur species might inadvertently be coupled with some carefully parameterized deposition velocity, and a completely erroneous dry deposition estimate results. In all cases, care must be taken not to conclude that dry deposition is occurring when the concentrations that are measured in the air are in fact a consequence of local emissions from the surface.

Two alternative approaches are presently used to derive estimates of appropriate deposition velocities for interpreting field measurements of air concentrations. For application in numerical models, the surface is typically categorized in terms of its land use, on the basis of which estimates of appropriate surface resistances are computed by using available field observations as guidance (32–34). For field programs and for possible extension to routine monitoring applications, the surface can be considered as a composite of individual biological species, each having its own physiological behavior controlling such factors as stomatal resistance and cuticular uptake (35, 36). The focus of the following discussion is on the field experimental requirements.

The three major resistance components in the multiple resistance approach are an aerodynamic resistance (R_a), a near-surface boundary resistance (R_b), and a resistance representative of surface reactions and canopy uptake (R_c).

Aerodynamic resistance is fairly well-understood for uniform, relatively flat terrain. For rough surfaces, the formulation of R_a is affected by uncertainty about how to consider the consequences of source and sink distributions that differ from those for momentum. In daytime, values of R_a tend to be relatively low, and thus the precise value is often not important. (Exceptions are very reactive species such as HNO_3, HCl, and HF, which are believed to deposit efficiently to any surface with which they come in contact.) At night, R_a is usually much greater, except over extensive water surfaces.

Diffusive effects strongly influence the resistance associated with transfer across the boundary layer immediately adjacent to a surface. Thus, gas speciation and particle size and density become important factors. The results of modeling studies and theory tend to agree for simple surfaces. There remains considerable doubt about the best representation for forests and for other surfaces with large roughness elements. Problems appear to be greatest for the case of particle exchange.

Canopy resistances are usually controlled by biological factors, most of which vary with the state of the plant canopy, water stress, and soil conditions, as well as with radiation, humidity, temperature, and other meteorological factors. Any single evaluation should be viewed as only one representation of a widely varying phenomenon.

The NOAA, DOE, and EPA CORE research program (28) has developed the guidelines in the box on page 308 for measurements to be made in support of air concentration sampling for dry deposition measurement. These guidelines identify environmental factors that should be quantified to permit extrapolation from research-grade CORE sites to other locations where intensive research measurements are not made. Many of these variables are being measured at the regular dry deposition monitoring (concentration monitoring) sites presently being set up under EPA and EPRI programs.

Guidelines for Measurements Made in Support of Air Concentration Sampling for Dry Deposition Measurement

Meteorological Measurements

Variables. In addition to measurements of the atmospheric concentrations of the chemical species of interest, measurements are needed of the quantities presently used in dry deposition algorithms (36) for deposition velocities: wind speed, wind direction standard deviation, incoming shortwave solar radiation, air temperature, air humidity, surface wetness, and precipitation. Quantities that might prove necessary in the future include net radiation, wind direction, and precipitation chemistry.

Time Resolution. Time resolution should be sufficient to resolve features of the diurnal cycle, typically 1 h. (Some 15-min averages are recorded in current work and are used in the quality assurance and quality control procedures to derive 1-h averages.)

Site Quality. Site quality should be adequate to assure that critical measurements are representative of the local surroundings (*see* the previous paragraphs), and unaffected by local topographical features imposing variability in either surface heating (e.g., nearby parking lots or buildings), roughness, moisture exchange (i.e., evaporation rates, perhaps from a nearby water body or irrigated crop), or slope. The matter of slope variation is especially important. Because flow separation often occurs when a slope changes by about 15%, sites should be selected so that no nearby surfaces exceed a slope of more than 8° from the horizontal, nor should the local slope change by more than this amount.

Surface Measurements

Variables. Quantities presently used in V_d algorithms include snow cover, plant species distribution, leaf area index, and canopy height. Quantities that are expected to be needed in the future include leaf water potential and soil moisture.

Time Resolution. Time resolution should be that of the basic dry deposition requirement (e.g., 1 week for the regular monitoring programs of the NTN).

Sampling Protocols

The question of sampling time arises frequently. Some workers argue that short-term concentration data are required in order to evaluate dry deposition. The perceived problem is the strong diurnal cycle that doubtlessly exists in many locations of interest. Two choices have statistical validity: either consider short-term data so that the long-term average can be constructed by integrating through the diurnal cycle, or consider sampling periods long with respect to 1 day and apply appropriately corrected interpretation procedures. Tests conducted so far have demonstrated that day–night sampling might be beneficial for ozone but not for sulfur species (27).

Siting Requirements and Representativeness

The inferential method as is being used in the dry deposition research programs of NAPAP (CORE and CORE satellite programs) evaluates the average dry deposition flux to an area surrounding the point of observation, which is typically several hundred meters in diameter. Hence, the observation site should be located within an area that is both spatially homogeneous and representative of the larger region of interest in which the site is located. ("Homogeneity" includes both topographical features and the distribution of plant species and other surface elements. For example, proximity to a body of water or to an area of different roughness should be avoided.)

For purposes of estimating dry deposition at some selected location, the appropriate deposition velocity must be computed. The deposition velocity is usually controlled by surface factors more than atmospheric ones, and hence the requirements associated with the operation of a normal meteorological reporting station can be relaxed somewhat. The extent to which such relaxation is permissible remains to be assessed fully. Some sites have been set up in forest clearings, for example, so that tests can be made of this kind of meteorologically imperfect exposure. These tests have not yet been completed.

The indirect estimation of dry deposition from air concentration data as collected in some concentration monitoring programs is not necessarily so constrained, but the results are correspondingly susceptible to greater errors and uncertainties. In essence, the philosophy of such methods is to characterize the atmospheric chemistry of some area of interest, and then to compute dry deposition by using some estimated deposition velocity derived from consideration of such factors as land-use categories.

Data Verification

The regular dry deposition monitoring network presently being set up under sponsorship of NAPAP is a modified concentration monitoring network, in which air chemistry and selected meteorological variables are recorded such that dry deposition can be computed once quality-assured deposition velocity routines are available. The accuracy of the results obtained in such an operation is assured, at least in concept, by the use of verified deposition velocity algorithms driven by on-site measurements of critical controlling variables.

In parallel with this regular monitoring activity, a nested-network approach is being tested in which the inferential methods applied at a wide array of locations are tested and results verified at a subset of more intensive CORE research stations. In this network operation, the accuracy of routine measurements is assured by direct reference against higher grade measurements made at a subset of more intensive stations: the CORE stations of the nested-network array (28). The various pieces of these various networks and the goals of each component are summarized in the appendix at the end of this chapter.

The network presently in existence consists of stations and satellite stations. CORE stations, designed to provide reference "benchmarks" data, are located at Oak Ridge, TN (servicing southern forested surfaces); Argonne, IL (servicing midwestern cropland); and State College, PA (servicing eastern cropland). CORE satellite stations, used for extrapolating results obtained at CORE sites, are located at West Point, NY (variable surface); Whiteface Mountain, NY (sloping terrain); Champaign–Urbana, IL (cropland); Panola, GA (comparison site for calibrated watershed); Sequoia, CA (comparison site for throughfall and stemflow); Pawnee, CO (comparison site for micrometeorology, U.S. Forest Service); Borden, Ontario (comparison site for micrometeorology); and Shenandoah, VA (comparison site for throughfall, stemflow, and watershed). The CORE stations at Oak Ridge and State College, as well as the satellite stations at West Point and Whiteface Mountain, are also sites of the EPA Concentration Monitoring, which is a regular dry deposition monitoring network. This EPA network will soon expand to more than 30 locations.

Conclusions

Improvement in the state of the art of sampling atmospheric concentrations for deriving dry deposition information is urgently needed. In particular, the research programs are in need of improved sensors having fast response or high precision, and routine monitoring programs require relatively simple but reliable devices suitable for year-round deployment in the entire range of environmental conditions.

Sampling designed to provide information relative to dry deposition must contain strong meteorological and surface components because for many chemical species, the deposition velocity is as variable a quantity as the concentration.

No well-defined optimal method for sampling dry deposition has been presented. Instead, a range of experimental methods are still being tested and developed. In this presentation, the focus has been on those methods that rely on the measurement of concentrations in air. Consequently, little attention has been given to many alternative methods that are of obvious promise. A more complete examination of the various alternatives is available (3).

Abbreviations and Symbols

ADOM	Canadian Acid Deposition and Oxidant Model
AESC	Atmospheric Environment Service of Canada
C	concentration in air (kg/m^3)
CORE	Core research establishments (of the dry deposition nested-network research program)
d	zero-plane height (m)
D	air density (kg/m^3)
DOE	U.S. Department of Energy
E	evaporation rate ($kg/m^2/s$)
EPA	U.S. Environmental Protection Agency
EPRI	Electric Power Research Institute
F	flux from the surface ($kg/m^2/s$)
G	heat transfer into the ground (W/m^2)
K	eddy diffusivity (m^2/s)
$L_w E$	latent heat flux (W/m^2)
NADP	National Atmospheric Deposition Program
NAPAP	U.S. National Acid Precipitation Assessment Program
NTN	National Trends Network of NAPAP
NOAA	U.S. National Oceanic and Atmospheric Administration
PAN	peroxyacetyl nitrate
R_a	aerodynamic resistance (S/m)
R_b	quasilaminar boundary resistance (S/m)
R_c	canopy resistance (S/m)
R_n	net radiation (W/m^2)
RADM	Regional Acid Deposition Model
S	canopy heat storage rate (W/m^2)
V_d	deposition velocity (m/s)
VOCs	volatile organic compounds
w	vertical wind velocity (m/s)
X	concentration mixing ratio (kg/kg)
z	height above the ground (m)

Appendix—Summary of Existing Networks Providing Data on Dry Deposition in the United States

The existing program of the NAPAP is largely directed toward provision of routine dry deposition measurements. The overall approach has been to address the problem at three groups of sites constituting a nested network. These groups of sites contain several levels of complexity and have substantially different goals.

The EPA Routine Air Concentration Monitoring Dry Deposition Sites

An EPA NAPAP program is being initiated to monitor variables related to dry deposition at locations identified as requiring dry deposition data for modeling purposes. Presently there are five sites, but the program will expand to at least 30.

Weekly data are desired and are initially to be obtained by using day–night sampling. However, the equipment will apparently be capable of operation over shorter time intervals as well in order to test shorter term predictions of models. Similar concentration monitoring activities are being conducted by EPRI and AESC.

Goal.   As a first step, the goal is to provide data for testing transport and deposition models, probably by using 12-h time resolution.

Chemical Focus.   The focus is on species important in modeling such as SO_2, HNO_3, NO_2, H_2O_2, O_3, NH_3, volatile organic compounds (VOCs), and peroxyacetyl nitrate (PAN), plus fine particulate SO_4^{2-}, NO_3^-, and NH_4^+. Coarse particles are not addressed.

Siting.   Initial emphasis is on the Northeast, at locations required for testing the Regional Acid Deposition Model (RADM) and Canadian Acid Deposition and Oxidant Model (ADOM).

The NOAA Trial Program (CORE and Satellite Stations)

Methods for extrapolating from research-grade CORE stations to other locations of simpler measurement are being tested at an array of satellite sites associated with the CORE program. The intent is to develop methods suitable for routine measurement of dry deposition of a few key chemical species (especially SO_2 and HNO_3) in a manner compatible with the operating procedures of the existing wet deposition monitoring networks; the installation and operating costs should be no more than 5 times that of the wet deposition operation. Measurements of air chemistry and supporting atmospheric and surface data are obtained with equipment that is relatively inexpensive, rugged, and suitable for long-term operation. The necessary techniques for inferring dry deposition from such field data are still being developed fully, and present capabilities are such that only a limited number of chemical species and surface configurations can be addressed. Nevertheless, the approach appears sufficiently practical that some limited expansion is planned.

All sites are selected to be in regionally representative areas where appropriate deposition velocities can be estimated from field data by using current capabilities. The measurements focus on SO_2, HNO_3, and particulate anions and cations as reported by the NTN. Weekly averages of concentrations are measured with simple filter packs (initially), and supported by hourly meteorological and surface data for computing deposition velocities. The method relies on the ability to deduce appropriate deposition velocities provided measurements are made of controlling surface and atmospheric properties. The approach is known as the *inferential method*, in order to differentiate it from the more conventional *concentration monitoring approach*, which focuses on air quality alone, although often with supporting measurement of a selected set of meteorological variables. Estimated deposition fluxes derived at the satellite sites are referenced against more direct methods at the subset of special CORE sites.

Goal. As a first step, the goal is to develop a routine method for providing dry deposition data that would parallel the wet deposition data of NTN.

Chemical Focus. The measurements will ultimately focus on fine-particulate anions and cations as reported by NTN, plus SO_2 and HNO_3. In general, these are species for which inferential methods are appropriate. Coarse particles are not yet being considered.

Siting. The siting is a regionally representative subset of NTN.

CORE Network

The first goal of the NAPAP CORE sites is to provide reference data for assuring the quality of results obtained by using simpler methods elsewhere (e.g., at the CORE satellite stations). A second goal has been to develop and assess state-of-the-art methods for measuring dry deposition fluxes. Several methods are explored concurrently (e.g., eddy fluxes, gradients, foliar extraction, throughfall, and watersheds). For those cases where these methods can operate side by side, rigorous intercomparisons of the mature methods are used to assess the reliability with which dry deposition fluxes can be measured. Strengthened by such assessments, the different methods can be separately applied in regions of their special applicabilities (e.g., hillsides, flat terrain, canopy, or open areas).

However, these special methods are usually more complicated than is feasible for routine application (e.g., in a monitoring network) and are hence applied regularly, not necessarily continuously, but normally, in short intensive case studies.

The DOE, NOAA, and EPA NAPAP network of CORE research stations is intended to provide the basis for the deposition velocity parameterizations to be used in routine monitoring operations, and also to provide benchmark verifications of routine methods. This network is presently made up of only three stations. All chemical, meteorological, and surface data are recorded hourly or more frequently and augmented by routine application of more direct dry deposition measurement methods (e.g., eddy correlation for 1 week every month) to provide the desired CORE site comparisons of both deposition velocities and dry deposition fluxes.

Goal. The goal is to provide detailed data on dry deposition at a limited number of regionally representative sites, and to test and improve methods used in routine measurement programs.

Chemical Focus. The emphasis is on species for which more direct measurement methods are appropriate (e.g., SO_2, O_3, NO_2, HNO_3, and fine particles).

Siting. Emphasis is on regionally representative sites having uniform surface and having existing research capabilities for application of different measurement methods.

Acknowledgments

This work was carried out under the sponsorship of the U.S. Department of Energy and the National Oceanic and Atmospheric Administration (NOAA) as a contribution to the National Acid Precipitation Assessment Program (NAPAP) (Task Group II: Atmospheric Chemistry).

References

1. Hicks, B. B. *Water, Air, Soil Pollut.* **1986**, *30*, 75–90.
2. Voldner, E. S.; Barrie, L. A.; Sorois, A. *Atmos. Environ.* **1986**, *20*, 2101–2123.
3. Hicks, B. B.; Wesely, M. L.; Lindberg, S. L.; Bromberg, S. M. *Proceedings of NAPAP Workshop on Dry Deposition, Harpers Ferry, VA, March 25–27, 1986*; NAPAP Report; National Technical Information System: Oak Ridge, TN, 1987.
4. Hicks, B. B. *Boundary-Layer Meteorol.* **1972**, *3*, 214–228.
5. Galbally, I. E.; Garland, J. A.; Wilson, M. J. G. *Nature (London)* **1979**, *280*, 49–50.
6. Nestlen, M. G. *Proceedings of the NATO Advanced Research Workshop on Acid Deposition Processes at High Elevation Sites, Edinburgh, Scotland, September 8–12, 1986*, *4*, 429–444.
7. Shepherd, J. G. *Atmos. Environ.* **1974**, *8*, 69–74.
8. Delany, A. C.; Fitzjarrald, D. R.; Lenschow, D. H.; Pearson, R., Jr.; Wendel, G. J.; Woodruff, B. *J. Atmos. Chem.*, in press.

9. Wesely, M. L.; Cook, D. R.; Williams, R. M. *Boundary-Layer Meteorol.* **1981**, *20*, 459–471.
10. Wesely, M. L.; Eastman, J. A.; Stedman, D. H.; Yalvac, E. D. *Atmos. Environ.* **1982**, *16*, 815–820.
11. Hicks, B. B.; Matt, D. R. *J. Atmos. Chem.*, in press.
12. Eastman, J. A.; Stedman, D. H. *Atmos. Environ.* **1977**, *11*, 1209–1211.
13. Lenschow, D. H.; Delany, A. C.; Stankov, B. B.; Stedman, D. H. *Boundary-Layer Meteorol.* **1980**, *19*, 249–265.
14. Wesely, M. L.; Eastman, J. A.; Cook, D. R.; Hicks, B. B. *Boundary-Layer Meteorol.* **1978**, *15*, 361–373.
15. Galbally, I. E.; Roy, E. R. *Proceedings, Quadrennial International Ozone Symposium, August 4–9, 1980, Boulder, CO*; National Center for Atmospheric Research: Boulder, CO, 1980; pp 431–438.
16. Huebert, B. J.; Robert, C. H. *J. Geophys. Res.* **1985**, *D1(90)*, 2085–2090.
17. Sievering, H. In *Precipitation Scavenging, Dry Deposition, and Resuspension*; Pruppacher, H. R.; Semonin, R. G.; Slinn, W. G. N., Eds.; Elsevier: New York, 1983; pp 963–978.
18. Fairall, C. W. *Atmos. Environ.* **1984**, *18*, 1329–1337.
19. Wesely, M. L.; Cook, D. R.; Hart, R. L.; Hicks, B. B.; Durham, J. L.; Speer, R. E.; Stedman, D. H.; Trapp, R. J. In *Precipitation Scavenging, Dry Deposition, and Resuspension*; Pruppacher, H. R.; Semonin, R. G.; Slinn, W. G. N., Eds.; Elsevier: New York, 1983; pp 785–793.
20. Wesely, M. L.; Cook, D. R.; Hart, R. L.; Speer, R. E. *J. Geophys. Res.* **1985**, *90*, 2131–2143.
21. Hicks, B. B. In *The Forest Atmosphere Interaction*; Hutchison, B. A.; Hicks, B. B., Eds.; Reidel: Boston, MA, 1985; pp 631–644.
22. Hicks, B. B.; Wesely, M. L.; Coulter, R. L.; Hart, R. L.; Durham, J. L.; Speer, R.; Stedman, D. H. *Boundary-Layer Meteorol.* **1986**, *34*, 103–121.
23. Hicks, B. B.; McMillen, R. T. *J. Clim. Appl. Meteorol.* **1984**, *23*, 637–643.
24. Speer, R. E.; Peterson, K. A.; Ellestad, T. G.; Durham, J. L. *J. Geophys. Res.* **1985**, *90*, 2119–2122.
25. Businger, J. A. *J. Clim. Appl. Meteorol.* **1986**, *25*, 1100–1124.
26. Wesely, M. L.; Hart, R. L. In *The Forest Atmosphere Interaction*, Hutchison, B. A.; Hicks, B. B., Eds.; Reidel: Boston, MA, 1985; pp 591–612.
27. Meyers, T. P.; Yuen, S. *J. Geophys. Res.* **1987**, *D6(92)*, 6705–6712.
28. Hales, J. M.; Hicks, B. B.; Miller, J. M. *Bull. Am. Meteorol. Soc.* **1987**, *68*, 216–225.
29. Brimblecombe, P. *Tellus* **1978**, *30*, 151–157.
30. O'Dell, R. A.; Taheri, M.; Kabel, R. L. *J. Air Pollut. Control. Assoc.* **1977**, *27*, 1104–1109.
31. Wesely, M. L.; Hicks, B. B. Ibid., 1110–1116.
32. Sheih, C. M.; Wesely, M. L.; Hicks, B. B. *Atmos. Environ.* **1979**, *13*, 1361–1368.
33. Walcek, C. J.; Brost, R. A.; Chang, J. S. *Atmos. Environ.* **1986**, *20*, 949–964.
34. Voldner, E. S.; Sirois, A. *Water, Air, Soil Pollut.* **1986**, *30*, 179–186.
35. Baldocchi, D. D.; Hicks, B. B.; Camara, P. *Atmos. Environ.* **1987**, *21*, 91–101.
36. Hicks, B. B.; Baldocchi, D. D.; Hosker, R. P., Jr.; Hutchison, B. A.; Matt, D. R.; McMillen, R. T.; Satterfield, L. C. "On the Use of Monitored Air Concentrations to Infer Dry Deposition"; NOAA Technical Memorandum; National Oceanic and Atmospheric Administration: Boulder, CO, 1985; pp 1–65; ERL ARL-141.

RECEIVED for review January 28, 1987. ACCEPTED April 6, 1987.

Chapter 22

Quality Control Infusion into Stationary Source Sampling

James A. Peters

The infusion of quality control (QC) procedures is discussed for emission testing of vents and stacks for hazardous air pollutants (HAPs). The concept of predefinition of data quality objectives (DQOs) enables many different levels of QC to be used. A representative sample collection strategy is outlined for several types of gas flow and composition scenarios. The U.S. Environmental Protection Agency's (EPA's) Method 18, an in-field gas chromatographic procedure for organic pollutants that has numerous options available to the tester, is reviewed, and QC suggestions are presented. Data and results presentation are examined.

S TATIONARY SOURCE EMISSION TESTING is the sampling and analysis of stacks and vents of air pollution sources. The reasons for emission testing may be quite varied and include testing for compliance with a specific set of regulations, responding to a regulatory request for emission data, acquiring information for control equipment design specifications, determining control equipment efficiencies, and acquiring emission inventory data for historical, contaminant reduction, or process control purposes. The types of sources to be sampled may also be highly variable, ranging from large stacks to small vents, and could include process gas streams; control device inlet ducts; area ventilation sources; storage tank vents; material handling systems; surface impoundments and landfills; and pumps, valves, flanges, and other devices with seals. The needs for data will help determine the methods of sample collection and analysis, the choice of sample points, and the number of samples collected. After the strategy for determining the number of samples collected is formulated, the choices of sample volumes and intervals, methods and equipment, and quality control (QC) procedures and goals can be determined.

1173–6/88/0317$06.00/0 © 1988 American Chemical Society

The U.S. Environmental Protection Agency (EPA) develops and publishes sampling and analytical procedures for pollutants from stationary sources in the *Code of Federal Regulations*, Title 40, Part 60 (40 CFR 60). In the last few years, QC procedures have been added to most of the 40 CFR 60 methods to establish the traceability of measurements and to define acceptable quality performance goals. In addition, EPA has published a book (*1*) that provides guidelines and procedures for achieving quality assurance (QA) for the more commonly used EPA reference methods, namely EPA Methods 2–10, 13A and B, and 17.

Stationary source evaluation is undergoing a quiet transformation and modernization as a result of two factors: technological change, and public and regulatory focus. Technological change has now enabled numerous pollutants to be sampled and analyzed on-site, continuously, and instrumentally rather than by manual, discrete sample collection and wet chemical or gravimetric analysis. These newer instrumental methods include EPA Method 3A for CO and O (*2*), EPA Method 6C for SO by pulsed fluorescence (*2*), EPA Methods 7E and 20 for NO_x by chemiluminescence (*3*), EPA Method 10 for CO by nondispersive IR (*4*), and EPA Method 25A for total hydrocarbons by flame ionization (*5*).

Continuous emission monitoring systems (CEMS) collect much more information than discrete sampling, and the use of digitized recorder data or data loggers can provide data for any time interval or averaging time desired. QA and QC procedures for instrumental measurement of stationary source emissions have been defined by EPA through the use of published performance standards (PS-1–5) (*6*).

Public pressure and regulatory focus has moved from criteria air pollutants (e.g., particulate matter, SO, NO_x, CO, and O) to hazardous air pollutants (HAPs) (i.e., both chronic and acute toxicity compounds and elements). This chapter will focus on infusing QC into the sample collection and measurement of HAPs.

Because much of modern source evaluation includes on-site analysis as well as sample collection, the total QC procedures that may be used will be discussed. Long before testing begins, the intended use of the data must be defined, and QC procedures for each method must be incorporated into the specific assignment plan.

Data Quality Objectives

Data quality objectives (DQOs) are qualitative and quantitative statements developed by data users to specify the quality of data needed from a particular data collection activity. Once the user has specified the quality of data needed, then the degree of QC necessary to ensure that the resultant data satisfy the user's specifications must be determined.

All data are subject to some error. Different types of error may be introduced at different stages of data collection. For instance, the error or uncertainty involved in sampling should determine to a large extent the variability that can be tolerated in the analysis, when the analytical methodology is not the limiting factor. When sampling error is already more than two-thirds of the total error, reduction in analytical error is of only marginal importance (7).

Some types of error can be controlled, whereas others cannot be controlled but can be recognized and described. Some types of errors can be quantified, whereas others can be only described qualitatively. The overall purpose of preparing detailed plans for data collection and QA is first to make sure that an appropriate level of control is exercised over sources of error that can be controlled, and second, to make sure that sufficient information is obtained to describe all known sources of error to the extent possible.

The magnitude of error associated with a particular data set is called *data quality*. As the magnitude of the error associated with a data set increases, the quality of data decreases. Collecting data with small levels of error may not always be possible or necessary; data of the very best quality are not always needed. Depending on the particular decisions that will be made with the data and the way in which the data will be used to support those decisions, specified levels of error may be tolerable or cost-effective. However, the quality of data must always be known. Data without error bars are not worthy of the name data.

The quality of a data set is represented in terms of five characteristics: precision, accuracy, representativeness, completeness, and comparability. For a particular data set, data quality may be expressed in quantitative terms for some of the characteristics and in qualitative terms for others. The relative significance of the five characteristics may differ for different types of data and for different uses of data.

The development of DQOs is a complex, iterative process involving both decision makers and technical staff. Both parties are responsible for defining how they intend to use the data and determining the quality needed to support that use. Defining the intended use depends on first understanding the types of decisions that will be made with the data. Information may be available from previous experience in the program, experience in related programs, or specific provisions of regulations. In any environmental data collection activity, multiple data users may be found having differing data quality requirements. When this is the case, DQOs should be developed with a balanced consideration of the needs of the primary user, the source of funds for data collection, and the level of quality required for the most demanding use of the data. Most often, the funds available dictate the amount and quality of data to be collected.

Source Types and Compositions

The purpose of emission testing (i.e., source or stack sampling) is to extract from the stack or duct a sample that is representative of emissions from that source during a time period in which the process is under a desired operating condition. When possible, the normal operation of a process should be determined during a pretest survey. If a process varies with time over a defined cycle, monitoring personnel should determine the variation in emission parameters during the cycle as a basis for deciding whether to sample during a part of the cycle, during an entire cycle, or during several cycles. If the process involves steady-state operation, the level of operation to be sampled should be determined. Any seasonal variations in process conditions should be noted, as should variations in feedstream composition or control device operation.

Selection of a suitable sampling location at a stack or duct should follow the procedure in EPA Method 1 whenever possible for the best quality sample and most accurate determination of gas flow rate (8). Lack of adherence to Method 1 location criteria can result in lower quality flow measurements due to undeveloped flow profiles, and can make sample point selection more complicated.

The sampling site or location and the number of traverse points designated for sample collection will affect the quality of the sample. Emission tests are based on the assumption that the sample obtained at a given point is representative of the concentration at that point. Therefore, a system in which concentrations are nonuniform with respect to the stack cross-sectional area will require more sampling points than will a system having uniform concentrations. One of four possible conditions can exist over the cross-sectional area where sampling points can be located, as shown in Table I.

Usually, gaseous concentrations are fairly uniform across a duct's cross section, unless low flow stratification conditions exist, and a single sampling point is sufficient. Cases 1 and 2 (Table I) are similar in that they both contain streams of constant composition, and representative samples can be obtained for each by employing proportional (to the flow rate or cycle period) sampling at one point over the cross section of the duct. The only difference between cases 1 and 2 in obtaining a representative sample is that the sample should be collected over a longer period of time in case 2 to account for the fluctuation in flow rate.

Table I. Stack Sampling-Point Decision Strategy for Cross Sections
at a Suitable Location

Case	Description	Sampling Strategy	
		Gases	Particulate Matter
Case 1	Constant composition and constant flow rate	One point	One point
Case 2	Constant composition and fluctuating flow rate	One point	One point
Case 3	Variable composition and constant flow rate	Traverse	Traverse and sample isokinetically
Case 4	Variable composition and fluctuating flow rate	Traverse	Traverse and sample isokinetically

Cases 3 and 4 (Table I) are similar in that both contain streams that vary in composition over the cross section of the duct, which is typically the case for particulate matter (i.e., particles greater than 4 μm in diameter). This condition can also exist when the sampling location is not sufficiently downstream of where two gas streams join to allow for complete mixing. For the effluent stream containing particulate matter, a method known as *isokinetic sampling* is employed to obtain a representative sample. For the case of unmixed gas streams, *proportional sampling* is employed. To account for the variable composition over the cross section of the duct, several traverse points that are located in the centroids of equal area segments of the cross section must be employed in both situations to obtain representative samples. Longer sampling times can again be employed to compensate for flow rate fluctuations.

The concentration level of a constituent in a gas stream will affect the choice of the sampling and analysis (S&A) method. For example, a compound may constitute 0.1% (1000 ppm) in a plant vent, and a particular S&A method will be most applicable for that concentration range. When testing for a compound at parts-per-billion levels, another S&A method is probably more appropriate. Figure 1 illustrates the general concentration range that emission testing must address and compares that range to the ranges most often tested for in workplace and ambient atmospheres. Stack sampling can range from trace analysis for compounds in combustion gases [e.g., dioxin and polychlorinated biphenyls (PCBs)] to percent levels of major constituents (e.g., CO_2, H_2O, NH_3, and SO_2).

One of the most useful concepts of data quality and error bars is the *heteroscedastic curve*, which relates interlaboratory coefficients of variation to concentration. This curve is also known as the *envelope of uncertainty* (9). During the establishment of DQOs and for subsequent decision making, uncertainty increases as the measured concentration levels decrease. This increase in uncertainty is due primarily to technological limitations of the methods used and the spatial variability inherent in the processes measured. Adding more QC to the measurement process as the expected analyte concentration decreases is prudent.

Hazardous Air Pollutants

Pressure from legislators and regulatory agencies has increased for industry to provide emission inventory data for a large list of air pollutants. Each state regulatory agency and congressman seems to have their own list of air pollutants, independent of EPA's priority list of 41 HAPs developed under section 112 of the Clean Air Act (10). The most recent new list includes the 329 compounds deemed as extremely hazardous substances under 40 CFR 372, which were legislated under section 313 of the Superfund Amendments and Reauthorization Act (SARA) of 1986 (11).

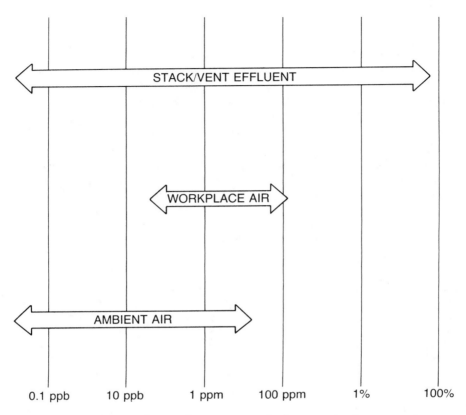

Figure 1. Comparison of expected concentration of pollutant analyzed in air sampling and analysis.

High quality monitoring data are most possible when detailed, standard-ized, validated S&A methods are used to make the measurements. Validated methods and detailed procedures are scarce for sampling and analyzing the large number of HAPs. For example, only 14 out of 41 EPA Office of Air Quality Planning and Standards (OAQPS) HAPs can be routinely measured in stacks at this time. Only four HAPs (mercury, beryllium, vinyl chloride, and polonium-210) have reference methods published for determining emission rates from stationary sources (*12*). Three years from now, EPA's air monitoring strategy indicates that they would like to be at a point where S&A methods for 25–30 compounds are available along with associated QA and data handling procedures (*10*). Five years from now, EPA would like to have methods for most of the 41 compounds (*10*). In contrast, current Comprehensive Environmental Response, Compensation, and Liability Act (CERCLA) legislation calls for emission inventory information from industry for 329 substances (*11*). In the meantime, the burden of selecting an acceptable S&A method lies with industry.

Nearly all HAPs will be in one of two forms: (1) particulate or particulate-associated, or (2) gaseous. The types of HAPs that will be particulate matter are metals, elemental species, and high vapor pressure organic compounds that are sorbed to particulates. EPA Method 5 (*13*) sampling trains for particulates and the proposed modified Method 5 (MM5) sampling train can collect particulate-associated HAPs for analysis (*14*).

The large majority of HAPs are gaseous compounds that will most likely be sampled by using EPA Method 18 techniques, which involve gas sample collection and delivery to a gas chromatograph, which separates the major components of the gas mixture, and measurement of the separated components with a suitable detector (*15*). Trace gas S&A methods that are developed for hazardous waste incineration and combustion gas characterization include the volatile organic sampling train (VOST) and the MM5 (*14*). The VOST procedure could be argued to be basically a subset of the many options of EPA Method 18.

EPA Method 18 is a set of "loose" procedures for measurement of gaseous organic compound emissions by gas chromatography (GC). Nearly all other EPA stack S&A methods contain very detailed procedures for sample collection, recovery, preservation, handling, analysis, and data reporting. Also, equipment design and specifications are often spelled out in detail. Detailed QA and QC procedures and performance specifications are now also added to these methods (*16*). EPA Method 18 has been called EPA's only nonmethod: "Do what you want to do, only do it right".

EPA Method 18 is certainly not without QA. Rather than specify equipment and procedures in detail, a set of QA and QC performance standards is required for generating valid data. A precision goal of 5% relative standard deviation (RSD) for replicate analyses and an accuracy goal of 10% deviation from prepared audit samples are given. Calibration curves for gas standards must be prepared at three concentrations per GC attenuation range, and a linear regression analysis must be performed. If gas dilution systems are used, then a test of the dilution system in meeting the performance goals must be demonstrated. If sorbent sampling tubes are used, QA goals for desorption efficiency, sample collection efficiency, and audit gas accuracy are provided.

The range of options for conducting an EPA Method 18 emission test is incredibly diverse. For example, the relationship between sample collection and analysis can range from direct interface GC with the stack (provided that the moisture content of the gas does not interfere with the analytical procedure) to discrete sample collection, transport, storage, and analysis. The time between sample collection and analysis can range from milliseconds to days, as long as efficient storage can be demonstrated.

Sample collection techniques can include gas bags (no type or material specified), charcoal tubes, sorbent tubes (no type specified), heated-line direct interface, and heated-line dilution system direct interface. The operator may use any GC system, any type of GC column, any GC detector type, any GC operating conditions, and any data reduction and recording system that is deemed appropriate for the task by the operator. Sample recovery can range from keeping a bag sample heated until analysis, to solvent or thermal desorption for sorbent samples. Gas standards for external calibration can be cylinder gases, diluted cylinder gases, prepared in gas bags from assay liquid injections, prepared by boiling semivolatile liquid materials into gas bags, or prepared by dynamic permeation devices.

The infusion of QA and QC procedures into a Method 18 emission test becomes an unbounded problem until the DQOs have been established and the specific S&A techniques have been defined. Once defined, an exercise through a QA and QC checklist such as that shown in Table II can guide the emission tester toward development of a QA project plan, whether it be formal or informal. The planning stage of an emission test is where the QA and QC procedures must be defined and where steps must be taken to obtain the necessary documentation for reporting on the QA and QC procedures and performance.

Proper sampling requires the use of the correct method, the equipment designated by the method, and competent personnel. Because EPA Method 18 rarely designates equipment as other EPA emission test methods do, the tester is left with conducting the method correctly and with competent personnel.

Table II. QA and QC General Checklist for HAP Emission Testing

Category	Sampling	Analysis	Reporting
Technical responsibility	Yes	Yes	Yes
QA responsibility	Yes	Yes	Yes
Data Quality Objectives	Yes	Yes	Yes
Quality control procedures			
Equipment calibration	Yes	Yes	No
Sample custody chain	Yes	Yes	No
Measurement traceability	Yes	Yes	No
Blanks	Yes	Yes	No
Control checks	No	Yes	No
Calibration standards	No	Yes	No
Storage time	Yes	Yes	No
Sample spikes	Optional	Yes	No
Internal standards	No	Optional	No
Surrogates	No	Optional	No
Replicates	Optional	Yes	No
Linearity	No	Yes	No
Leak testing	Yes	Yes	No
Data review	Yes	Yes	Yes
Quality Assurance Procedures			
System audits	Optional	Optional	Optional
Performance audits	Optional	Yes	No
Written SOP[a]	Yes	Yes	Yes
Documentation	Yes	Yes	Yes

[a]The abbreviation SOP denotes standard operating procedure.

Prior to the test date, the tester should determine that the proposed test methods comply with appropriate testing regulations, expectations of the customer or agency, and the DQOs. In many instances, deviating from proposed methods may be necessary. In such cases, the tester should make an engineering analysis of the test site and immediate problems, and then proceed only after obtaining the approval of the regulatory authority or customer, depending on DQOs. This determination should be recorded in the field notes or logbook. An after-the-fact site analysis may suffice in some instances, but good QA techniques dictate that this analysis be made prior to spending the many work hours required to extract and analyze the sample.

Method 18 Sampling Equipment

Sampling equipment, such as flow meters and gauges, must be properly calibrated and maintained. As standard practice for EPA source test methods, the monitoring team should check and record the dates of calibration or servicing. Gas sampling equipment that requires maintenance and calibration includes pitot tubes, nozzles, manometers, thermometers and thermocouples, flow meters, orifices, and dry gas meters. The same principles also apply to the analytical equipment used in an EPA Method 18 test. Routine "absolute" calibration is not the same as chemical calibration, where the relationship between instrument response and concentration is established. Absolute calibration ensures that the perceived instrument response corresponds to the correct physical signal that should produce that response. Examples of equipment that must be absolutely calibrated include, but may not be limited to, balances, thermometers, other temperature sensors and controllers, flow controllers, autoinjectors and gas sample loops, and recorders.

Balances. Balances are the clearest examples of equipment requiring calibration. National Bureau of Standards-certified (NBS-certified) weights are used to establish the traceability and accuracy of measurements.

Thermometers. NBS-certified thermometers are used to verify the accuracy and to establish traceability of measurements.

Other Temperature Sensors and Controllers. For analytical equipment that incorporates temperature sensing or control, the accuracy of the sensors and controllers will affect method performance. When a method specifies an injector temperature of 100 °C, the analyst must be sure that the instrument settings for 100 °C actually correspond to that temperature. Oven temperatures (e.g., GC ovens) must be accurately known. Equipment manufacturers describe procedures for temperature calibration with either NBS-calibrated thermometers or measured electrical signals. Documentation of these calibrations will establish traceability for these measurement subparts.

Flow Controllers. Measuring and controlling gas and liquid flow are integral parts of many instrument analysis systems. The devices used to measure or control flow must also be calibrated to ensure that the actual flow corresponds to instrument readings or settings. GC, gas chromatography–mass spectrometry (GC–MS), dynamic permeators, and gas dilution systems are examples of systems that must be calibrated for flow.

Autoinjectors and Gas Sample Loops. For autoinjectors and gas sample loops, the actual volume injected into the analytical system must correspond to the instrumental settings for the intended volume. This calibration is particularly critical when absolute analyte response (e.g., peak height) is used for quantification (as opposed to the ratio of analyte peak height to internal standard peak height).

Recorders. When physical records (e.g., strip charts) are used for quantification, the recorder response must correspond to the electronic signal received. If the basis of quantification is a linear relationship between response and concentration, the recorder must exhibit linear response to linear changes in electrical signals.

QA and QC Features of Method 18

The key QA and QC features of an EPA Method 18 emission test are to establish traceability for each component of the measurement system to a primary or secondary measurement standard (e.g., NBS materials) and to show that the measurement systems used are "in control" with established limits, that estimates of the bias and precision errors are possible, and that measurement uncertainty can be quantified. The setting of DQOs prior to the emission test will dictate the amount and level of necessary calibrations, documentation to be collected, and quality of the reference and calibration materials. This level can range from very little for an in-house survey to find the emission rates range to order-of-magnitude quality, to a full-blown high quality emission test needed to establish regulatory compliance with permitted levels of pollution. Practical concerns to consider when planning an EPA Method 18 test are discussed in the next section.

Practical Concerns for Method 18 Test Plan and Execution

General Method. All equipment used in S&A activities should have written calibration procedures. Procedures and documentation of the most recent calibration should be available. Post-test calibration of some equipment used in S&A activities may be warranted. Chain-of-custody procedures should be established to ensure sample integrity.

Precision. The precision can range from replicate analyses for 1 in every 10 samples to replicate analysis for each sample if time permits. A whole method precision estimate is possible by collecting duplicate gas samples. Valid data should be collected when duplicate analyses are within 5% of the mean value. A table of replicate precision results should be prepared for the final report.

Accuracy. External standards should be prepared at three concentrations per attenuation range. The linearity of external standards should be established by using regression analysis, and documented. Audit gas samples should be obtained and analyzed to within 10% agreement. The frequency of audit gas analysis should be established. Analysis of actual samples should be suspended if check or audit samples are outside the desired agreement range.

External Standards. Prepared cylinder gases should be traceable to a secondary or primary NBS standard. The dilution apparatus should be calibrated for preparation of diluted standards, and the Method 18 procedure should be followed and documented. Dynamic permeation devices require calibration of rotameters, oven temperatures, and the analytical balance used to weigh the permeation or diffusion tube. Intervals for weighing permeation or diffusion tubes should be established within the testing period. For standards from volatile liquid materials, the balance used for density measurement, the dry gas meter, and the syringe volume must be calibrated, and the material assay purity must be estimated. Standards injections must be repeated until area counts are within 5% of the mean. Dilution calibration checks are possible with prepared gas cylinders. Dilution gas standards must be within 10%.

Blanks. Consider tiers of blanks and documentation of analyses: laboratory blanks to demonstrate lack of contamination in the laboratory (these blanks may be very appropriate for field trailer labs), trip blanks to demonstrate lack of contamination between the collection point and the lab, field blanks to demonstrate lack of contamination from sample collecting activity exposure to the lab, and reagent and solvent blanks to determine the background of materials used in the analysis.

Gas Bags. The bag material selected should be best suited for the sampling application [e.g., poly(vinyl fluoride) (Tedlar) or a copolymer of tetrafluoro-ethylene and hexafluoropropylene (Teflon FEP)]. A storage time study should be performed to assess sample loss or degradation with time for the compounds analyzed. The time between sample collection and analysis should be documented for each sample. Any bag reuse must have a documented procedure for cleaning and for blank determination. Because new bags often have off-gassing problems, documentation that the blanks are acceptable is needed. When gas bag standards are used, the acceptable storage time must be documented before repreparation of standards. The bag water vapor content must be documented. The vapor content can be found by using an ambient temperature, pressure, and humidity determination.

Reporting and Documentation

Stack samplers are generally also the report writers, so data reduction, presentation, and documentation responsibilities in the report fall on the sample collector's shoulders. In Table II, numerous QA and QC checklist activities were shown to involve data documentation. The most helpful hint to achieving the proper QA and QC documentation in a final report is to communicate that list of required documentation before sampling begins. The operator should never assume that the analysts have their QA and QC shop in order. The operator should find out what they do and how they do it, and what limitations are involved.

Manual recording of data is required for source tests, especially for extractive sampling that leads sample recovery for later analysis. Standardized forms should be used to ensure that all necessary data are obtained. These forms are typically designed to clearly identify the process tested, the date and times, the test station location, the sampling personnel, and the person who recorded the data. During the actual test period, the meter readings, temperature readings, and other pertinent data are recorded in the spaces immediately upon observation. These data determine the accuracy of the test and should not be erased or altered. Any error should be crossed out with a single line; the corrected value is then recorded above the crossed-out number. The raw field data sheets are the first wave of documentation in a source test report.

The format of a stationary source test report has been formalized over the years (17). Including a separate section on QA and QC procedures (*see* box on page 329.) will facilitate the reader's understanding of the data quality. Results from this section should tie into the discussion of errors in the summary of results. If the QA project plan was written before the project began, then somewhere in this section the reader should be able to find the match of results versus project goals for precision, accuracy, and completeness. Similarly, a match of the DQOs with results should be provided.

In this modern era of hand-held calculators and computerized data reduction, the conventions of significant figures and rounding have been "lost". Significant figures in an emission test are usually well-defined by means of the field data sheets, on which the most reliable last digit of a measurement has been documented. The data reduction and calculation activities often cause the reporting of excessive numbers of digits, or insignificant digits.

Conventions or rules for significant digits and rounding exist. Conversion of quantities should be handled with careful regard to the implied correspondence between the accuracy of the data and the given number of digits. In all such conversions, the number of significant digits retained should be such that accuracy is neither sacrificed nor exaggerated.

Suggested Source Testing Report Format

Cover
1. Plant name and location
2. Source(s) sampled
3. Testing company or agency, and name and address

Certification (if required)
1. Certification by team leader or technical management
2. Certification by technical reviewer (e.g., P.E.)

Introduction
1. Test purpose
2. Test location(s) and type of process(es)
3. Test dates
4. Pollutants tested
5. Observer's names and affiliations (industry and agency)
6. Any other important background information

Summary of Results
1. Emission results, with tables and figures (wherever possible, these should contain coefficients of variation, standard deviations, and error bars)
2. Process data as related to determination of compliance
3. Allowable emissions
4. Description of collected samples
5. Discussion of errors, both real and apparent

Source Operation
1. Description of process and control devices
2. Process and control equipment flow diagram
3. Process data and results, with example calculations
4. Representativeness of raw materials, products, and operation
5. Any specially required operation demonstrated

Sampling and Analysis Procedures
1. Sampling port location(s) and dimensioned cross section
2. Sampling point description, including labeling system
3. Sampling train(s) description
4. Brief discussion of sampling procedures, including discussion of deviations from standard methods
5. Brief discussion of analytical procedures, including discussion of deviations from standard methods

QA and QC Procedures
1. DQOs
2. QC results for precision and accuracy
3. Comparison of QC results with goals
4. Discussion of QA and QC documentation provided in the appendix to establish traceability to primary standards

Appendix
1. Complete results including example calculations
2. Raw field data (copies of originals)
3. Laboratory reports, including chain of custody
4. Raw production data signed by plant official
5. Calibration procedures and results
6. Test logs
7. Project participants and titles
8. Related correspondence
9. Standard procedures (excluding 40 CFR 60 methods)

Significant Digits

Any digit that is necessary to define the specific value or quantity is said to be significant. When measured to the nearest 1 m, a distance is recorded as 157 m; this number has three significant digits. If the measurement had been made to the nearest 0.1 m, the distance may have been 157.4 m; this number has four significant digits. In each of these cases, the value of the right-hand digit was determined by measuring the value of an additional digit and then rounding to the desired degree of accuracy. In other words, 157.4 was rounded to 157; in the second case, the measurement in hundredths, 157.36, was rounded to 157.4.

Zeroes may be used either to indicate a specific value like any other digit, or to indicate the magnitude of a number. The 1970 U.S. population figure rounded to thousands was 203,185,000. The six left-hand digits of this number are significant; each measures a value. The three right-hand digits are zeroes that merely indicate the magnitude of the number rounded to the nearest thousand. The identification of significant digits is only possible through knowledge of the circumstances. For example, the number 1000 may be rounded from 965, in which case only one zero is significant, or it may be rounded from 999.7, in which case all three zeroes are significant.

Usually, data required for a calculated value in a report must be drawn from a variety of sources where they have been recorded with varying degrees of refinement. Specific rules must be observed when such data are to be added, subtracted, multiplied, or divided.

The rule for addition and subtraction is that the answer shall contain no significant digits farther to the right than occur in the least precise number. Consider the addition of three numbers drawn from three sources, the first of which reported data in the millions, the second in thousands, and the third in units:

$$
\begin{array}{r}
163{,}000{,}000 \\
217{,}885{,}000 \\
96{,}432{,}768 \\
\hline
477{,}317{,}768
\end{array}
$$

The total indicates a precision that is not valid. The numbers should first be rounded to one significant digit farther to the right than that of the least precise number, and then the sum taken as follows:

$$
\begin{array}{r}
163{,}000{,}000 \\
217{,}900{,}000 \\
96{,}400{,}000 \\
\hline
477{,}300{,}000
\end{array}
$$

The total is rounded to 477,000,000 as called for by the rule. If the second of the figures to be added had been 217,985,000, the rounding before addition would have produced 218,000,000, in which case the zero following 218 would have been a significant digit.

The rule for multiplication and division is that the product or quotient shall contain no more significant digits than are contained in the number with the fewest significant digits used in the multiplication or division. The difference between this rule and the rule for addition and subtraction should be noted; the latter rule merely requires rounding of digits that lie to the right of the last significant digit in the least precise number. The following examples highlight this difference for multiplication, division, addition, and subtraction.

$$113.2 \times 1.43 = 161.876, \text{ rounded to } 162$$

$$113.2 \div 1.43 = 79.16, \text{ rounded to } 79.2$$

$$113.2 + 1.43 = 114.63, \text{ rounded to } 114.6$$

$$113.2 - 1.43 = 111.77, \text{ rounded to } 111.8$$

The preceding product and quotient are limited to three significant digits because 1.43 contains only three significant digits. In contrast, the rounded answers in the addition and subtraction examples contain four significant digits.

Numbers used in the preceding illustrations have all been estimates or measurements. Numbers that are exact counts are treated as though they consist of an infinite number of significant digits. More simply stated, when a count is used in computation with a measurement, the number of significant digits in the answer is the same as the number of significant digits in the measurement. If a count of 40 is multiplied by a measurement of 10.2, the product is 408. However, if 40 were an estimate accurate only to the nearest 10, and hence contained only one significant digit, the product would be 400.

Rounding Values

When a figure is to be rounded to fewer digits than the total number available, the procedure should be as follows:

When the first digit discarded is less than five, the last digit retained should not be changed. For example, 3.46325, if rounded to four digits, would be 3.463; if rounded to three digits, 3.46.

When the first digit discarded is greater than five, or if it is a five followed by at least one digit other than zero, the last figure retained should be increased by one unit. For example, 8.37652, if rounded to four digits, would be 8.377; if rounded to three digits, 8.38.

When the first digit discarded is exactly five, followed only by zeroes, the last digit retained should be rounded upward if it is an odd number, but no adjustment made if it is an even number. For example, 4.365, when rounded to three digits, becomes 4.36. The number 4.355 would also round to the same value, 4.36, if rounded to three digits.

Abbreviations

CEMS	continuous emission monitoring systems
CERCLA	Comprehensive Environmental Response, Compensation, and Liability Act
CFR	*Code of Federal Regulations*
DQOs	data quality objectives
EPA	U.S. Environmental Protection Agency
GC	gas chromatography
GC–MS	gas chromatography–mass spectroscopy
HAPs	hazardous air pollutants
MM5	modified Method 5
NBS	National Bureau of Standards
OAQPS	Office of Air Quality Planning and Standards
PCBs	polychlorinated biphenyls
PS	performance standards
QA	quality assurance
QC	quality control
RSD	relative standard deviation
S&A	sampling and analysis
SARA	Superfund Amendments and Reauthorization Act
SOP	standard operating procedures
VOST	volatile organic sampling train

References

1. *Quality Assurance Handbook for Air Pollution Measurement Systems: Volume III, Stationary Source Specific Methods*; U.S. Environmental Protection Agency: Research Triangle Park, NC, 1977; EPA–600/4–77–027b.
2. *Fed. Regist.* **1986**, *51(112)*, 21164–21172.
3. *Code of Federal Regulations*; Title 40, Chapter I, Part 60, Appendix A, Method 20: Determination of Nitrogen Oxides, Sulfur Dioxide, and Oxygen Emissions from Stationary Gas Turbines; U.S. Government Printing Office: Washington, DC, 7–1–85.
4. *Code of Federal Regulations*; Title 40, Chapter I, Part 60, Appendix A, Method 10: Determination of Carbon Monoxide Emissions from Stationary Sources; U.S. Government Printing Office: Washington, DC, 7–1–85.
5. *Code of Federal Regulations*; Title 40, Chapter I, Part 60, Appendix A, Method 25A: Determination of Total Gaseous Organic Concentration Using a Flame Ionization Analyzer; U.S. Government Printing Office: Washington, DC, 7–1–85.
6. *Code of Federal Regulations*; Title 40, Chapter I, Part 60, Appendix B: Performance Specifications; U.S. Government Printing Office: Washington, DC, 7–1–85.
7. Horwitz, William. *IUPAC Commission Sampling Monograph*; International Union of Pure and Applied Chemistry: 1986; Vol. 3.
8. *Code of Federal Regulations*; Title 40, Chapter I, Part 60, Appendix A, Method 1: Sample and Velocity Traverses for Stationary Sources; U.S. Government Printing Office: Washington, DC, 7–1–85.

9. Horwitz, W. *Anal. Chem.* **1982**, *54(1)*, 67A–76A.
10. Hauser, T. R.; Scott, D. R.; Midgett, M. R. *Environ. Sci. Technol.* **1983**, *17*, 86A–96A.
11. *Fed. Regist.* **1987**, *52(107)*, 21152–21208.
12. *Code of Federal Regulations*; Title 40, Chapter I, Part 61, Appendix B: Test Methods; U.S. Government Printing Office: Washington, DC, 7-1-85.
13. *Code of Federal Regulations*; Title 40, Chapter I, Part 60, Appendix A, Method 5: Determination of Particulate Emissions from Stationary Sources; U.S. Government Printing Office: Washington, DC, 7-1-85.
14. *Test Methods for Evaluating Solid Waste*, 3rd ed.; *Physical/Chemical Methods*; U.S. Environmental Protection Agency: Washington, DC, 1987.
15. *Code of Federal Regulations*; Title 40, Chapter I, Part 60, Appendix A, Method 18: Measurement of Gaseous Organic Compound Emissions by Gas Chromatography; U.S. Government Printing Office: Washington, DC, 7-1-85.
16. *Fed. Regist.* **1987**, *51(33)*, 5105–5112.
17. *Industrial Guide for Air Pollution Control*; U.S. Environmental Protection Agency: Research Triangle Park, NC, 1978; EPA–625/6–78–004.

RECEIVED for review January 28, 1987. ACCEPTED May 14, 1987.

SAMPLING BIOTA

Chapter 23

Coping with Sampling Variability in Biota

Percentiles and Other Strategies

Richard Albert and William Horwitz

Principles of good sampling should be followed regardless of the population sampled. Especially important is the requirement that the sample taken be representative. A complication arising in sampling biota is the large variability encountered and the corresponding uncertainties in the estimates. Also, the distribution of values (e.g., mercury concentrations in fish) may not be normal (i.e., bell-shaped, or Gaussian). To reduce the uncertainties, such strategies may be resorted to as increasing the number of units sampled or breaking down the population into distinct subgroups. A naked, unqualified estimate of an average is useless. Associated with any average should be an indication of how the individual values are distributed. For biota, the various percentile points can be reported along with the confidence limits derived from the binomial distribution. The final results of sampling biota may well be an average plus key percentiles of the distribution values.

S TATISTICAL MANIPULATION ALONE cannot reduce variability to manageable proportions; therefore, nonstatistical strategies must also be brought to bear. The more that is known about the target biota (i.e., plants and animals) before sampling, the more readily can the effects of variability be mitigated and biased estimates be guarded against. The emphasis of this chapter is on coping statistically with the inherently large variability encountered in sampling biota.

Example 1. The contamination of the bluefish by polychlorinated biphenyls (PCBs) off the Atlantic Coast of the United States is investigated. To check each bluefish is an absurd impossibility, so only some are checked, and the few that are caught will somehow represent all the bluefish. On the basis of the PCB concentrations measured in the bluefish sample, what can be said about the PCB concentrations in all Atlantic bluefish?

Example 2. The amount of aflatoxin in the U.S. peanut crop is investigated. The investigator goes to a nearby grocery store and buys all the peanuts therein. The aflatoxin content of each peanut is meticulously measured. From these measurements, what can be said about the contamination of the peanut crop?

Example 3. Having heard the famous rhyme about purple cows, an investigator wants to find out if any purple cows really do exist. While traveling through a farm district, the investigator spots 20 cows, but none are purple. What can be concluded about the number of purple cows in existence?

These three scenarios share something in common: Data are desired about all members of a certain large group (i.e., the population is Atlantic bluefish, U.S. peanuts, or cows), but only a minuscule fraction of the large group is inspected. From this fraction, information must be extrapolated to the entire population. The goal is to make estimates and inferences that are valid for the entire population, although data are contained on only a sample of the population. The fundamental question is this: What can be asserted, and how certain can the investigator be in these assertions? This chapter focuses on the certainty, or lack thereof, in estimating specific statistical quantities associated with various biota.

In assessing certainty or uncertainty (i.e., in trying to quantitate confidence or diffidence), a fundamental role is played by the variability of the population. If all peanuts were identical, then the sampling of just one peanut would suffice to monitor the entire peanut crop. Alas, in the real world, the pervasive existence of variability of all sorts compels us to sample more than just one unit: more than one bluefish, more than one peanut, and more than one cow.

Confronting variability and taming it so that valid estimates can be made with adequate precision is the burden of the working statistician. By clever use of the variability found in the sample itself, the statistician can typically associate with an estimate a *confidence interval*, which is a range of values that can be declared with a specified degree of confidence to contain the correct or true value for the population. One quality control manual defines confidence interval as "the interval between two values, known as *confidence limits* [italics added], within which it may be asserted, with a specified degree of confidence, that the true population value lies." The more variable the population, the broader are these confidence intervals.

Assuming the 20-cow sample to be representative, the statistician can be 95% confident that the actual percentage of cows that are purple is somewhere between 0.0% and 13.9% ($1 - (1 - 0.139)^{20} = 0.95$). For a 40-cow negative sample, the 95% confidence interval would be from 0.0% to 7.2% ($1 - (1 -$

$0.072)^{40} = 0.95$). A user of statistics should be critical and alert in making decisions based on sample estimates and corresponding confidence intervals. These statistical artifacts are only as sound as the assumptions and procedures that went into creating them. The validity of the confidence interval is inextricably linked to the validity of the sampling; coordination and quality control are called for throughout the whole complex integrated process that begins with a question and ends with a confidence interval.

In the following portions of this chapter, some mathematical "cookbook recipes" will be presented, but the essential points to remember are as follows:

Point 1. Sampling should not be undertaken until the questions to be answered have been determined and properly framed.

Point 2. The individuals included in the sample must have been chosen at random, or, more generally, by probability sampling from a population that is well-defined. If this procedure is not done, experienced statisticians have warned that the investigator can be 100% confident that the final confidence interval estimates will be catastrophically wrong.

Point 3. The distribution of values in a population is often Gaussian (i.e., "bell-shaped", or "normal"), so that the estimates of the average and of the standard deviation are all that is needed to characterize the entire distribution. Confidence intervals are readily computed from a sample of the distribution if the parent population is Gaussian.

Point 4. Non-Gaussian populations can be sampled by empirical estimates based on samples of the 10-, 25-, 50-, 75-, and 90-percentile points. Confidence intervals for these percentile-point estimates are much broader (i.e., less precise) than corresponding confidence intervals of percentile-point estimates for Gaussian populations. (The P-*percentile point* is that value at or below which P percent of all population values lie.)

Proper Sampling

The order of evolution is from raw data to information, from information to knowledge, and from knowledge to wisdom. The crucial first step is to intelligently gather the appropriate raw data. What is appropriate is determined by what sort of knowledge is sought. A campaign must be mapped out where all available resources are inventoried, and the scope of the questions is guided in part by the extent of these resources. After careful framing of the questions, the relevant population must be sought. Here, as elsewhere, the more that is known, the more that can be learned. For example, in a real-life bluefish survey, PCBs were known a priori to concentrate in the larger adult fish. Such knowledge served to direct the sampling procedure.

After question posing and population identifying, the actual sample is selected (e.g., the specific individuals, the specific bluefish, the specific peanuts, or the specific cows). Such selection must be done in a probabilistic way (i.e., every individual in the population has a prior known probability of being included in the sample). Only with such a probabilistic selection can certain potential biases be avoided.

The final phase in getting the raw data is to reduce the sample to manageable portions, aliquots, or pieces so that the chemical or other measurements can be made. The homogeneity of the test portion at this stage is crucial in minimizing the overall variability or uncertainty in the ultimate estimates of the average and associated confidence intervals. Obviously, cows can just be looked at for being purple, but a fish has to be skinned, deboned, and subjected to extraction of its fatty portion in order to estimate PCB content.

Further details on these consecutive links in the chain leading from the problems (i.e., the questions) to the answers (i.e., the estimates of averages and confidence intervals) are provided in the rest of this chapter.

Proper Questions

The toil and expense of a survey can be justified only if the questions answered have a value. Vague, unstructured exploratory surveys are wasteful. A need to answer certain questions must be present. Then the price of getting the answers should be looked at. The cost per datum can be disconcertingly high.

The routine analysis of aflatoxin concentration in peanuts at the parts-per-billion level costs perhaps $100 per data point. The moral of the story is to make sure the questions are worth asking in terms of dollars and cents and resources.

Questions should be couched in quantitative terms. In the real-life bluefish survey, the preliminary work established over two dozen questions: "What is the average PCB concentration in bluefish?" and "Are there significant differences in PCB concentrations based on geographical, sex, or seasonal differences?" Even "significant" should be assigned a more quantitatively precise meaning. Otherwise, the investigator would be violating the principle of posing only those questions that allow definite, clear-cut answers.

Basically, at this stage the investigator should decide what should be measured and how accurately it should be measured.

Proper Population

Coupled with the issue of what is measured is the issue of what it should be measured in. Needs include what is to be included as a potential candidate to be sampled, and, equally as important, what is not a candidate to be sampled.

Basically the right answer should not be obtained to a wrong question. Are PCB raw data desired from all bluefish at all levels in all regions? Definitely not. The purpose of one recent real-life bluefish survey was to assess human exposure to PCBs; accordingly, the field workers for this survey sampled only from those bluefish that were actually "landed" (i.e., brought to an Atlantic Ocean dock for sale or personal use). Only those bluefish entering the human food chain were of interest.

Proper Selection and the Importance of Randomness

Sufficiently large sample sizes can provide very "tight" estimates of averages (i.e., estimates bracketed by very narrow 95% confidence limits). The smaller the confidence interval, the more precise (and expensive) the estimated average. However, the danger always lurks that a precise estimate of the wrong average may be obtained. The persistent difference between the true or population average and the average obtained from repetitions (actual or potential) is a manifestation of bias. (In other words, the difference between the true population average and the estimate from the particular survey of the average may be called the *error of the estimate*. *Bias*, on the other hand, is a theoretical construct, and is equal to the algebraic average of such errors obtained from an indefinitely large number of repetitions of the survey.)

The sovereign remedy against bias is proper sample selection by a random process. The use of randomness helps preclude the presence of a persistent bias. A random sampling scheme can help ensure that in the long run, representative, nonbiased samples are obtained. In a random sampling scheme, every member of a population—every bluefish, every peanut, or every cow—has an equal probability of being included as part of the sample. In a random sampling scheme, the proportion of "positives" selected would in the long run be equal to the actual proportion of positives in the population (i.e., a representative sample would be achieved).

Knowledge is power, and foreknowledge of certain features in the target population can help reduce the variability in the sample, which in turn can lead to more precise estimates (i.e., to narrower confidence intervals). In particular, if before a survey is taken, a certain subpopulation (e.g., positives) is known to constitute a certain fraction of the target population, and the quantity to be estimated is especially variable among that subpopulation, then the investigator could randomly sample more frequently among the more variable population (i.e., larger random samples could be taken from the more variable subpopulations). Trivial recipes exist for combining the averages and variances of the nonoverlapping subpopulations (strata) into an overall average with an associated confidence interval surrounding this average. The resulting estimates are more precise but are also more liable to bias should the investigator be mistaken in the preliminary information on the population breakdown into strata.

Whether or not the investigator has prior information on the population subgroups, sampling should be done via random selection within each subpopulation or within the total population. However, in addition to knowing how to sample, the investigator needs to know how much to sample. Formulas exist to calculate the number (N) of items or individuals to be included. This number is directly proportional to the variability of the population expressed as the variance, but how can the investigator know this variability before the survey? Experience with previous similar surveys or results from a crude preliminary survey may give an adequate estimate of the variance, which in turn will help to estimate the sample size in the full-blown survey. The required sample size is also inversely proportional to the square of the a priori desired width of the confidence interval.

This section can be summarized as follows: Sampling must be random, either overall over a population or within its well-defined subpopulations, in order to ensure representativeness and lack of bias.

Final Link: Proper Test Sample and Measurement

A chain, being only as strong as its weakest link, is a good metaphor to describe the progression in estimating the relevant characteristics of a population. The investigator starts off by framing the questions in an answerable format and by defining in detail the population for which answers are desired. A random-sampling plan is then prepared that includes specifications as to how many individuals are to be included in the sample and from which strata these individuals are to be selected. Careful design and control up to this point help to cope with the intrinsic variability of the population.

The researcher is now faced with a hopefully representative sample and must make measurements on the individuals or units in the sample. How is the PCB content of a sample fish measured? How can the researcher ascertain the amount of aflatoxin in a mound of peanuts? [In fact, especially dramatic contributions to variability arising from sample inhomogeneity have been uncovered in the measurement of aflatoxin in peanuts (1).] The analytes whose concentrations are to be analytically measured are not uniformly distributed throughout the individual. This within-sample unit variability contributes its undesirable share to the width of confidence intervals obtained.

Homogenization or common sense can help remove this variability contribution. The U.S. Department of Agriculture, in its inspection of certain antibiotic residues in cattle, examines just the liver instead of the total carcass. Putting a liver into a high-speed blender is easier than putting in a whole cow. Similarly, often only after a fish has been gutted or fileted, certain chemical measurements are taken. The overall survey scheme can focus on just that portion of the individual where the analyte of interest is usually found.

Even with the best homogenization and the best selectivity of sites to remove within-sample unit variability, the analytical method itself has an irreducible, inevitable variability. Horwitz (2) has found that this variability, expressed as a relative standard deviation [RSD, where %RSD = 100 × (standard deviation/concentration)], often may be empirically given by the following equation:

$$\text{RSD}_{\text{method}} = 2^{1 - 0.5\log_{10}C} \qquad (1)$$

where the concentration C is expressed as a decimal fraction ($1.0 = 100\%$ and $0.000001 = 1$ ppm), and the $\text{RSD}_{\text{method}}$ is expressed as a percentage. For biota, this analytical source of variation would be expected to be negligible compared with those large variations from all the other sources such as sampling.

But what about bias? The analytical method may be adequately precise, but what if it is persistently off the mark? A classic horror story of bias in analytical methods comes from the nonbiotic realm of mineral composition. The details can be found in an article by Abbey (3). The author's conclusion is worth quoting: "There is no such thing as a bad method—only bad analysts who fail to allow for its limitations."

The lesson to be learned here is that even in the final stage, due care must be taken to eliminate avoidable variability and bias.

Variance and Confidence Intervals

Any distribution may be characterized by a variance, even if that variance equals infinity for certain U-shaped distributions. *Variance* is a measure of the variability in a population and can be used to compute confidence intervals about estimates of population averages. Distributions in biota might be expected to have especially large variances and hence especially large confidence intervals (i.e., estimates of averages would tend not to be very precise). However imprecise an estimate, at least statistical computations will alert the investigator to the magnitude of the imprecision. This section applies to all sampling, including sampling biota. However, the expected large variance and the possibility of non-Gaussian distributions in biota make more relevant the section on non-Gaussian distributions.

Measures of Variability

Consider a set of values, such as 244 PCB concentrations from 244 bluefish. The most convenient single statistic to characterize this set of values is the average, $\bar{X}$, given by $\bar{X} = (\Sigma X_i)/n$, where X_i is the PCB concentration measured in the ith bluefish, and n is the number of values (here $n = 244$).

A supplement to the average would be some measure of the spread or of the variability of the values. A simple measure of spread is the *range*, which is equal to the highest value minus the lowest value:

$$\text{range} = X_{(n)} - X_{(1)} = X_{\text{highest}} - X_{\text{lowest}} \qquad (2)$$

where the integer subscripts in parentheses conventionally indicate the rank of the value when the values are arranged in increasing order from lowest to highest. Thus, $X_{(1)}$ is the lowest PCB value, and $X_{(244)}$ is the highest PCB value.

A statistic that is more stable than the range is the variance (V) of the set of values, which is simply the average of the squared deviation from the average of a set of values:

$$V = \frac{\Sigma(X_i - \bar{X})^2}{n} \tag{3}$$

Usually, however, the variance of the set of values is not what is wanted, but rather the variance of the population from which these values are merely samples. No matter what the population, the estimate of the population variance is conveniently obtained via multiplication of the sample variance by $n/(n-1)$:

$$\text{population variance} = \frac{\Sigma(X_i - \bar{X})^2}{n-1} \tag{4}$$

The average of a sample is a good estimate of the corresponding average of the population, but the variance of a sample must be multiplied by the $n/(n-1)$ factor to get a good estimate of the population variance. For $n = 100$, no palpable differences between the population variance and the sample variance exist.

Other measures of variability come immediately to mind, such as the range divided by n, the average magnitude of deviation from the average, and the 75-percentile point minus the 25-percentile point. However, the variance remains the statistic of choice to indicate the variability. The prime reason is that variances are additive. This additivity means that the total variance can be expressed as a sum of the variance of assorted factors (e.g., within individual, within geographical region, and among geographical regions).

Moreover, the square root of the variance, which is the *standard deviation*, is the crucial quantity used in calculating confidence intervals. In Gaussian distributions, knowledge of the average and the variance (or equivalently, knowledge of the average and the square root of the variance) is enough to describe the whole distribution. In particular, with Gaussian distributions, the investigator can state a priori what percentage of the values in a population lie at or below such and such a number of standard deviations away from the average (*see* Table I in the appendix at the end of this chapter).

Once and For All: Standard Deviation versus Standard Error

"The *standard error* [italics added] is the standard deviation of the average and is equal to the standard deviation based on the individual values divided by the square root of the number of individual values involved." This definition is the recipe for the standard error. A typical question is "But do you use the standard error or the standard deviation in calculating the confidence intervals?" The correct answer is yes. Simply put, when confidence intervals are desired for single values in a population, the standard deviation (as calculated from the sample by using the $n/(n - 1)$ factor) should be used; when confidence intervals for averages are desired, the standard error (i.e., the standard deviation based on the sample divided by the square root of the number of individuals in the sample) should be used.

Qualitatively, taking averages leads to a tighter distribution: In any set of values being averaged, even an extreme value is likely to be neutralized and driven toward the population average. Nature is kind in that this tightening—this reduction in variance—is proportional to $1/n$, or for standard deviation, proportional to $1/n^{1/2}$.

The Magnificent Gaussian Distribution

Gaussian distributions, or at least near-Gaussian distributions, occur naturally. These distributions arise when many small independent fluctuations in components tend to cancel each other out to yield a stable average. Little subcomponents that add up and vary are what generate a Gaussian distribution. A more classic example is the distribution of IQs among a fixed population, where a wide variety of factors (e.g., genetic, economic, and social) combine to generate a final IQ. (IQ is a measure of how well one does on an IQ test.) Such small independently contributing components that fluctuate exactly match the model of a Gaussian distribution.

Consider the distribution of analytical results from an interlaboratory study of a given method applied to a given concentration of analyte in a given material. The roster of small, independently contributing factors might include sampling time, background noise, ambient temperature, and inhomogeneity.

Statisticians relish the Gaussian distribution because it is easy to handle: Knowledge of the average and standard deviation completely specifies the whole distribution. As Table II in the appendix at the end of this chapter shows, the location of every percentile point is known a priori. The fraction of the population that falls between any two specified values is known also.

Besides being handy, the assumption of a Gaussian distribution is often realistic. Moreover, rigorous statistical demonstration shows that even if a population is not Gaussian, its averages tend to become Gaussian-like. This point shows the true magnificence of the Gaussian distribution. The larger the sample size, the more Gaussian-like is the population of averages. Therefore, statisticians agonize over large sample sizes for two reasons: (a) the population gets more Gaussian-like, and the validity of the confidence-interval formulas increases; and (b) the standard deviation is diminished by a factor of $1/n^{1/2}$, which leads to tighter estimates (i.e., narrower 95% confidence intervals).

Another reason for advocating a large n is that the 1.96 in the 95% confidence interval formula

$$X \pm (1.96S/n^{1/2}) \tag{5}$$

is, strictly speaking, valid only for infinite n ($n = 30$ is acceptable), and the coefficient before the standard deviation (S), denoted $t_{0.025,n}$, is a function of n. For $n = 2$, the t coefficient is 12.71, and for $n = 30$, the t-coefficient is 2.05. Remembering 1.96 is easier than looking up the appropriate value for the t-coefficient, but the looking up should be done even if the refinement is only minor.

Confidence Intervals for More General Distributions

Gaussian distributions apply to continuous-valued variables, such as PCB or aflatoxin concentration. Obviously, an indefinitely large number of distributions for continuous values can be contrived, and many such distributions are well-known and fully described in the literature. Among such candidate distributions are the log-normal distribution, Pearson-types I–IV, the F-distribution, and the t-distribution ($t_{0.025,n}$ is a value in this distribution category). If the investigator knows a priori what distribution the population of values will have, then specialized tables can be looked at to calculate appropriate confidence intervals. Many times, Gaussian is a good guess, for which researchers should be thankful.

Biota distributions often fall into the dreaded "none-of-the-above" category: neither fish nor fowl. Particularly pernicious would be a bimodal distribution consisting of the superposition of two Gaussian distributions, each having its own average and its own standard deviation. Nevertheless, a confidence interval formula exists that is applicable to any continuous variable distribution, even a bimodal one. The price paid for the generality and the universal applicability of these so-called *Chebyshev intervals* is that the confidence intervals are quite broad.

The Chebyshev confidence interval is given by the following equation:

$$X \pm (ZS/n^{1/2}) = 100 \times [1 - (1/Z^2)]\% \text{ confidence interval} \qquad (6)$$

where X and the standard deviation S are calculated from a large sample, and Z is any positive value greater than one. The price of ignorance is explicit: for $Z = 1.96$ with a Gaussian distribution, a 95% confidence interval is obtained; for an unknown distribution, for $Z = 1.96$, at least a 74% confidence interval is obtained, but the researcher does not know how much more confident he or she can be. To construct a 95% confidence interval about an average in the most general case, use $X \pm 4.47 \, S/n^{1/2}$.

Even if the true population distribution is not known, this interval will be at least a 95% confidence interval. Had the population been Gaussian, this interval would be a 99.9+% confidence interval. Therefore, the way to tighten the interval—the way to get a more precise estimate—is to take a larger number of samples.

Confidence Intervals for Percentile Estimates

By applying due care and attention to details, the researcher can get a serviceable estimate of the average and the standard deviation. In fact, with sufficient skill and extraordinarily large samples (i.e., large *n*), very good estimates can be made of the population average. But often after all that labor, the mere average is not enough.

The researcher must be aware of how misleading an average alone can be, because the average tells nothing about the underlying spread of values. Think about the old joke about the statistician who drowned in the river whose average depth was 5 in.; the statistician just happened to be trying to walk on water 10 ft deep. The moral is this: Often the overall distribution is a crucial feature to assess. Especially in regulatory situations, great importance is attached to knowing what percentage of the population has values lying above the maximum level permitted by law. In short, the task of the statistician can also include that of estimating the distribution.

This section is devoted to a very simple, straightforward procedure for estimating the population distribution: reporting for the sample the 10-, 25-, 50-, 75-, and 90-percentile points, and then reporting the confidence interval for each of these percentile points. The appendix at the end of this chapter presents the confidence intervals for these percentile points as a function of sample size (*n*) in Table III. Statisticians sometimes use *box plots*, which are geographical depictions of percentiles.

Customer's Favorite: Two-Item Sample

Cost considerations lead users of statistics to minimize the sample size. Two-item samples are the ideal. (Even the most cost-conscious manager would grant that a one-item sample is too small, because at least two items are needed to estimate variability.) This section serves to show the quantitative consequences of picking just two individuals.

If the investigator is estimating some continuous quantity such as PCB concentration, then the 95% confidence interval is

$$[(X_1 + X_2)/2] \pm (12.706) \times \sqrt{(X_1 - X_2)^2/(2)(2)} \qquad (7)$$

$$[(X_1 + X_2)/2] \pm [6.353 \times (X_1 - X_2)] \qquad (8)$$

(For the connoisseur: The number 12.706 is the *t*-value for $n = 2$ at the 95% confidence interval, and the estimated variance from two values is $\Delta^2/2$, where $\Delta = X_1 - X_2$.)

Besides estimating continuous values, often the problem of estimating fractions is confronted (e.g., what fraction of the cow population consists of purple cows?). With fractions like this, any particular individual either belongs or does not belong to the class under investigation.

The crucial trick that provides a mechanism to convert continuous data to fractional or proportional type data is this: The statistician should mentally break down the population of values (e.g., actual PCB concentrations) into those values at or below the 90-percentile point and those values above the 90-percentile point. By definition, the probability of any value being at or below the *P*-percentile point is *P*% (or *P*/100 when probabilities are expressed as a decimal fraction). Therefore, if any item (e.g., a fish) is taken at random, the probability is 0.1 that its value (e.g., PCB concentration) is at or below the 10-percentile point. The probability is 0.9 that the value for any item is at or below the 90-percentile point. Associated with this probability is the probability that the value is above the 90-percentile point. This probability is 0.1 (= 1 − 0.9) because any value must be in one of the two classes, and therefore, the total probability must equal one.

For any two-item sample, the probability that both values are at or below the 90-percentile point is $0.9 \times 0.9 = 0.81$. The probability that both are above the 90-percentile point is $0.1 \times 0.1 = 0.01$. The probability that one value is at or below this point and that the other is above this point is $(0.9 \times 0.1) + (0.1 \times 0.9) = 0.18$. All possibilities have been exhausted, which is indicated by the fact that $0.81 + 0.01 + 0.18 = 1.00$.

For the 97.47-percentile point, the probability is $0.9747 \times 0.9747 = 0.9500$ that both item values lie below this point. This value of 0.9500 is also the probability that the higher of the two values is at or below the 97.47-percentile point.

Therefore, for the higher of the two-item values, the investigator can be 100% certain that it lies at or below the 100-percentile point of the population, 95% certain that it lies at or below the 97.47-percentile point of the population, 81% certain that it lies at or below the 90-percentile point of the population, and only 1% certain that it lies at or below the 10-percentile point of the population.

Before undertaking sampling, the researcher should know that the higher (or highest) value in the sample will be equal to or less than the 100-percentile point of the population (i.e., the population's maximum value). With only two items, the researcher can be 95% confident that the higher value is at or below the 97.47-percentile point. Not much improvement is obtained in going from 100% to 97.47%. But what would the researcher expect with only two items?

Bigger Samples

A more meaningful sample size would be 28 items, for then the researcher could be 95% confident that the highest value in the sample lies above the 90-percentile point of the population. (For the connoisseur, $0.9^{28.4} = 0.05$.)

For any either/or, yes/no, belongs/doesn't belong variable, the characteristic distribution to be used is known unequivocally as a *binomial distribution*. Use of the binomial distribution is not a mere approximation as is the case so often with the Gaussian distribution. The shape of the distribution depends only on two parameters: f, which is the fraction of the population that consists of "yes" items, and n, which is the number of items in the sample. By using conventional statistical techniques that are simple and obvious but tedious extensions of the work done for the two-item sample, the confidence intervals for percentile estimates can be obtained. Table III in the appendix at the end of this chapter reveals the results of some of these extensions. Specifically, for the indicated sample size (n) and the indicated sample percentile, the 95% and 99% confidence intervals are given for the true population fraction (f).

The following tactics are recommended when the population of values is so non-Gaussian that knowledge of the average and standard deviation is not enough to characterize the distribution of values. First, the sample values should be arranged from smallest to largest. Then the values for the $(n/10)$th, $(n/4)$th, $(2n/4)$th, $(3n/4)$th, and $(9n/10)$th items in this series should be reported. These five values are, respectively, the 10-, 25-, 50-, 75-, and 90-percentile points of the sample. These five values serve as estimates of the corresponding percentiles in the population. Next, after the computations are done, the confidence intervals for the estimates should be reported. For example, with a sample size (n) of 20, the 10th smallest value should be reported as the estimate of the 50-percentile point. (For the connoisseur: A subtle distinction, of no relevance for this exposition, exists between the *median*, which has 50% of the values below it and 50% of the values above it, and the *50-percentile point*, which has 50% of the values at or below it.) This estimate for the 50-percentile point could in reality (i.e., in the population) be reported as anywhere from the 27.2-percentile point to the 72.8-percentile point, with a confidence of 95%. These confidence intervals for percentile estimates are painfully broad, but they do become narrower with increasing sample size. These intervals are the best that can be done if no knowledge of the population distribution is available. These intervals have the virtue of being applicable to any distribution; they have the vice of being broad.

A fruitful comparison can often be made between the observed sample percentile points and the percentile points predicted on the basis of a Gaussian distribution having the average and standard deviation calculated on the basis of the sample. Large discrepancies reinforce the fears that the population is not Gaussian. Concordance among the percentile estimates, on the other hand, might indicate that the population is Gaussian after all. (*See* Table III in the appendix at the end of this chapter.)

Conclusion

The four key points are as follows: (1) frame the questions properly, (2) obtain representative samples, (3) estimate confidence intervals about estimates of population average under the assumption of a Gaussian distribution for the population of averages, and (4) report certain percentile points and their confidence intervals especially if the population has an unknown or non-Gaussian distribution.

If the sample is representative, the variability in the sample will reflect the variability in the population. Large variability leads to large confidence intervals encompassing the average. This problem is often remedied by adequate sample size. If the values are not part of a Gaussian population, estimates of the percentile points may suffice to answer relevant environmental and regulatory questions. Huff (4) stated this point:

> By the time the data have been filtered through layers of statistical manipulation and reduced to a decimal-pointed average, the result begins to take on an aura of conviction that a close look at the sampling would deny... To be worth much, a report based on sampling must use a representative sample...from which every source of bias has been removed... Even if you can't find a source of demonstrable bias, allow yourself some degree of skepticism about the results.

The investigator never knows for certain and can only play the odds. Statistics helps the investigator play these odds in a rational way.

Abbreviations and Symbols

Δ	difference between two values
Σ	summation
C	concentration
f	fraction of a population
n	sample size
N	number of items or individuals
P	percent of population values
PCBs	polychlorinated biphenyls
RSD	relative standard deviation
S	standard deviation ($S = V^{1/2}$)
t	Student's t-value, usually accompanied by subscripts indicating the probability level and number of degrees of freedom involved
V	variance
$\overline{X}$	average
$X_{(i)}$	the ith value when the n values of X are arranged in increasing order from lowest to highest
Z	a value, X, expressed in standard deviation units ($Z = X/S$)

Appendix

Table I. Confidence Intervals for Gaussian and Chebyshev Conditions

Z	Gaussian Conditions	Chebyshev Conditions
0.1	8.0	NA[a]
0.2	15.9	NA
0.3	23.6	NA
0.5	38.3	NA
1.0	68.3	NA
1.1	72.9	17.4
1.2	77.0	30.6
1.3	80.6	40.8
1.4	83.9	49.0
1.5	86.6	55.6
1.645	90.0	63.0
1.8	92.8	69.1
1.96	95.0	74.0
2.0	95.5	75.0
2.1	96.4	77.1
2.5	98.8	84.0
3.0	99.7	88.9
4.0	100.0	93.8
5.0	100.0	96.0

NOTE: The confidence intervals are presented as the percentage of values between $X \pm Z \times S$, where X is the average, Z is a value expressed in standard deviation units, and S is the standard deviation. These are confidence limits for single values.
[a]NA indicates not applicable.

Table II. Percentiles for a Gaussian Curve

Z	Percentile
−1.282	10
−0.674	25
0.000	50
0.674	75
1.282	90

Table III. Sample Size Dependence of Confidence Intervals for Percentile Estimates

n	Sample Percentile	95% Confidence Interval		99% Confidence Interval	
		Lower Value	Upper Value	Lower Value	Upper Value
10	10	0.3	44.5	0.1	54.4
10	50	18.7	81.3	12.8	87.2
10	90	55.5	99.8	45.6	100.0
20	10	1.2	31.7	0.5	38.7
20	25	8.7	49.1	5.8	56.0
20	50	27.2	72.8	21.8	78.2
20	75	50.9	91.3	44.0	94.2
20	90	68.3	98.8	61.3	99.5
40	10	2.8	23.7	1.7	28.3
40	25	12.7	41.2	10.0	46.1
40	50	33.8	66.2	29.5	70.5
40	75	58.8	87.3	53.9	90.0
40	90	76.3	97.2	71.7	98.3
50	10	3.3	21.8	2.2	25.8
50	50	35.5	64.5	31.6	68.5
50	90	78.2	96.7	74.2	97.8
60	10	3.8	20.5	2.6	24.1
60	25	14.7	37.9	12.3	41.8
60	50	36.8	63.2	33.1	66.9
60	75	62.1	85.3	58.2	87.8
60	90	79.5	96.2	75.9	97.4
80	10	4.4	18.8	3.3	21.7
80	25	16.0	35.9	13.7	39.4
80	50	38.6	61.4	35.4	64.7
80	75	64.1	84.0	60.7	85.3
80	90	81.2	95.6	78.3	96.7
100	10	4.9	17.6	3.8	20.2
100	25	16.9	34.7	14.8	37.7
100	50	39.8	60.2	36.9	63.1
100	75	65.3	83.1	62.3	85.2
100	90	82.4	95.1	79.8	96.2

References

1. Whitaker, T. B. *J. Am. Oil Chem. Soc.* **1972**, *49*, 590– 592.
2. Horwitz, W. *Anal. Chem.* **1982**, *54*, 67A–76A.
3. Abbey, S. *Anal. Chem.* **1981**, *53*, 528A–534A.
4. Huff, D. *How To Lie with Statistics*; Norton: New York, 1954; p 18.

Bibliography

Bennet, C.; Frankin, N. *Statistical Analysis in Chemistry and the Chemical Industry*; Wiley: New York, 1966.

Cochran, W. G. *Sampling Techniques*; Wiley: New York, 1977.

Huff, D. *How To Lie with Statistics*; Norton: New York, 1954.

Huntsberger, D.; Billingsley, P. *Elements of Statistical Inferences*; Allyn and Bacon: Boston, 1973.

Langley, R. *Practical Statistics Simply Explained*; Dover: New York, 1970.

Zervos, C.; Fringer, J. J. *Toxicol. Clinical Toxicol.* **1984**, *21* (special symposium issue on "Exposure Assessment: Problems and Prospects").

RECEIVED for review January 28, 1987. ACCEPTED April 24, 1987.

Chapter 24

Sample Size

Relation to Analytical and Quality Assurance and Quality Control Requirements

John B. Bourke, Terry D. Spittler, and Susan J. Young

Biota present unique challenges to valid sampling because of the vast size differences between species, variations within a study population, and tissue differentiation. Several considerations dictate the number of specimens involved and place limits because of growth stage, habitat, and availability. Subsample size selection must consider the limitations of the analytical method, sensitivity requirements, and statistical requirements. Variation in sampling and subsampling methods for a study of grapes is discussed. No differences were found over a three-fold subsample size range. Preliminary results from a study of subsample homogeneity over the range of 2–100 g are presented. These field-treated produce reflect nonuniform pesticide residue distributions requiring careful preparation.

Preliminary Considerations

Samples are taken for a number of reasons, including monitoring, regulatory activities, quality control, scientific study, disaster assessment, or just idle curiosity. Whatever the reason, this first collection of any material for the analysis of one or more analytes is considered the *primary sample*. The method of collection and sample size are described by the sampling plan, which in turn is designed to ensure that this primary sample will be representative of the whole. In most instances, this primary sample will be sent to the laboratory or some intermediate location for subdivision into manageable-sized secondary samples.

No one set procedure or sample size is appropriate for all sampling events. Many factors must be considered before sampling; the most important factor is the objective of the study for which the sample is being taken. The size and method of sampling for monitoring studies are considerably different than those employed when attempting to describe some cause-and-effect relationship as is usually the case in either applied or basic research projects. The precision and accuracy required by the experimental protocol has an influence on sample frequency and size. The degree of validation needed to ensure the integrity of results will also influence sample frequency and replication. Data destined for litigation are produced under conditions more stringent with regard to sample identity and custody then are results collected for incorporation into a scientific data base.

Therefore, before a sample is taken, a sampling plan or protocol must be developed; the purpose for undertaking the study will form the basis of this plan. If, for instance, monitoring is the objective and the investigator wishes to know only if some analyte is present, a single grab sample will be all that is required, and the sample may be relatively small. However, the concentration of analyte found is representative of the sample only and not the environment from which it is taken.

In other chapters (Chapters 6, 11, 20, and 28), problems were presented regarding sampling and the determination of sample size necessary to ensure that a primary sample is representative when studying analytes in water and air. Some of the same sampling philosophy applies with respect to the sampling of plants, animals, microorganisms, or other parts of the biota. Obviously, when sampling biota, a number of factors must be included such as purpose of the study, homogeneity of the matrix to be sampled, concentration of the analyte, efficiency of the methods of extracting and concentrating, and sensitivity of the method to be employed. Nature of the organism and size of the population under consideration as well as availability and cost of the material to be studied are factors more unique to biosamples. In addition, analyte physical characteristics such as phase, volatility, oxidation state, chemical properties, and biological activity will have pronounced influences on sampling, and these characteristics specify when, where, and how much material must be removed from the environment to ensure sample validity.

Distribution

When the homogeneity of the matrix to be sampled is referred to, not only the matrix itself must be considered, but also the distribution of the analyte throughout that matrix. For instance, if investigators wish to determine the residues of a pesticide applied to plants in a field, they need to know something about how the pesticide was applied. If the application was done by an accurately calibrated and designed ground boom sprayer, where care was taken to avoid skips and oversprays, then the number of sampling points required to describe the area will be determined by the number and uniformity of the plant population. If, however, the pesticide was applied by a sprayer having poor nozzle design and positioning or with improper control over swath width and placement, a nonuniform distribution of pesticide will result, and considerably more sampling sites will be required to adequately assess the coverage. The resulting primary sample may then become large. Although this composite sample may accurately describe the average analyte concentration on the plants, it will not tell us anything about the variability. If that profile is desired, produce from a large number of sampling points must be kept separate and analyzed independently.

This problem can be illustrated by visualizing the results of spring-applied nitrogen to winter wheat. If each pass across the field is separated by a distance greater than the application swath width, alternating light and dark green streaks will be seen. These streaks are the result of an uneven distribution of nitrogen. A single sample from such a field will represent only that part of the field from which it was taken. Walking down the field parallel to the direction of application and taking a number of samples will result in analytical data representative of that particular swath only. Walking across the swath paths, however, and taking samples at a spacing unequal to the swath width will better represent the average analyte content of the field. Keeping the original objectives in mind for the sampling is important. If application uniformity is the subject of interest, each sample must be analyzed individually, and the sample size will approach that of a subsample. If the mean application is the subject of interest, for example, as in yield studies, then all the individual samples can be composited, the composite made as homogeneous as possible, and a subsample taken for analysis. Duplicate subsamples in this case tell something about the homogeneity of the composite, and the total composite sample size will be dependent on the total number of field samples taken.

Sample Size

The size of each sample will depend on a number of factors but should be kept at a minimum. For biological materials, the limiting factor may often be the availability of substrate. The collection of uncommon or even rare and endangered organisms often limits the sample to the material that suggested the need for analysis (i.e., a dead organism). In another biological material, the ability to capture organisms may be impaired by the population's habitat, size, or life cycle. The cost of some biotic materials may make the acquisition of large samples impossible; thus, resources can limit sample size. In these cases, the data may represent only the individual, primary sample and not the extant population. Extrapolations from these results to describe the world at large may be tenuous at best. When working with larger organisms, one must decide if it is the average occurrence within a population that is desired or the probability of a single organism containing the analyte. In the case of average occurrence, many organisms may be composited to give a single sample, and in the case of the probability of a single organism, each sample will, by definition, be derived from a single organism.

The concentration of the analyte may have a significant bearing on the size of the sample. If the analyte is in a very low concentration, the sample may have to be rather large to allow for the extraction and concentration of analyte. Occasionally, the entire primary sample may have to be used as the laboratory sample, and extraction carried out on the whole sample as obtained. The size of each unit within a sample also affects primary sample size. A primary sample of peppercorns will be much smaller than an equivalent sample of oranges, which will in turn be smaller than one sample comprised of watermelons. In regard to the distribution of analyte at very low concentrations within a sample, one orange in a boxcar represents 1 ppm, and one orange in 1000 boxcars represents 1 ppb. Such visualizations put in perspective some of the problems of sampling and analysis.

So far, the pulling or taking of the primary sample has been discussed. Such a sample may be made up of snakes or snails or even puppy dog tails. Whatever the composition, the primary sample must be of sufficient size to be representative of the system or population that the sampling plan wished to address and is generally far larger than can be used for individual chemical or physical analyses. These samples must then be reduced to replicates of a test size that can be analyzed in the laboratory. The manner in which this reduction takes place is critical to the eventual validity of the data.

Size Reduction

Air and water samples, once taken, are and generally remain homogeneous. Although this homogeneity can also be true of biological fluids, it rarely holds for other materials. Reduction can therefore proceed in two ways: nonhomogeneously or homogeneously.

Nonhomogeneous Reduction

First, a selected portion of the sample is isolated for analysis because that portion is either of special interest, is the known accumulation site, is the regulated commodity, or represents a limitation in the method. Arguments can be made that this separation process is merely generating more primary samples and does not really constitute a reduction. In instances where this takes place directly on-site, as in picking of fruit or foliage or in removing blood or tissue specimens from captured and released animals, the argument is probably valid. Other times, the segregation of the primary sample into differentiated tissues is an intermediate step performed either in the laboratory or at selected facilities. Fat is removed from fish samples for analysis of lipophilic chlorinated pesticides and polychlorinated biphenyls (PCBs); food commodities are reduced to the edible components for regulatory or nutritional testing. Frequently, a collection of excised parts will subsequently undergo a second or homogeneous reduction process.

Homogeneous Reduction

The sample (primary or secondary) can be made homogeneous by grinding, milling, blending, chopping, mixing, or any of a number of physical processes such that further subsampling by techniques of dividing, riffling, or aliquot taking will produce uniform portions of a size amenable to the analytical method and representative of the whole. The final subsample is called a test sample and its size will be primarily determined by the concentration of the analyte, the sensitivity of the method, and the capability of the analytical equipment. At any of these steps in which a solution is produced, the increment becomes homogeneous, and further reductions by aliquot are redundant and for convenience only. At this stage, replicate test samples are produced and checked, and storage and recovery spike material is prepared and incorporated into the analytical scheme. That topic has been thoroughly covered in prior chapters.

Case History

An interesting series of studies illustrating these concepts was conducted in cooperation with the Geneva laboratories over 25 years ago and pertained to the selection of grape berries for surface stripping of pesticides by solvent for residue determination (1–3). Analytical subsamples consisting of as many as 900 grape berries were stripped of DDT (1,1'-(2,2,2-trichloroethylidene)bis[4-chlorobenzene]) by tumbling in solvent. The average DDT residues for each sample size were essentially equal at two different residue levels for samples of 300 berries or more. The 300 count was thus retained as the standard sample size. Standardizing on a berry count rather than a weight was done because the increase in berry size by growth during the residue-weathering study's time frame resulted in a smaller arithmetic change in surface area. Had weight been the constant, progressive sample changes would have been much greater because of the geometric surface-to-mass relationship.

The selection method for the 300 berries was found to be critical. One experiment involved the designation of 20 consecutive vines from a 70-vine replicate plot. Three clusters were randomly picked from each vine, and five individual berries were snipped from random sections of each cluster; the total was 300 berries. Residues were found to be 9.2 ppm. When five berries were selected from each of 15 clusters derived from only four random vines in the row, the average residue was 28% lower (6.6 versus 9.2 ppm). As a check, all berries from all clusters on four random vines were separated and randomized; 300 berries randomly selected and analyzed from these were found to have a DDT concentration of 9.2 ppm. The efficient first method duplicated the very thorough but time-consuming third method. The four-vine shortcut employed in the second trial gave inadequate representative samples.

Analysis of the surface of the grapes only and the separation of the berries from row, vine, and cluster stems constitute nonhomogeneous reductions. Yet, the final objective was still the production of analytical samples representative of the entire crop.

Current Study

The limits of homogeneous reduction is the topic of a current cooperative study between the Food and Drug Administration and Cornell University, and some preliminary illustrative data are available (Young, S. J.; Newell, D. F.; and Spittler, T. D.; in press). Most applied pesticides, with the possible exception of some having systemic mechanisms, are unevenly distributed in and on the target. In these experiments, methoxychlor (1,1'-(2,2,2-trichloroethylidene)bis[4-methoxy-benzene]) is used as a foliar spray after the heads of cabbage are well-formed. Consequently, distribution on the leaf surfaces is not uniform and obviously heavier on the outer layers of the head. With the current trends toward microanalysis and solid-phase cartridge extraction, and the high costs of solvent purchase and disposal, a small subsample (e.g., 5.0 g) is more desirable than a large one (e.g., 500 g). But at what point do the fewer and fewer chopped pieces that make up size-reduced subsamples become nonrepresentative of the primary sample and give unacceptable replicate variation?

A large primary sample of methoxychlor-treated cabbage (ca. 10 kg) was thoroughly chopped, and 12 replicates of each of six subsample weights were obtained in a randomized manner. The subsamples were 100, 50, 25, 10, 5, and 2 g. Because this study was to test the representativeness of subsample size and not the miniaturization of the method, appropriate amounts of chopped, untreated check cabbage were added to bring the total test sample weight of each subsample to 100 g. Analytical results for this first series are shown in Table I along with the respective coefficients of variance (CV).

Although results for other commodities (e.g., green beans and apples) and full statistical analyses are incomplete, inspection reveals that at least in this instance, homogeneous reduction to 2 g is reasonable and representative of the primary sample. This statement does not mean that an analytical method based upon 2 g of extractable produce will give results of either this precision or

Table I. Means and Coefficients of Variation for Decreasing Subsample Weights

Sample Weight (g)	Mean (ppm)		CV[a]	
	Area	Peak Height	Area	Peak Height
100	8.882	8.874	4.638	4.421
50	9.066	9.048	3.251	3.146
25	9.062	9.063	3.650	3.703
10	9.399	9.401	5.116	4.978
5	9.525	9.510	4.784	4.915
2	9.883	9.899	6.011	5.725
	Overall mean[b] 9.296		Overall CV[b] 5.800	

[a]The abbreviation CV denotes the coefficients of variation.
[b]The overall value was calculated from 144 determinations.

accuracy, but rather that 2 g is a valid subsample size when employed with validated analytical procedures.

Size reduction is a determination that must still be made on the basis of each new set of circumstances, but eventually a reliable body of literature will accumulate to provide guidance and insight in establishing workable experimental designs.

References

1. Taschenberg, E. F.; Avens, A. W. *J. Econ. Entom.* **1960**, *53(2)*, 269–276.
2. Taschenberg, E. F.; Avens, A. W. *J. Econ. Entom.* **1960**, *53(3)*, 441–445.
3. Taschenberg, E. F.; Avens, A. W.; Parsons, G. M.; Gibbs, S. D. *J. Econ. Entom.* **1963**, *56(4)*, 431–438.

RECEIVED for review January 28, 1987. ACCEPTED April 20, 1987.

Chapter 25

Composite Sampling for Environmental Monitoring

Forest C. Garner, Martin A. Stapanian, and Llewellyn R. Williams

Guidance for selecting a plan to composite environmental or biological samples is provided in the form of models, equations, tables, and criteria. Composite sampling procedures can increase sensitivity, reduce sampling variance, and dramatically reduce analytical costs, depending on the exact nature of the samples, the analytical method, and the objectives of the study. The process of taking random grab samples and individually analyzing each sample for elements, compounds, and organisms of concern is very common in environmental, biological, and other monitoring programs. However, the process of combining aliquots from separate samples and analyzing this pooled sample is sometimes beneficial. The researcher must consider detection limits, probability of analyte occurrence, criterion level, sample size, aliquot size, analyte stability, the number of samples, analysis cost, sampling cost, monetary resources, biological interactions, chemical interactions, and other factors in order to make a wise decision of whether to composite or not, and how many sample aliquots to composite.

I N TYPICAL ENVIRONMENTAL, BIOLOGICAL, and other monitoring programs, each monitored site or individual is sampled, and each sample is analyzed individually. A chemical or biological measurement is thus obtained for each sample. This procedure is the optimal plan when a measurement is needed for every sample. However, combining aliquots from two or more samples and analyzing this pooled sample is advantageous at times. Such *composite sampling plans* can provide various advantages but should be used only when the researcher fully understands all aspects of the plan of choice. Composite sampling, as discussed in this chapter, applies to each of the following situations:

1173–6/88/0363$06.00/0 © 1988 American Chemical Society

1. when samples taken from varying locations or individuals need to be analyzed to determine if the component of interest is present or exceeds criterion limits in any of the samples;

2. when aliquots of extracts from various samples composited for analysis need to be analyzed to determine whether the component of interest is present or exceeds criterion limits in any of the samples;

3. when representativeness of samples taken from a single site, waste pile, product lot, household, community, or population need to be improved by reducing intersample variance effects;

4. when representativeness of random aliquots removed from a potentially heterogeneous sample needs to be ensured by reducing the effect of variance between aliquots;

5. when the material available for analysis in samples of necessarily limited size, such as blood samples, needs to be increased to achieve analytical performance goals; and

6. when the confidentiality of the individual donors of samples needs to be ensured.

Applications of composite sampling have involved sampling bales of wool (1–3); estimating plankton in freshwater (4–6); and determining pyrethrin levels in fruit (7), fat content in milk samples (8), pesticide levels in drinking water (9), and pH in soil samples (10). Sobel and Oroll (11) listed many industrial applications. Garrett and Sinding-Larsen (12) presented applications in geochemistry and remote sensing. Much of the pioneering theory of composite sampling was provided by Dorfman (13). Watson (2) derived equations for relative testing costs for some types of composite sampling plans. Watson discussed the problems of detection and false detection of factors and optimum group size in composite sampling. Connolly and O'Connor (8) compared random and composite sampling methods. Brown and Fisher (14) considered the problem of estimating the mean of the characteristic of interest from composited samples.

Potential Advantages

Composite sampling can be used to reduce the cost of classifying a large number of samples. Suppose the objective of the monitoring program is to identify samples of hazardous material, diseased individuals, or defective units of product when the a priori probability of hazard, disease, or defect is believed to be very low. For example, when checking human blood samples for syphilis, analysts may reduce cost by pooling aliquots from pairs of samples. On the rare occasion when a composite sample tests positive, the two original samples can be found and tested to identify the diseased individual. Over a large number of samples, the cost of such a plan would be little more than half the cost of analyzing every sample. Costs might be reduced further by pooling three or more samples. However, if too many samples are pooled, then expected cost may rise, and other limitations may be reached.

In any monitoring program, sample analyses may occasionally indicate the presence of analyte when none is actually present. Such *false positives* may be caused by random errors, instrument anomalies, or laboratory contamination. A false positive is less likely to occur in a composite sampling plan for classification than in ordinary sampling plans. If a composite tests positive, then each of the contributing samples is tested individually. A false positive will only occur in the unlikely event that both the composite and an individual sample both falsely test positive. Thus, composite sampling may also provide an added benefit by reducing the number of false positive results.

Composite sampling may also be used to reduce intersample variance due to heterogeneity of the sampled material. This reduction can be useful when the objective is to characterize a potentially heterogeneous material by estimating the mean concentration of analyte or total amount of analyte present in the material. For example, a hazardous waste site might be characterized by chemically analyzing 10 surface soil samples and averaging the result. However, the same value might be obtained by pooling the 10 samples and performing just one analysis. By pooling 10 sample aliquots and homogenizing the composite, the variance due to sample differences (site heterogeneity) is reduced by a factor of 10. If the variability between samples is much larger than the variability within samples, then one analysis of a composite of 10 samples will yield an estimate of the average concentration of the site that is of similar quality (equal mean and nearly equal variance) to that of averaging 10 sample analyses. However, the same amount of information is not obtained because no variance estimate is possible from one analysis and no unusual samples can be identified. Therefore, this composite sample plan is recommended only when funds are limited and analytical costs are much larger than sampling costs.

Composite sampling can also be used, with certain limitations, to increase the amount of material available for analysis. This procedure could potentially increase method sensitivity and achieve lower method detection limits, which would reduce the rate of false negatives. Additionally, composite samples may be used to ensure confidentiality of the donor. This purpose may be beneficial when social concerns may discourage certain classes of individuals from participating in a random survey because of fear of identification; composite sampling may help to preserve randomness and anonymity.

Potential Limitations

When the objective of the monitoring program is classification, sample compositing may result in dilution of the analyte to a level below the detection limit of the analyte. This dilution would result in an error in decision making known as a *false negative*. Suppose that the detection limit of a particular analyte in drinking water is 0.005 mg/L, and samples from five wells having true concentrations of 0.003, 0.015, 0.0000, and 0.001 mg/L are composited. Further, suppose that the criterion level for this analyte is 0.005 mg/L. In this example, the true concentration of the composite would be 0.004 mg/L. Therefore, the analyte may not be detected in the composite. Such a situation is unacceptable because one well has levels of the analyte 3 times the criterion level. This problem can occur when the number of samples composited exceeds the ratio of the criterion level to the detection limit. Care should be taken so that sample dilution does not substantially reduce the ability of the researcher to identify target analytes in the composite. Many definitions of detection limit exist, and the researcher must be certain that the detection limit used implies a high probability of detection in the sample matrices of interest.

If sampling costs are greater than analytical costs, then analyzing each sample individually may be more cost-effective. Garrett and Sinding-Larsen (12) presented statistical and cost-based approaches to optimal composite sampling plans that allow for nontrivial sampling costs.

When considering multiple analytes in a composite, information regarding the relationships among analytes in individual samples will be lost. The experimenter cannot be sure from which sample each analyte came or in what proportion with other analytes. Further, analytes or organisms in separate samples may be mutually destructive; therefore, after some time, the composite may not be representative of any of the individual samples.

Composite Sampling for Classification

Suppose the experimental objective is to classify each of N units into exactly one of two categories. These discrete categories might be detected and nondetected, diseased and nondiseased, or defective and nondefective. Each unit could be tested separately, which would require N tests. Alternatively, a group of units could be tested simultaneously, and the group would be classified by the test into one of the two categories. Such a composite sampling plan is appropriate for discrete measurement methods.

Other studies (2, 11, 13, 15) have considered the case where the number of defective units is a binomial variable. These studies used the expected number of tests for the entire population to determine the efficiency of composite sampling. In Dorfman's study (13), blood samples were pooled into groups. When a group was identified as defective, each blood sample in that group was tested individually. Li (16) proposed a multicycle version of Dorfman's method. Watson (2) derived equations for relative testing costs for discrete models and considered instances of zero and nonzero error variances. Watson discussed the problems of detection and false detection of factors and optimum group size. Hwang (3) proposed a method of composite sampling that does not require the assumption of binomial distribution and applies to the cases where either the distribution or an upper bound of the number of defective units is known. Hwang's algorithm for finding the optimum number of units to group into a composite sample that minimizes the total number of tests is also relatively efficient and simple to use. Mack and Robinson (*17*) researched the advantages of composite sampling in a national survey of pesticides in human adipose tissues and derived an approximate model of composite sample concentration from a model of individual sample concentration.

Now suppose that the objective of a sampling plan is to determine if the concentration of an analyte or organism exceeds a criterion level. For example, the criterion level may be the maximum concentration of iron deemed to be safe for drinking water. If each of N samples was tested individually, then N tests would be required. As for a discrete measurement method, use of a composite sampling plan for comparing continuous measurements to a criterion level may reduce the number of tests required. When determining the number of samples that can be pooled into the composite, researchers must consider the detection limit of the analyte. If the criterion level is near the detection limit of the analyte, then the number of samples that can be pooled into a composite is limited.

Thus far we have considered the measurement of only one analyte that will yield univariate data. The discrete and continuous measurement models can be extended to situations in which more than one analyte, organism, or attribute is measured. In such cases, a battery of tests is performed simultaneously. When a positive result occurs (discrete measurement model) or the level exceeds criterion (continuous measurement model) for any single test in the battery, then the n units are tested individually. If the multiple analytes are expected to be highly correlated, then a univariate approach may be useful.

In all of the classification models, the original samples are assumed to be homogeneous with respect to the analytes or organisms of interest.

Models for Discrete Distributions

The probability that the analyte or organism of interest occurs in a sample (p) may be estimated by the proportion of positive samples in a similar study, by presampling, or by any other reasonable procedure.

If the analyte or organism of interest is not found in the composite of n pooled samples, where n is the number of samples aliquoted and combined into each composite, then only one analysis needs to be performed. If the analyte or organism of interest is found in the composite, then $n + 1$ analyses must be performed. Hence, the total number of analyses (M) needed is 1 if the test of the composite is negative or $n + 1$ if the test of the composite is positive.

In order to design the sampling program that minimizes cost, the expected number of analyses, $E(M)$, must be calculated as follows:

$$E(M) = q^n + (n + 1)(1 - q^n) \tag{1}$$

where q is the probability that the analyte or organism of interest does not occur.

The relative cost factor (RCF) is the ratio of the expected number of analyses calculated by using composite samples to the number of samples that would have to be analyzed if samples were not composited. Therefore,

$$RCF = \frac{NE(M)}{nN} = \frac{E(M)}{n} = \frac{q^n + (n+1)(1-q^n)}{n} = 1 + \frac{1}{n} - q^n \qquad (2)$$

This result was also found by Dorfman (13). The optimal composite sampling plan is that which has minimum cost relative to a traditional plan in which all samples are analyzed. Table I provides the value of n that minimizes the RCF for a given value of q.

The model described by equation 2 (model 1) can be generalized to accommodate cases where more than one analyte or organism is tested in each composite sample. The resulting multivariate model (model 2) assumes independence among the various analytes.

Model 2 is a simple extension of model 1. In equation 2, the expected number of analyses is simply the product of the expected number of analyses for each of the individual analytes or organisms over the number of analytes or organisms of interest (m). Therefore, the relative cost factor is

$$RCF = \frac{E(M)}{n} = \frac{Q^n + (n+1)(1-Q^n)}{n} = 1 + \frac{1}{n} - Q^n \qquad (3)$$

where n is the number of samples aliquoted and combined into each composite, and Q is the product of q_i values, where $i = 1$ to m.

Table I. Optimal Values of n for Given Values of q or Q

p	q or Q	Optimal Value of n	Expected Savings[a] (%)
0.0001	0.9999	101	98.0
0.0002	0.9998	71	97.2
0.0004	0.9996	51	96.0
0.0008	0.9992	36	94.4
0.0016	0.9984	26	92.1
0.0032	0.9968	18	88.8
0.0064	0.9936	13	84.3
0.0128	0.9872	9	77.9
0.0256	0.9744	7	69.1
0.0512	0.9488	5	56.9
0.1024	0.8976	4	39.9
0.2048	0.7952	3	17.0
0.4096	0.5904	1	0.0

[a] Expected savings are the percent reduction of analytical costs expected when a composite sampling plan is used instead of when each sample is analyzed individually.

A limitation of model 2 is the assumption that the analytes of interest are not correlated. Extreme caution, therefore, should be exercised when this multivariate procedure is used. The optimal value of n can be obtained from Table I.

Models for Continuous Distributions

A model based upon a continuous distribution of measurements is appropriate when the concentration of the analyte or organism is measured and compared to a criterion level, rather than when the presence or absence of the analyte or organism is merely detected or recorded.

The criterion level for the composite sample should be determined so that the composite sample concentration will exceed the criterion level whenever any one sample exceeds the criterion level of the analyte or organism in the sample (c'). Therefore, the composite sample criterion (c) can be calculated as

$$c = \frac{c'}{n} \qquad (4)$$

Let q be the probability that $y < c$, where y is the concentration of the analyte or organism in the composite of n sample aliquots of equal mass or volume. The value of q can be estimated by integrating the probability density function of the composite sample results. This probability density function may be estimated by using some knowledge of the density function of results for individual samples, or by using the central limit theorem to justify normality of y. If the composite sample analysis yields a result less than c, then each of the n samples must have concentrations less than c'. If the composite sample analysis yields a result in excess of c, then each of the n samples must be analyzed to determine which samples contributed the analyte. If M is the number of analyses performed for each group of n samples, then $M = 1$ if $y \le c$, or $M = n + 1$ if $y > c$. Also,

$$\mathrm{RCF} = \frac{E(M)}{n} = \frac{q + (n + 1)(1 - q)}{n} = 1 + \frac{1}{n} - q \qquad (5)$$

The minimum RCF can be found by substituting various values of n. Restrictions on n do exist, however. If the method detection limit is D, then

$$n \le \frac{c'}{D} \qquad (6)$$

Model 3, which is described by equation 5, can be generalized to another model (model 4), where $m > 1$ analytes or organisms are measured. If aliquots are taken from n samples and combined for each composite, then the RCF is calculated according to equation 7.

$$\text{RCF} = \frac{Q + (n + 1)(1 - Q)}{n} = 1 + \frac{1}{n} - Q \qquad (7)$$

The probabilities q_i are estimated by integrating an estimated probability density function of the composite sample results. This density function may be estimated by using some knowledge of the probability density of the individual sample results or by using the central limit theorem assuming approximate normality of the composite sample results for each analyte. These steps can be simplified by assuming that the m analyte concentrations are independent, but this assumption may be violated in many applications. Such estimation is very involved mathematically. Therefore, a mathematical statistician should be consulted to ensure that good estimates are obtained for the probabilities q_i and the RCF. Such estimates could be used to obtain an optimal value of n, but a table of optimal values such as those in Table I, which applies to models 1 and 2 only, cannot be derived without specific knowledge of the statistical distribution of analyte concentrations.

The number of samples aliquoted and composited is restricted by detection limit considerations. The restriction is that n must be less than or equal to the minimum of the ratios of c'_i to D_i, where c'_i and D_i are the criterion level and detection limit, respectively, of the ith analyte from $i = 1$ to m.

Reducing Variance

The difference between the variances from individual and composite analyses can be illustrated with an example. If k samples are randomly taken from a hazardous waste site and each is quantitatively analyzed, then the mean of the k results is an estimator of the average concentration of the site. The variance of this estimator (σ^2_M) can be calculated as follows:

$$\sigma^2_M = \frac{\sigma^2_E + \sigma^2_H}{k} \qquad (8)$$

where σ^2_H is the variance between samples due to site heterogeneity, and σ^2_E is the random error variance of the analytical method.

If, however, equal aliquots are taken from each of the k samples, and these aliquots are pooled and homogenized, then one analysis of this composite sample will also be an estimator of the average concentration of the site. The variance of this estimator (σ^2_C) is

$$\sigma^2_C = \sigma^2_E + \frac{\sigma^2_H}{k} \qquad (9)$$

By comparing the estimator obtained by averaging k analytical results to that obtained by analyzing a composite of k samples, σ^2_M and σ^2_C will be very nearly equal if $\sigma^2{}_H$ is much greater than $\sigma^2{}_E$. On many occasions, the variance between samples at different locations of a waste site, or that between aliquots of a single sample, greatly exceeds the random error variance of the analytical method. In such cases, using a composite sample approach to estimate the average concentration at the waste site is cost-effective. Rohde (15) developed distributional properties of the analytical result of the composite sample. Rohde considered complex situations in composite sampling, such as the presence of more than one variance component.

Increasing Sensitivity

Composite sampling may also be useful in increasing analytical sensitivity and potentially reducing the rate of false negatives. Occasionally, an analytical method will require a larger sample than is possible from one individual. For example, if a method requires 2 L of blood to achieve the required limit of detection, then no living human could be adequately sampled. However, if four 0.5-L samples could be taken from four similar subjects, perhaps from the same household, then the desired sensitivity could be reached. Analysis of the composite sample would yield a result appropriate for the household as a whole, but not necessarily for each included individual. This type of plan is recommended only when the level of concern is below the limit of detection of practical individual samples, and a larger amount of material is needed for each analysis than is possible from a single individual.

Achieving Confidentiality

Occasionally, obtaining a random sample of individuals for testing may be difficult because of personal concerns about being identified as diseased. In such cases, composite sampling may be a useful technique when the objective of the study is to estimate the proportion of individuals that are diseased but not to identify these individuals. This objective may be beneficial when social concerns may discourage part of the population from participating in a random survey because of fear of identification. In such cases, composite sampling may help preserve randomness through assured anonymity.

For example, suppose that one wishes to estimate the proportion of individuals exposed to the acquired immune deficiency syndrome (AIDS) virus. Certain individuals, for social reasons, may be less willing to volunteer for individual testing; thus, a biased sample may result. By pooling groups of $n > 2$ samples for each test, no single individual can be identified as an AIDS carrier, and a less biased or an unbiased sample may result. The proportion of individuals that would have tested positive (p) can be estimated from the observed proportion of groups that test positive (P) by

$$p = 1 - (1 - P)^{1/n} \tag{10}$$

This estimator may also be useful to obtain a cost-effective estimate of the proportion of individuals possessing a characteristic (especially when sampling costs are much less than analytical costs) even when confidentiality is not a specified requirement.

Discussion

The prudent use of composite sampling can dramatically reduce analytical costs of classification experiments, reduce intersample variance in estimation experiments, improve sensitivity when individual sample size is limited, or improve the randomness of surveys through assured confidentiality. These benefits can be realized only under certain objectives and experimental conditions, and generally only after careful consideration on the part of the experimenter.

In classification experiments, where individuals are classified into exactly one of two groups, using composite sampling is typically cost-effective when there is a low probability of detecting analyte (models 1 and 2) or when determining analyte above the criterion limit (models 3 and 4). The optimal number, n, of samples to composite for models 1 and 2 is shown in Table I. Finding the optimal value of n for models 3 and 4 is more difficult because the solution must account for the exact statistical distribution of analyte concentrations. The experimenter must be careful that n does not exceed the ratio of the criterion level to the method detection limit.

In experiments where the objective is to estimate the mean concentration of analyte over a region or throughout a population, composite sampling will reduce the intersample variance. The result will be a more cost-effective estimate of the mean concentration. This procedure is especially useful when the cost of analysis substantially exceeds the cost of sampling. This procedure should not be used when an analytical result is needed for each sample.

Occasionally, achieving desired analytical performance, such as detection limits, with individual samples will be impossible because of sample size limitations. Using composite sampling to accumulate sufficient material may be beneficial to achieve the desired analytical performance. The experimenter must be cautioned that concentration estimates will not be available for each individual. Conclusions must be drawn for groups of individuals.

In random surveys, such as one to estimate the proportion of a population having a disease, composite sampling may be used to ensure anonymity of the sample donors. This use could help to achieve randomness of the sample when individual concerns of identification could cause a biased sample.

In any composite sampling plan, interactions among analytes must be carefully considered. Care must be taken to ensure that analytes or organisms from different samples will not be mutually destructive nor will create analytical interferences. If such problems are suspected, then the relevant procedures must be modified to eliminate them or render them irrelevant. Otherwise, composite sampling may create a material that is not representative of any of the original samples.

Abbreviations and Symbols

σ^2_C	variance of measurements of composite samples
σ^2_E	random error variance component
σ^2_H	heterogeneity variance component
σ^2_M	variance of the mean of k analyses
c	criterion level for a composite sample
c'	criterion level for a single sample
c'_i	criterion level for the ith analyte
D	method detection limit
D_i	method detection limit of the ith analyte
$E(M)$	expected value (mean) of the variable M
k	number of samples to be analyzed
M	number of analyses necessary to classify n composited samples
m	number of analytes measured in each analysis
N	total number of samples
n	number of samples composited
P	observed proportion of composites that test positive
p	probability of presence of analyte in one sample
Q	probability of simultaneous absence of all analytes
q	probability of absence of analyte in one sample
q_i	probability of absence of the ith analyte
RCF	relative cost factor, which is the expected cost of a composite sampling plan expressed as a proportion of the cost of a conventional sampling plan
X_i	concentration of analyte in the ith sample
y	concentration of analyte in the composite sample

Acknowledgment

The research described in this chapter was funded wholly or in part by the U.S. Environmental Protection Agency through contract number 68–03–3249 to Lockheed Engineering and Management Services Company, Inc. This chapter has not been subjected to Agency review and does not necessarily reflect the views of the Agency; therefore, no official endorsement should be inferred.

References

1. Cameron, J. M. *Biometrics* **1951**, 7, 83–96.
2. Watson, G. S. *Technometrics* **1961**, 3, 371–388.
3. Hwang, F. K. *J. Am. Stat. Assoc.* **1972**, 67, 605–608.
4. Cassie, R. M. In *Secondary Productivity of Fresh Water*; Edmondson, W. T.; Winberg, G. G., Eds.; International Biological Programme Handbook No. 17; Blackwell Scientific Publications: Oxford, England, 1971; pp 174–209.
5. Hrbacek, J. Ibid., pp 14–16.
6. Heyman, U.; Eckbohm, G.; Blomquist, P.; Grundstrom, R. *Water Res.* **1983**, 16, 1367–1370.
7. Ryan, J. J.; Pilon, J. C.; Leduc, R. *J. Assoc. Off. Anal. Chem.* **1982**, 65, 904–908.
8. Connolly, J.; O'Connor, F. *Ir. J. Agric. Res.* **1982**, 20, 35–51.
9. Bruchet, A.; Cognet, L.; Mallevialle, J. *Water Res.* **1984**, 18, 1401–1409.
10. Baker, A. S.; Kuo, S.; Chae, Y. M. *Soil Sci. Soc. Am. J.* **1981**, 45, 828–830.
11. Sobel, M.; Oroll, P. A. *Bell Syst. Tech. J.* **1959**, 38, 1179–1252.
12. Garrett, R. G.; Sinding-Larsen, R. *J. Geochem. Explor.* **1984**, 21, 421–435.
13. Dorfman, R. *Ann. Math. Stat.* **1943**, 14, 436–440.
14. Brown, G. H.; Fisher, N. I. *Technometrics* **1972**, 14, 663–668.
15. Rohde, C. A. *Biometrics* **1976**, 32, 273–282.
16. Li, C. H. *J. Am. Stat. Assoc.* **1962**, 57, 455–477.
17. Mack, G. A.; Robinson, P. E. In *Environmental Applications of Chemometrics*; Breen, J. J.; Robinson, P. E., Eds.; ACS Symposium Series 292; American Chemical Society: Washington, DC, 1985; pp 174–183.

RECEIVED for review January 28, 1987. ACCEPTED May 5, 1987.

Chapter 26

Considerations for Preserving Biotic Samples

Terry D. Spittler and John B. Bourke

Biological samples are noted for being much more susceptible to the decomposition of organic analytes than are other types. Even inorganic compounds can be altered by further reactions. Although this alteration may not affect the usefulness of many elemental inorganic determinations, anion determinations may be confounded by oxidation–reduction unless proper preservation measures are taken. The emergency situations under which some biomaterial is gathered frequently require that available rather than preferred measures be employed, and the method selected may preserve one component but destroy or compromise others. Several illustrative cases are presented.

SELECTION OF THE PRIMARY SAMPLE(S) is influenced by the objectives of the investigation (presuming that these objectives are known), and the whole series of protocol operations is essential to the production of uniform test samples (i.e., samples representative of the whole and homogeneous with respect to all analytes). Analyses, researchers trust, will proceed smoothly by well-documented methods under uncompromising quality assurance and quality control protocols to yield results of splendid accuracy and unblemished precision. However, a missing factor is that at all successive stages, from the removal of a biotic sample from its environment to its controlled disintegration in the first analytical steps, the integrity of the analytes of interest must be protected and preserved.

In a well-defined study, an appropriate sample preservation scheme will be incorporated. An appropriate sample preservation study is one that is within the physical limitations of the available equipment and compatible with the information or analytes sought. Selection of this scheme should include the following special considerations when dealing with biomaterials:

1173–6/88/0375$06.00/0 © 1988 American Chemical Society

- timing of sample movement,
- visual clues,
- segregation,
- necessary homogeneous reductions,
- knowledge of the analyte,
- containers and stability,
- storage stability assessment, and
- packing and transport.

Timing of Sample Movement

All or part of the sample must eventually reach the laboratory, but a first consideration is if the site should be marked or photographed for future reference before being disturbed. For most monitoring or field research, a site map or coordinate designation is sufficient. In emergency response situations, where suspected biotic victims are involved, complete descriptions of the circumstances at the site are essential. This close attention to prevailing conditions can be instrumental in distinguishing between a relevant sampling point and a coincidental event. For purposes of this discussion, all samples are assumed to be relatively inanimate and need not be taken alive.

Visual Clues

Will the sample have to be examined macroscopically or microscopically for deposits, wounds, burns, lesions, or pathology? Many of these conditions are indicative of chemical contamination, or, conversely, are symptomatic of diseases similar in appearance but unrelated to chemical exposure. Not only are chemists misled by biological phenomena or physical causes, but biologists also frequently ascribe pathological significance to symptoms that are wholly chemical in origin. Many honeybee kills occurring in the late 1960s and early 1970s were presumed to be the result of known viral or protozoal infections: these conclusions were based on the hive die-out pattern and absence of recent pesticide application actively in the vicinity. Only much later was it determined that the bees were collecting microencapsulated insecticides that they were mistaking for pollen grains, and returning to the hives (1, 2). Once placed in pollen storage cells, the insecticide eventually killed not only the existing colony but also destroyed subsequent colonies introduced to repopulate the hive after the "disease" had run its course. Suspicions were aroused when pollen taxonomists endeavoring to identify the plants from which pollen had been gathered started reporting the presence of unidentifiable 50-μm spheres, some full, some empty (3, 4). A great deal of apicultural literature had to be reevaluated.

Component Segregation

Are any nonhomogeneous reductions to be performed that would be complicated by an untimely or inappropriate procedure? Frequently, plant parts are separated in the field but at a point removed from the actual collection site; for example, whole carrot plants may be pulled, tagged, and moved to a service vehicle or field shed. There, tops and bottoms are severed and treated as two different commodities (i.e., a nonhomogeneous reduction). Had the carrot plants immediately been placed on dry ice in the field, the separation would have been complicated later in the laboratory.

Necessary Homogeneous Reductions

Homogeneous reductions may be similarly facilitated when large units (e.g., melons or squash) are held temporarily at ambient temperature and then sectioned, as discussed in other chapters (Chapters 9, 14, 24, and 29), before increments are bagged and preserved. This separation would be more difficult if the units had been frozen before this reduction, so anticipation of these useful procedures in the protocol can save later aggravation. Whole organisms having much less tissue homogeneity than a spherically symmetrical melon may also have to be handled. Beached whales, large game kills in areas of limited access, woodland defoliation, and California's entire watermelon crop are situations left to the imagination.

Knowledge of the Analyte

Knowing the identity and the nature of the analyte allows the luxury of having to be concerned with only one compound's stability. When this situation is not the case, either a compromise is made, or replicate samples are preserved by two or more means. The preservation scheme should be kept as simple as possible. Freezing is a logical first choice for retarding decomposition of most analytes. Elemental analyses do not strictly require that biomaterials be frozen, but it is probably a good idea: acid digestion and subsequent atomic absorption or inductively coupled plasma emission cation determinations are blind to the biomaterial's organic integrity (the analyst's nose may not be); however, anion measurements, particularly the multispecies schemes developed for ion chromatography, frequently are not. Nitrogen-, phosphorus-, and sulfur-containing analytes can undergo numerous redox transformations. In addition to freezing, pH adjustments can be made. Acidification is generally preferred unless NaCN or something forming a volatile acid is to be measured. Organic solvents can also act as preservatives, with the more volatile solvent being readily removed prior to digestion or extraction. Ideally, if the solvent selected is to be used later in the analytical scheme, complications are minimized.

Containers and Stability

Unless specifically prohibited for a particular analyte, biota samples should be placed and held in closed glass containers having inert [e.g., polytetrafluoroethylene (Teflon)] seals or cap liners. Instances have occurred of analytes decomposing on glass surfaces or being adsorbed from dilute solutions onto the vessel surface. Rumors of organic materials disappearing into plastic storage vessels also abound. These possibilities will not be discounted, but usually the migration of analytes from the matrix to or into the container will be slow at low temperatures. Decomposition of organic analytes is slowed by freezing, and the majority of preservation schemes consist of freezing alone. (The effectiveness is determined in a judicious program of freezer spikes wherein standards are added to uncontaminated check materials and held along with the field samples for later analysis to assess analyte disappearance.) Care must still be taken that containers are sealed. Some organic compounds sublime to the container sides or sublime out if the container is improperly sealed. If glass vessels are used, the first extraction solvent should rinse the biotic specimen and possible translocated analyte from the container into the first step of the analytical scheme. Depending upon the analyte and the detection method employed, other container materials such as polyethylene and polypropylene can also be solvent-rinsed. Chlorinated polymer extractives from plastic materials used to interfere disastrously with several chromatographic detectors, but newer laboratory products show little interference, and the practical advantages of plastic are considerable. Analysts still occasionally find phthalate plasticizers or halogenated scraps in samples; usually, these substances are derived from plastic laboratory tubing exposed to solvent or from household plastic materials wandering into a laboratory situation.

Many older analytical schemes for organic pesticide analyses recommend that samples be extracted with organic solvents as soon as possible and that the organic solutions be stored pending chromatographic or spectrophotometric analysis. For the stable (i.e., persistent) chlorinated hydrocarbon pesticides, this recommendation was reasonable. Before this option is considered, however, the analyte's storage stability in the selected solvent should also be ascertained. An instance of unexpected analyte (and standard) instability in solvent is currently annoying our laboratory. Ethyl carbamate is a suspected byproduct of fermentation and thus a contaminate in wines, cordials, and liquors, and has been found in bottled products aged several years (5). Yet, its stability in the laboratory in 10% ethanol/water is frequently only a few days, and in ethyl acetate, ethyl carbamate deteriorates overnight. In methylene chloride, standards have held steady for months. An ethyl carbamate standard's disappearance might be explained by transesterification in another alcohol/ester solvent, but in ethyl acetate or ethanol, all pathways lead back home. Also puzzling is the compound's apparent persistence for months and years in 10%–50% ethanol-containing beverages. Solving this problem is not necessary, but being aware of it is.

Direct collection into a preserving solvent is occasionally practiced, either because of a known rapid degradation (e.g., sodium azide-treated produce is harvested and dropped directly into jars of aqueous Na_2CO_3), or to prevent losses occurring because of delays in samples reaching frozen storage or a laboratory. Our laboratory participated as the analytical facility in a program where dead bees were field-collected by the beekeepers into jars containing 100 mL of acetone (200 bees/jar = 20 g), and sent to our laboratory several days later. Because an acetone blending was the first step in our multiresidue scheme, this step worked quite well (6). By insisting on supplying the jars and solvent, we were able to establish background purity of our materials before they went out; some control was therefore maintained. Allowing field operators to furnish analytical or collection components can lead to disaster. In the long run, money can be saved if the laboratory buys, tests, and supplies everything possible. One such disaster is to watch what happens to an electron-capture detector when a sample is injected that was collected in paint-store acetone and stored in a red-rubber-sealed mason jar. Hours or days later, the signal may come back down scale to zero.

Storage Stability Assessment

When uncontaminated check material is available (frequently it is not), analytes are known (they frequently are not), and standards of the analytes are readily obtainable (they frequently are not), then a storage stability study should be implemented as soon as possible after the primary sample collection, preferably when subsampling takes place. Standard analyte in a suitable solvent is added to replicate check subsamples and intimately mixed. Whereas the primary samples are probably being chopped or ground at this time, the mixing is not too difficult. On occasion, when the subsample contains undisrupted grains or small fruits, spikes are only surface-applied, and this application should be noted. A spike should never be subsampled! Aliquots should be taken only from a step in which the spike is in solution.

In the absence of untreated, uncontaminated check material, attempts at storage stability approximations are made by spiking surrogate material of similar matrix, or replicates of an actual sample are spiked at several levels in the hopes that a plot of analyte found against analyte added will go through the origin or intercept an axes at the original, but unknown concentration. This procedure is much like a standards addition determination except that the expression for degradation with time is unobtainable. This procedure may give a semiquantitative assessment of stability.

Sometimes storage stability approximations can't be made at all: An analytical scheme for the fungicide captafol (3a,4,7,7a-tetrahydro-2-[(1,1,2,2-tetrachloroethyl)thio]-1*H*-isoindole-1,3(2*H*)-dione) warns against chopping or disrupting the cellular structure of plant tissue samples until immediately before extraction and analysis (Helrich, K., Rutgers University, personal communication, 1983). The caution was issued because of suspected enzymatic breakdown of analyte. Storage assessment is thus precluded, unfortunately, in a situation where stability is most uncertain. A solvent-rinsable jar is always preferred for spiked checks and storage stability samples because any added analyte standard inadvertently placed on the walls is recoverable. Plastic bags, although handy, do not always offer this option, although it has been done.

Quality assurance and quality control protocols are frequently compromised during primary sample reduction when sample temperatures are allowed to rise unnoticed during chopping or grinding, or when merely sitting on the bench thawing in anticipation of the sample reduction. Usually, frozen materials must be softened somewhat before maceration, but they then should be immediately weighed into test-size replicates and refrozen. Grinding in dry ice is an often-cited option, but, inconveniently, accurate aliquots cannot be weighed until the dry ice completely dissipates, which usually means the bulk lot is warmed. Moisture also condenses on cold samples with or without dry ice and cannot only confound accurate sample weights but may have undesirable chemical effects.

Packing and Transport

Much of this topic is self-explanatory or so subject to local availability that few options are available. Most commercial carriers are fast and reliable as long as care is taken in the custody scheme to ship to recipients when they are available to receive the material. One common school of thought is that commodities, analytes, and items that common sense tells us are stable and traditionally held at ambient temperature can be simply boxed and sent by bus or common carrier. Potatoes fall under this designation, along with nuts and grains, for example. Packing can be more adventuresome, particularly when a crisis response places the investigator far from a supply of clean containers, insulated cases, dry ice, and workable permanent marking pens. The earlier-scorned household plastic can be pressed into service as can the expedient method of emptying available food containers such as glass jars. At least one full glass jar should be kept for background determinations in the event the cleaning out was less than thorough. Multiple layers of newspaper and cardboard make excellent insulation for ad hoc transporting of items.

A recent plane crash on the East Coast of Africa put 2000 gal of malathion ([(dimethoxyphosphinothioyl)thio]butanedioic acid diethyl ester) into a Senegal bay. The technical materials at hand were limited to several sizes of plastic bags, a freezer, and a picnic chest. No dry ice was available. A clever visiting scientist recognized the potential problem and instead of bagging small marine samples and freezing them, he placed them in a bag of the water from which they were obtained and froze the whole block. By so doing, two primary samples were well-protected. Others, like sand and sediment, were in small bags frozen inside large water samples. Upon their arrival at our laboratories 48–60 h later, the water samples were still partially frozen (Schaefers, G. A.; Spittler, T. D., Cornell University, unpublished data, 1986). The proper handling of biomaterial samples can come down to good judgment and ingenuity.

References

1. Stevenson, J. H.; Walker, J. *Bee World* **1974**, *55*, 64–67.
2. Stevenson, J. H.; Needham, P. H.; Walker, J. Rothamsted (Experiment Station) Report for 1977; Harpenden: 1977; pp 55–72.
3. Rhodes, H. A.; Wilson, W. T.; Sonnet, P. E.; Stoner, A. *Environ. Entomol.* **1979**, *8*, 944–948.
4. Burgett, M.; Fisher, G. *Am. Bee J.* **1977**, *117*, 626.
5. Ough, C. S. *J. Agric. Food Chem.* **1976**, *24*, 323–328.
6. Spittler, T. D.; Marafioti, R. A.; Helfman, G. W.; Morse, R. A. *J. Chromatogr.* **1986**, *352*, 439–443.

RECEIVED for review January 28, 1987. ACCEPTED April 20, 1987.

SAMPLING SOLIDS, SLUDGES, AND LIQUID WASTES

Chapter 27

Sampling Variability in Soils and Solid Wastes

Elly K. Triegel

Solids, sludges, and liquid wastes frequently exhibit wide variability in their properties and chemical compositions. Variations in the sampling program results may be a consequence of the inherent properties of the materials or an artifact of the sampling process itself. To design a sampling program to adequately characterize these variations, the goals of the project must be delineated, the most significant parameters selected, and an appropriate sampling plan and methods chosen. Interpretation of the data may involve statistical methods to quantify variations and relationships among variables. Confidence in the conclusions may be increased by more thorough characterization of variations, by the use of alternate means of data evaluation, or by the use of backup systems such as continued monitoring.

VARIABILITY IS INHERENT in naturally deposited solids (i.e., soils and rock) and disposed waste materials. This variation has a significant impact on the reliability of data generated and conclusions drawn from a sampling program. This chapter considers several aspects of the environmental sampling process: (a) the identification of the sources of variability and their significance, (b) the development of a plan to increase the level of confidence in the data, (c) the interpretation of the data collected during the sampling program, and (d) the impact of variability in the decision-making process.

In general, consideration is given to the broader definitions of level of confidence and the significance of data, rather than the more specific statistical tests. Examples in which sampling variability was a significant factor in the outcome of the program are given to illustrate some of the issues involved in each step of the sampling effort. These examples are followed by more general

1173–6/88/0385$06.00/0 © 1988 American Chemical Society

discussions of the various aspects of designing, implementing, and interpreting data on solids and waste materials.

Sources of Variability

In the initial stages of planning a sample collection program, identification of the potential sources of variability is critical. The nature of the variability may affect the number of samples to be collected, the method(s) of collection and analysis, and the overall design of the sampling program. The identification of the sources of variability and bias before starting field operations may eliminate the use of inappropriate collection and analytical methods or sampling intervals. Other parameters may be added to the program at this initial stage to facilitate the interpretation of the data variation. Finally, a phased approach, in which the sources of the variability are identified and characterized in the first phase, may be required.

Example

Soils at a proposed landfill site are being sampled to evaluate their ability to adsorb potential contaminants (chiefly metals) that may be present in the leachate. Soil borings are drilled, and samples of the soil are collected. The soils consist of partially weathered bedrock that has a weathered zone of highly variable thickness and rock fragment content. Recovery of samples from the rocky horizons is difficult and results in an undersampling of that horizon during drilling of the borings.

In a separate portion of the sampling program, soils are collected from test pits that allow better estimations of the rock fragment content and more complete sampling of the material. The soils are analyzed to determine the *cation exchange capacity* (CEC), which is a test performed on the fine-grained (<2 mm) fraction of the soils. Results of the analysis of the fine-grained fraction produce fairly consistent results. When the highly variable rock fragment content (10%–65% of the total by weight, with a CEC of essentially zero), is included, however, the CEC is found to be much lower than indicated in the analysis of the fine-grained fraction and is found to vary by over an order of magnitude.

Discussion

In any sampling program, variability may arise from (a) the nature of the solids, (b) the adequacy of the sampling population in representing the total population, and (c) the bias resulting from the methods of sampling and analysis. Frequently, the variations due to the heterogeneity of the materials exceed the deviations that result from the analytical method. Characterization of the nature and significance of the data variability may be difficult because of nonreproducibility of measurements, technical and cost considerations in collecting an adequate number of representative samples, and the potential environmental importance of outlier values.

In a sampling program, the mean and range of values is generally better defined when a large number of samples are collected (1-3). Eventually, the mean of the population approaches a limiting mean, or *statistically true value* (4), as the number of samples increases. The statistically true value is characteristic of the parameter being measured and the measurement process used. The usefulness of the statistically true value as an indicator of the actual value depends on the appropriateness and accuracy of the measurement process.

Even if determined precisely, the statistically true value may deviate from the true value, or *scientifically true value*, because of the failure of the sampling method. The method may exclude portions of the relevant population, include data not relevant to the assessment, use inappropriate methods of sampling and analysis, or bias the collection method such that certain parts of the population are systematically excluded. Examples of such processes include (a) the inability to collect coarse-grained materials in a borings program, as illustrated in the example; (b) the difficulty in distinguishing the boundaries of the population to be sampled, which results in oversampling or undersampling; and (c) the inappropriate analysis of samples composited from solids having significantly different properties.

Goals of the Sampling Program

Soils and waste materials frequently vary in a large number of their properties. Characterization of the variation in a cost-effective manner requires the identification of the goal(s) of the sampling program and the required level of confidence. The amount of data collected may then be reduced to that amount needed to answer the questions under consideration, excluding extraneous (though possibly interesting) information.

Example

A sampling program is undertaken at a former waste lagoon to sample the underlying soils and to analyze the extent of contaminant migration. The groundwater beneath the site has not been contaminated because of the long migration times of the contaminants. The principal goal of the program is to collect enough data to determine the depth of excavation required to remove contaminated material.

A preliminary analysis of existing data indicates that the wastes are strongly attenuated in the soil, and that the soil concentrations drop to background levels rapidly. In order to characterize the sorption process and the rate of decrease in the contaminant, sampling at 2-in.-deep increments is required. However, because excavation of the soils cannot be practically controlled to depth increments of less than 1 ft, the sampling intervals are changed to conform to the goal of the program (i.e., to determine the depth of excavation, not to quantify the migration process).

Discussion

Data collected during a sampling and analysis program may ultimately be used for a range of purposes. These purposes may vary widely in their requirements for confidence in the data. An essential part of planning the sampling program is to identify the goals of the program and the confidence levels needed. The data may be needed (a) to determine if a problem exists; (b) to monitor site conditions to determine if remedial action is required; (c) to design and implement remedial measures; (d) to assess the total exposure and risks to the population; (f) to determine appropriate treatment levels or monitoring criteria; or (e) to predict future conditions used in remedial design, litigation, or settlement procedures. Quantification of site conditions, in and of itself, is not generally a goal but may be a requirement in implementing one of the goals.

Ranking of Variables

Once the potential sources of variation are identified and the goals of the program determined, the parameters to be measured must be chosen. Frequently, the number of parameters that may be important exceeds the ability of the sampler to adequately characterize all the variables. A ranking of the variables may then be used to identify the most critical or sensitive parameters in order to narrow the sampling program to a manageable number of measurements or analyses.

Example

A sampling and testing program is being considered to evaluate the potential for desorption of mercury from a sediment and the possible contamination of an estuary. The site is characterized by sediments that range from a well-sorted sand to an organic muck. Redox conditions vary from oxidized surface water to reduced organic sediments. The goals of the site assessment are to determine if the current risk to the environment is significant and to recommend an appropriate remedial program.

The factors that might affect mobility of the mercury include the redox conditions; adsorption on colloidal iron, aluminum oxyhydroxides, and organic matter; pH; the surface area of the soil particles; the concentration and form of the mercury in the soil; the solubility reactions; coprecipitation with iron as an oxyhydroxide; potential volatilization; and biologically induced reactions (5).

Analysis of all the variables for all relevant soil types is thought to exceed the needs of the remedial program, to be excessive in cost, and to be subject to a great deal of uncertainty in the prediction of the net effect.

A ranking of the variables is undertaken to narrow the scope of the sampling plan. A model of the geochemical reactions indicates that two parameters significantly affect the mercury solubility and its environmental impacts. These factors are the concentration of the mercury in the sediment and the redox conditions. Sampling efforts are concentrated on delineating the spatial variability of these parameters, coupled with sparcer sampling to characterize a number of less significant parameters.

Discussion

In an assessment such as in the preceding example, a large number of factors may affect the environmental behavior of the contaminants. Ranking of the variables in terms of their influence on the mobility of the contaminant provides a means for reducing the number of parameters. Because the number of analytical parameters is reduced, the total number of samples can then be increased to provide a more representative sampling of all soil types.

Two considerations in ranking the importance of these variables are the range of values expected and the influence of this range on the outcome of the analysis. The ranking process may include a formal sensitivity analysis, where the values of relevant parameters are varied over the observed or expected ranges. The impact of their variation is assessed through a modeling effort. In the absence of a reliable model of the processes, ranking the parameters may be accomplished by past experience in similar cases, laboratory studies (e.g., leaching tests), or a more simplified analysis of each variable.

Design of the Sampling Plan

Several sampling designs may be considered to characterize the nature and variability of the soils or wastes. These designs depend on the nature of the variation and goals of the program. The principal factors include the number and distribution of sampling points and the measurements or analyses to be performed.

Example 1

Groundwater contamination has been found in the aquifer beneath an uncontrolled waste disposal area containing volatile organic compounds. A sampling program is undertaken to evaluate the extent of contamination in the wastes. No previous studies of the site wastes have been undertaken, but the wastes are viewed as a potential source of future contamination of the groundwater. The goal of the investigation is to assess the need and feasibility of removing or containing the waste contamination.

No data are available on the composition and concentrations of the contaminants in the waste at the site prior to sampling. Hence, no straightforward statistical method is available for determining the expected variations or the sampling density required to obtain a representative picture of the contamination. Instead, a sampling grid is established on arbitrarily chosen 50-ft centers to provide a random distribution of borings. A portable organic vapor analyzer with a gas chromatograph (GC) attachment is used to provide data on the types and order-of-magnitude concentrations of volatile compounds in each of the split-spoon samples taken.

The areal extent of the contamination is found to vary in a consistent and predictable manner, but the composition of the waste materials, as detected by the GC attachment, varies in the proportion of fuel oil products and chlorinated hydrocarbons. The sampling program for laboratory analysis of the wastes is modified in the field to incorporate the observed range of compositions and concentrations.

Example 2

A sampling program is designed to measure the annual variation of fertilizer loss from soils in a large section of the southeastern United States. The goal of the sampling and testing is to evaluate the overall risk of fertilizer use to the aquatic environment. A ranking of the factors affecting concentration indicates that the most significant parameter is the timing of a runoff event after application of the fertilizer. Historical records of the stream discharges are used to assess the variability of occurrence of a peak discharge soon after fertilizer application. The number of watersheds to be sampled is then calculated by using the criterion of a 90% chance that at least one watershed will experience a runoff event soon after application.

Discussion

In general, the design of the sampling plan should consider the statistical characteristics of the population to be sampled so that a large enough sample size is obtained (6, 7). In many cases of sampling solids or waste materials, however, little or no information is available on the expected deviation of the data. To some extent, this lack of information can be overcome by the modification of the sampling plan in the field such that the entire range of site conditions is represented in the final data set. The modifications may be based on field testing results or field observations by trained personnel.

The level of sampling effort to adequately characterize the site is also influenced by the type of data needed. Calculations may require an estimate of the mean or median of the population, or the distribution of values to be expected. Estimation of the total flow from a waste impoundment or landfill having an essentially homogeneous liner is an instance in which a value characteristic of the hydraulic conductivity of the liner is necessary. In an effort to reduce the consequences of overestimating or underestimating critical parameters, a *worst case* may be considered by using values that produce the most conservative results. In general, these values are chosen from the existing data set or are presumed from the commonly encountered range of values (e.g., the greatest expected hydraulic conductivity, if the type of liner material is given).

In contrast, determination of outlier values that are infrequent in their occurrence, but may have significant impacts on the conclusions, requires more extensive sampling to identify the anomalous values. Although these values may be infrequent, they may exert a disproportionate influence on the conclusions. Examples of these outlier values include the presence of vertical channelways for contaminant migration (e.g., root channels or vertical fractures in an otherwise low-permeability soil), and small volumes of highly toxic materials in a larger waste mass.

Sampling Method

Part of the design considerations in a sampling program is the selection of an appropriate sampling method. Sampling may be done on a random basis, or in deliberately selected areas. Both methods have specific advantages and disadvantages. The choice of the method depends on the overall goals of the sampling program.

Example

A field plot is being analyzed to determine the potential for movement of metals downward in the profile. The soils consist of a fine-grained, structureless lake sediment overlying a truncated B-horizon of an oxidized silt loam. A test pit exposes some of the vertical and lateral variations in the soil profile that are not easily observed in soil borings. In particular, a number of vertical channels are coated by brightly colored iron oxyhydroxides. This coating is indicative of preferential pathways of movement of iron and possibly other metals. An analysis of soils indicates a wide range of conditions, such as an Eh of 130–615 mV, where Eh is a measurement of the redox state, and a pH of 4.2–6.4. Sampling from specific locations of interest in the soils (e.g., channelways, oxidized subsoil, or reduced lake sediment) permits a more complete representation of the overall variability of the soil. Such sampling would also result in a more complete characterization of the less frequent vertical channels than would composite samples from soil borings. Care should be taken, however, to provide information on the frequency and magnitude of such outlier values, such that their relative importance can be evaluated.

Discussion

Selection of a sampling method to adequately characterize the nature of the soil or water materials is a function of the goals of the investigation, the variability of the parameter being measured, and the impact of the variation on the conclusions of the study. Samples may be collected randomly, or from specific areas deliberately chosen to represent the range of conditions observed in the field or unusual conditions thought to be of particular interest. In general, randomly chosen samples or samples composited from randomly selected locations are appropriate (a) for overall assessments of site conditions, (b) where variations are essentially random in nature, (c) where enough samples are available to sufficiently characterize the range of values, or (d) where outlier samples are considered either unlikely or unimportant.

Significance of the Observed Variation

After the sampling program is implemented and data collected, the significance of the observed variation in the data must be evaluated. Interpretation of the results may require the use of statistical methods.

Example

Methylene chloride is detected at low concentrations during several sampling quarters in a number of monitoring wells at a waste disposal facility. The principal question is whether the detection of the compound is indicative of a contaminant plume or a result of cross-contamination during the sampling and analysis program. Methylene chloride is thought to be potentially the most mobile contaminant at the site and should theoretically appear first at the leading edge of a plume.

Two groups of wells, upgradient and downgradient of the disposal areas, are evaluated to determine the frequency of detectable levels of methylene chloride. The frequency of occurrence of the compound is found to be the same in both groups of wells (i.e., the probability of finding the compound upgradient of the site is similar to the likelihood of detecting it downgradient of the site). In addition, the occurrence over time in each well is generally nonrepeating and random. Hence, the compound is probably the result of cross-contamination. Although trip blanks are included in the analyses, the number of blanks is considerably less than the total population of analyses, and the conclusions drawn from the blanks have a lower level of confidence.

Discussion

Interpretation of data that are variable in nature requires identification of the source of the variation, an assessment of the adequacy of the characterization of the variation, and the evaluation of the significance of the range of values obtained. A number of statistical methods are available to quantify the variation or trends (*8–10*), though care must be taken to ensure that the data meet the requirements of the statistical test used. A common example is the misapplication of a test requiring a normally distributed data set (e.g., Student's *t*-test) to an abnormal data set. The data can be checked for normality by the useful graphical technique, the Lilliefor's test (*11*). Another example is the pooling of errors (e.g., sampling and analytical errors) without testing the validity of the procedure.

Outlier values (i.e., anomalous data) are probably the most difficult to analyze and evaluate (*12*). Such data may be a result of (a) measurement error, (b) actual site conditions that are not of particular importance in the evaluation of site conditions, or (c) real and significant conditions at the site that affect data interpretation. Criteria used to distinguish these possibilities may or may not include reproducibility, frequency of occurrence, magnitude of the deviation from the norm, results of sensitivity analyses, impacts on the goals of the sampling program, and relevance to the on-site conditions or operations. Because many environmental analyses are concerned with low levels of contamination, the effects of anomalous conditions may be of considerable significance from an environmental standpoint. Examples of significant but rare occurrences include the presence of occasional highly permeable vertical channelways (e.g., old tree roots), or "hot spots" of very high concentrations or very toxic materials in a waste mass.

Increasing the Level of Confidence

A number of approaches can be used to increase confidence in the conclusions drawn from the results of the sampling program. These approaches include (a) increases in the number of samples or tests conducted; (b) adoption of a multifaceted approach that examines the site by using several methods of analysis; and (c) use of backup systems, such as monitoring, to reduce any potential negative effects of inadequate or incorrect sampling, analysis, or data interpretation.

Increasing the number of samples collected or the number of tests conducted is a straightforward method of achieving a statistically true value. The method does not necessarily result in a more scientifically true value or a more reliable evaluation of the data, particularly if the uncertainty arises from method bias. However, many cases occur in which the field observations or the results of a phase-one sampling program make the expansion of the sampling program the logical choice (*13*).

A multifaceted approach to the analysis of the existing data consists of using all available tools to evaluate the data and the reasonableness of the results. These tools include field observations, laboratory results, historical data, consideration of theoretical relationships, results of published laboratory models, mathematical modeling, statistical analyses, and past experience at similar sites or under similar conditions. Of necessity, comparison of conclusions generated by using these different tools must be qualitative to some extent. The comparison is most useful in identifying potential errors in the data, assumptions, or conclusions that may not be obvious if only one approach is used. The confidence in the results of an analysis using only one or two of these techniques is generally low, even if the data are reliable.

In certain cases, such as in the settlement of cost allocations and remedial performance standards, the level of confidence required is high. The costs and time required to more completely understand the variation in site conditions should be considered in terms of the need to proceed with facility design or remedial efforts and the appropriate allocation of the resources available. Confidence in the adequacy of the site assessment and the prediction of future conditions can be increased by the continued monitoring of the site such that violations of health or other criteria are detected and the performance criteria of the remedial program are achieved.

References

1. Greminger, P. J.; Sud, Y. K.; Nielsen, D. R. *Soil Sci. Soc. Am. J.* **1985**, *49*, 1075.
2. Albrecht, K. A.; Logsdon, S. D.; Parker, J. C.; Baker, J. C. *Soil Sci. Soc. Am. J.* **1985**, *49*, 1498.
3. Ameyan, O. *Soil Sci. Soc. Am. J.* **1986**, *50*, 1289.
4. Kempthorne, O.; Allmaras, R. R. In *Methods of Soil Analysis*; Black, C. A., Ed.; American Society of Agronomy: Madison, WI, 1965; Chapter 1.
5. Triegel, E. K. In *Attenuation by Soils of Selected Potential Contaminants from Coal Conversion Facilities: A Literature Review*; Oak Ridge National Laboratory: Oak Ridge, TN, 1980; p 29; ORNL/TM-7249.
6. Holcombe, L. J.; Eynon, B. P.; Switzer, P. *Environ. Sci. Technol.* **1985**, *19*, 615.
7. Schweitzer, G. E.; Black, S. C. *Environ. Sci. Technol.* **1985**, *19*, 1026.
8. Addiscott, T. M.; Wagenet, R. J. *Soil Sci. Soc. Am. J.* **1985**, *49*, 1365.
9. Davidoff, B.; Lewis, J. W.; Selim, H. M. *Soil Sci. Soc. Am. J.* **1986**, *50*, 1122.
10. Yeh, T.-C. J.; Gelhar, L. W.; Wierenga, P. J. *Soil Sci.* **1986**, *142*, 7.
11. Iman, R. L.; Conover, W. J. In *A Modern Approach to Statistics*; Wiley: New York, 1983; p 153.
12. Dixon, W. J. In *Methods of Soil Analysis*; Black, C. A., Ed.; American Society of Agronomy: Madison, WI, 1965; Chapter 3.
13. Barcelona, M. J.; Gibb, J. P.; Helfrich, J. A.; Garske, E. E. In *Practical Guide for Groundwater Sampling*; U.S. Environmental Protection Agency: Ada, OK, 1985; Section 4; EPA/600/2-85/104.

RECEIVED for review January 28, 1987. ACCEPTED April 29, 1987.

Chapter 28

Relations of Sampling Design to Analytical Precision Estimates

Larry J. Holcombe

The frequency of sampling or the spatial distribution of samples required to obtain a representative sample depends on the variability of the waste composition and the precision of the required waste analysis. An example study conducted at a coal-fired power plant is described to illustrate the relation of sampling designs to precision requirements. To achieve at least 50% precision in a 30-day average coal ash composition, the optimal sampling design required seven samples, each analyzed twice, for a total of 14 analyses over the 30-day period. The approach presented in this chapter can be used for other waste sampling programs to assist in sampling design.

COMPARED TO ANALYTICAL RESEARCH, little research has been conducted on sampling design and its relation to the quality and precision of a waste composition analysis. This chapter briefly addresses some of the work underway on waste sampling requirements, namely, work sponsored by the American Society for Testing and Materials (1), the Electric Power Research Institute (EPRI) (2), and the U.S. Environmental Protection Agency (3).

The frequency of sampling and the number of composites required to obtain a descriptive sample of a waste depend on the following:

- the time or spatial variability of the waste composition,
- the time span or spatial area that the samples are to represent, and
- the precision of waste analysis that is required (e.g., if a hazardous constituent is present in the waste at levels near the regulatory

1173–6/88/0395$06.00/0 © 1988 American Chemical Society

limit or another limit of concern, then greater precision will be
required than if the levels are well below the limits of concern).

To obtain a representative sample of a waste, the concentration levels and
approximate variation in the constituents of concern should first be estimated.
In some cases, an estimate of the characteristics of a waste can be made on the
basis of knowledge of the processes that produce the waste. This method of
designing a sampling strategy is commonly referred to as engineering judgment.
Engineering judgment is often sufficient if the goal is simply to establish whether
a waste meets some set criteria, such as the Resource Conservation and Recovery
Act (RCRA) hazardous criteria. However, a more exacting approach to sampling
design is needed for wastes produced from complex and variable processes, or
if precision requirements are more demanding.
    A more exacting approach to sampling design first involves defining the
time period or volume of the waste for which a composition estimate is desired.
If the waste is not homogeneous, then suitable strata may be defined, and each
is handled separately. Differences in levels of chemical constituents and in
variability of the constituents may provide a basis for stratification.
    Next, a sampling unit is defined. This unit is the *primary sampling unit*,
which may be a unit of time if the waste is a flowing stream, or a volume if the
waste is in an impoundment.
    Finally, by using the estimated sample composition and variance either
from a pilot sampling effort or from engineering judgment, the number of
samples required to achieve the desired precision in waste composition can be
estimated by using fundamental statistical concepts as follows (financial
constraints are not considered):

$$n = t_{0.80}^2 S^2 / d \tag{1}$$

where n is the appropriate number of samples to be collected, and $t_{0.80}^2$ is the
square of the tabulated value of Student's t-statistic for a two-sided confidence
interval and a coverage probability of 0.80 for the unknown mean, where the
degrees of freedom are defined for S^2, which is used to estimate the population
variance δ^2. (S^2 is a preliminary estimate of the population variance (δ^2)
obtained from previous samplings, from a pilot sampling effort, or from other
information such as the likely range of the population values.) The symbol d
denotes the deviation from the mean, which is exceeded only in two cases out
of 10 in repeated sampling (i.e., the precision requirement). For environmental
sampling, the precision requirements may depend on how close the waste
composition is to a regulatory threshold.
    Although the use of Student's t-distribution is based on an underlying
normal distribution for the measurements, the robustness of the t-statistic may
be relied upon for this use. If information seems to indicate that normality may
not be assumed, then a goodness-of-fit test should be performed to determine
if the assumption of a normal distribution is reasonable. The Lilliefors'

goodness-of-fit test is an example (4). If the goodness-of-fit test shows that a normal distribution does not adequately fit the data, then further sampling and analytical considerations will be required.

A recent sampling study (2) is presented as an example of the relation of sampling design to precision requirements. EPRI sponsored this study to determine the variability of elemental concentrations in coal ash and thus assist in the creation of sampling design for power plants (2). EPRI recognizes the potential magnitude of the effects of coal ash disposal because over 70 million tons of coal ash are disposed of each year (5). A 500-MW power plant may generate between 150 and 300 tons of coal ash per day. Because of the large volumes of coal ash generated at a given plant, reliable and efficient means of obtaining representative samples are important. The representative samples may then be used to determine average chemical and physical properties for purposes of marketing the ash or for regulatory purposes.

A slightly more rigorous approach to determining the number of samples, n, required to achieve a desired precision was used than that just presented in equation 1. In this case, maximizing the sampling efficiency was accomplished by determining both the optimum number of samples required and the number of analyses to be performed per sample. To determine these numbers, the components of variability associated with sampling and analysis had to be estimated. These components include the following: S_T^2, the time variation in the mean concentration of samples taken on different sampling days; S_S^2, the sampling variation between samples taken at the same sampling point at the same time; and S_A^2, the variation between sample results due to analysis. The sum of these components of variability equals the total sample variability S^2, which is defined as

$$S^2 = S_T^2 + S_S^2 + S_A^2 \tag{2}$$

Given this information, the most efficient sampling design can be established to meet the precision requirements. The first step in the EPRI-sponsored study was to conduct a pilot sampling effort to collect and analyze coal ash samples over a set time period. Then, a variance component analysis was performed to estimate the variance in measured sample compositions attributable to time, sampling, and analysis. Finally, sampling designs were established for meeting the desired level of precision.

Pilot Study

Site Description

The coal-fired power plant used in the study is located on the east bank of a lake in the southeastern United States. The plant has one unit with a maximum generating capacity of 950 MW and was placed in commercial operation in 1967.

Figure 1 presents a simplified schematic diagram of material flow through the plant and identifies locations used for collecting samples of the feed coal and precipitator ash. Coal is transported by train, and the plant uses about 8200 tons of coal per day. From the coal pile, the coal is conveyed by belt to the powerhouse, where it is crushed and loaded into bins that hold a 1–2-day supply. The crushed coal passes through the pulverizer, where it is reduced to a fine powder. From the pulverizer, the coal is mixed with hot air and blown through burners into the boiler furnace, where the mixture ignites. The heavy ash drops to the bottom of the boiler and is disposed of. Fly ash is carried out of the boiler by the flue gas and subsequently removed by electrostatic precipitators (ESP). The precipitator ash is collected in hoppers beneath the ESP and drawn by vacuum to be slurried with water. The slurried ash travels through a glass-fiber pipe to the ash ponds.

Sampling Procedures

Sampling was conducted from January 9, 1982, to October 19, 1982, with some down time for scheduled plant maintenance during February, March, April, and May 1982. Precipitator ash was collected from the ash sluice line at a point about one-quarter mile from the ESP. The sluice line at this point was under a positive head pressure of approximately 20 psig; therefore, pumping was not required to obtain a sample of the slurry.

A probe was permanently fixed to the sluice line. Figure 2 illustrates the probe assembly. All parts were constructed of high-pressure poly(vinyl chloride) to avoid metal contamination. To sample, a polypropylene line was attached to the faucet, and the faucet was turned on for 1 min to purge the probe and line. The end of the line was then inserted into an 8-L linear polyethylene (LPE) bottle, and a sample was taken. This sample size was necessary in order to obtain between 200 and 400 g of dry precipitator ash. The LPE bottle was taken back to the laboratory, where the ash mixture was allowed to settle for 1 h. The liquid was filtered off (Whatman no. 41 filter paper) by vacuum, and the solids were dried overnight in a 35 °C (95 °F) convection oven. The dried sample was sealed in a bottle and labeled for analysis.

Sampling Design

The pilot study sampling design allowed for calculations of time variability between samples over lag times ranging from 24 h to approximately 3 months. Other components of sample variability besides time—sampling, sample preparation, and sample analysis—were also calculated from this design as follows: The sampling episodes were grouped in 3-day periods. Within a 3-day period when samples were taken, a specified number of composite and replicate samples also were prepared. The composite sample consisted of equal portions of day 1, 2, and 3 samples mixed thoroughly by riffling. This composite was

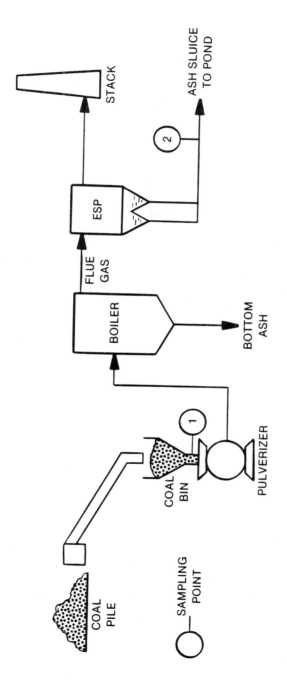

Figure 1. Locations of coal and precipitator ash sampling points.

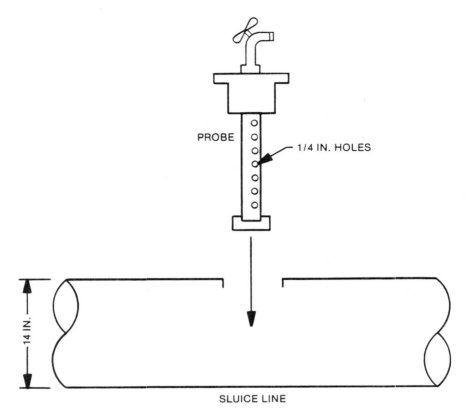

Figure 2. Precipitator ash sampling probe.

then split for duplicate analysis. On day 2 of each 3-day period, a replicate sample was collected to determine time-zero variability, or the variability associated with the sampling procedure. This replicate was collected in a separate bottle immediately after the primary sample was taken on day 2.

Analytical Procedures

The variability over time of the coal and ash compositions was calculated on the basis of total chemical concentrations and on concentrations measured in water extracts. The elements analyzed in the ash digests (6) and extraction procedure extracts (3) were the eight trace elements regulated under RCRA (arsenic, barium, cadmium, chromium, lead, mercury, selenium, and silver) and additional elements that are minor and major mineral constituents of the ash. Analytical techniques used in this project included inductively coupled argon plasma emission spectrometry (ICAP), atomic absorption (AA) spectrometry, and ion chromatography. Detection limits by ICAP for four elements (lead, arsenic, selenium, and mercury) were too high to provide quantitative results. Therefore, samples were analyzed by AA for these elements.

Variance Component Analysis

The replication design for the sampling plan in this study allowed estimation of three components of variability for the noncomposite samples: S_T^2, S_S^2, and S_A^2 as defined in equation 2.

These three parameters plus the overall mean, m, were estimated by maximum likelihood methods (2). [This approach wouldn't be useful when dealing with trace anthropogenic compounds (e.g., volatile organic compounds).] Situations with a large number of detection limit values (i.e., >50% of the data) were not analyzed for variance components. In other cases, the detection limit observation was included in the analysis with a value that was the median probability point of the data below the detection limit, according to the fitted normal-probability-density function.

The results of this analysis are given in Tables I and II, which list for each element the mean concentration and measurement units, and the three components of variability expressed as a proportion of the total variance.

$$100(S_i^2/S^2) \tag{3}$$

S_i^2 is the variance for each individual component, and S^2 is the total variance as defined in equation 2.

In order to further quantify the effect of time on the correlation between trace element compositions in samples, a variogram analysis was performed on the data. This analysis investigates the persistence of element concentrations over time. In a previous study conducted at a different power plant (7), the variability in ash composition was found to be time-dependent, and a variogram analysis was required to provide a rigorous fit of the data. In this study, the majority of elements exhibited constant variability with time. And, although the variance components were fitted by a variogram function in this study, the variability can be reasonably described by a constant function.

Sampling Designs

The data on variance components presented in Table II could be used to estimate the number of samples to achieve a required precision by using equation 1. However, the construction of a representative sampling plan involves a trade-off between precision and cost. The objective is to estimate from sample data the average composition of a waste stream for a given time. The average composition should be within a specified precision at least cost.

Table I. Percentage of Variance from Each Source
of Daily Samples, Where Preparation Equals Digestion

Element	Mean Conc.	Total Variance[a]	Percentage of Total Variance		
			Time	Sampling	Analysis and Preparation
Aluminum	15.27 wt %	0.0007	13.0	71.8	15.0
Arsenic	100.7 μg/g	0.0043	32.1	41.9	25.8
Barium	566.3 μg/g	0.0115	85.6	13.8	0.5
Beryllium	22.74 μg/g	0.3403	97.5	0.5	1.9
Calcium	4124 μg/g	0.0060	79.3	4.5	16.0
Chromium	167.7 μg/g	0.0009	0.0	75.9	24.0
Cobalt	83.32 μg/g	0.0052	65.6	33.2	1.0
Copper	186.9 μg/g	0.0032	56.2	36.3	7.4
Iron	2.305 wt %	0.0035	47.6	50.3	2.0
Lead	154.6 μg/g	0.0005	0.0	83.1	16.8
Magnesium	4553 μg/g	0.0038	76.2	20.8	2.9
Manganese	74.61 μg/g	0.0196	63.9	1.7	34.2
Mercury	0.2423 μg/g	0.1341	1.5	91.1	7.2
Molybdenum	32.33 μg/g	0.0075	54.2	0.0	45.7
Nickel	145.1 μg/g	0.0038	45.2	41.5	13.1
Potassium	2.059 wt %	0.0036	86.6	11.3	2.0
Selenium	14.33 μg/g	0.1118	92.9	5.9	1.1
Silicon	21.96 wt %	0.0035	0.0	94.9	5.0
Sodium	2167 μg/g	0.0235	0.0	0.0	0.0
Sulfur	412.9 μg/g	0.0223	8.5	0.0	91.4
Titanium	1.12 wt %	0.0021	64.0	30.4	5.5
Vanadium	262 μg/g	0.0020	35.3	55.3	9.2
Yttrium	80.59 μg/g	0.0049	45.3	51.1	3.5
Zinc	255.8 μg/g	0.0040	23.2	67.1	9.5
Gross alpha	39.68 pCi/g	0.0654	0.0	0.0	0.0
Gross beta	68.18 pCi/g	0.0149	0.2	0.0	99.7
Radium 226	3.57 pCi/g	0.1486	0.0	51.7	48.2

[a]The total variance is computed as the sum of all variance components divided by the square of the mean.

Mathematical Description

Let C_D equal the cost of daily sampling; C_A equal the cost of sample splitting, preparation, and analysis per split (costs of compositing sample material are assumed to be relatively negligible); and d_1 equal the specified statistical precision requirement expressed as a percentage of the estimated concentration. Also, let the waste stream variability characteristics for each chemical equal S_A^2, which is the variance between determinations from splits of a single homogenized sample or composite; let $S_S^2 + S_A^2$ equal the variance between determinations from simultaneous samples of unit volume V_0; let $S_D^2 + S_S^2 + S_A^2$ equal the variances between determinations on different sampling days from samples of unit volume V_0, where S_D^2 is the time variance in sample concentrations; and let m equal the concentration mean value.

Table II. Percentage of Variance from Each Source
of Daily Samples, Where Preparation Equals Extraction

Element	Mean Conc. (µg/mL)	Total Variance[a]	Percentage of Total Variance		
			Time	Sampling	Analysis and Preparation
Aluminum	0.3945	0.2421	51.1	40.0	8.7
Arsenic	0.0488	0.4595	58.4	30.2	11.3
Barium	0.4021	0.1742	70.7	21.8	7.4
Calcium	5.144	0.1754	71.1	26.7	2.1
Copper	0.0091	1.5445	1.5	85.5	12.8
Lithium	0.0063	1.4100	0.0	18.2	81.7
Magnesium	0.9537	0.1164	77.6	14.9	7.4
Manganese	0.0163	0.2948	33.2	62.9	3.7
Molybdenum	0.0856	0.3832	49.8	36.3	13.7
Nickel	0.0039	0.4849	8.5	23.0	68.4
Potassium	0.4296	1.7302	0.0	91.5	8.4
Selenium	0.2193	0.1609	82.8	12.6	4.4
Silicon	0.863	0.2008	71.3	26.4	2.2
Sodium	0.6217	0.0779	0.0	43.9	56.0
Sulfate	7.22	0.1086	77.4	9.3	13.2
Sulfur	2.565	0.1072	57.4	23.5	19.0
Vanadium	0.0666	0.3237	60.7	20.0	19.2
Zinc	0.0079	0.5451	47.6	43.5	8.8

[a]The total variance is computed as the sum of all variance components divided by the square of the mean.

For a constant variability with time, the optimal sampling design is defined by the following parameters: T is the specified length of the time interval for which average concentrations need to be estimated; N_D is the number of sampling days spread evenly over the time interval of length T (i.e., N_D subintervals each of length T/N_D having a sampling point at the center of each subinterval); N_A is the total number of sampling splits analyzed; and V is the total "sample" volume per analysis, which is to be made as large as practical. The analysis and sampling variance components can be combined to yield $S_{AS}^2 = S_A^2 + S_S^2/(V/V_0)$, where V_0 is the sample volume used in this study (1 gal). Because the cost structure is relatively insensitive to sample volume, the total cost (C) may be written approximately as

$$C = N_D C_D + N_A C_A \qquad (4)$$

The relative precision of estimation may be expressed as

$$d_1 = (1.645/m)[(S_D^2/N_D) + (S_{AS}^2/N_A)]^{1/2} \qquad (5)$$

where m is the mean concentration. This equation has the same form as equation 1. In this case, d_1 has been used to define the precision requirement: $d_1 = d/m$. [An upper 95% confidence limit for the mean is approximately $m(1$

$+ d_1$).] Also, the value 1.645 is used in place of Student's t-statistic because in this case, the distribution is normal and has greater than 30 samples. For specified precision, the total cost is minimized mathematically when

$$N_D = \frac{S_D{}^2 \, [(C_A/C_D)^{\frac{1}{2}}(S_{AS}/S_D) + 1]}{(md_1/1.645)^2} \qquad (6)$$

$$N_A = \frac{S_{AS}{}^2 \, [(C_D/C_A)^{\frac{1}{2}}(S_D/S_{AS}) + 1]}{(md_1/1.645)^2} \qquad (7)$$

If $N_A \geq N_D$, then on each sampling day, N_A/N_D samples are drawn, each of maximum volume V. If $N_A < N_D$, then on each sampling day, one draws a single sample of volume $V(N_A/N_D)$ and composite samples from N_D/N_A consecutive sampling days. From each sample or composite, a single split is taken for analysis. The minimized cost (C^*) will then be

$$C^* = \frac{[S_{AS}(C_A)^{\frac{1}{2}} + S_D(C_D)^{\frac{1}{2}}]^2}{(md_1)^2} \qquad (8)$$

The mathematical calculation will need to be rounded up because both N_A and N_D need to be integers. Also, either N_A/N_D or N_D/N_A needs to be an integer. This requirement results in precision slightly better than d^2 and a total cost slightly higher than C^*. Because the variability parameters for different chemical species are likely to be different, it may be necessary to choose a sampling plan that meets estimation precision requirements for the most variable chemical species of interest.

The precision d_1 does not depend on the length T of the sampled time interval. This independence is an approximation that is reasonable provided the time variability of the chemical species in question is constant over such a time range. If T is either very short or very long, then this approximation could break down. For many of the chemicals in this study, a fairly constant variability was observed over a range from a few weeks to a few months.

Examples

To demonstrate the application of these formulas, examples of sampling designs are presented. Assume that the object of interest is measurement of the average extract element concentrations over a 30-day period for the set of elements found in the ash extracts. The sampling cost is estimated to be about $60 per visit, and the analysis cost per sampling for the whole battery of elements is estimated to be about $240 (i.e., $C_A/C_D = 4$). A modest increase in the sample volume was assumed from that used in this study (i.e., from 1 to 3 gal). The two levels of desired precision considered were for $d_1 = 0.30$ and $d_1 = 0.50$.

For each element, the cost equations were used to find the minimum cost sampling plan meeting the desired precision. The minimum was obtained over all plans with an integral number of samples and analyses and an integral ratio of either N_D/N_A or N_A/N_D. Tables III and IV give the optimal values of N_D and N_A for the plan for each individual element, the cost of that plan, and the achieved precision of the plan.

The analysis was then extended to consider all elements simultaneously to produce an overall sampling plan. The sampling plan that achieves the desired precision for all elements is listed at the bottom of each table, along with the cost. The achieved precision of this plan for each element is also tabulated. Most elements actually achieve a precision somewhat better than the target. If lithium were dropped from the list of elements, these plans would become cheaper by about a factor of 2.

If even less precision is needed (i.e., if the observed concentrations are well below the regulatory limit), then even less sampling would be necessary.

Conclusions

The statistical methods presented can be used to estimate components of variability in waste composition. With this information and knowledge of the

Table III. Optimal Sampling Designs, Where Preparation Equals Extraction and Delta = 0.30

Element	Mean Conc. ($\mu g/mL$)	Number of Samples	Number of Analyses	Total Sampling Cost ($\$$)	Achieved Precision of Indiv. Plan	Achieved Precision of Overall Plan
Aluminum	0.3945	12	3	1440	0.28	0.19
Arsenic	0.0488	8	4	1440	0.30	0.15
Barium	0.4021	6	1	600	0.30	0.11
Calcium	5.144	6	1	600	0.29	0.13
Copper	0.0091	3	15	3780	0.30	0.19
Lithium	0.0063	6	36	9000	0.30	0.30
Magnesium	0.9537	2	1	360	0.27	0.08
Manganese	0.0163	1	2	540	0.27	0.06
Molybdenum	0.0856	10	5	1800	0.30	0.19
Nickel	0.0039	7	14	3780	0.30	0.19
Potassium	0.4296	3	18	4500	0.28	0.20
Selenium	0.2193	3	1	420	0.27	0.10
Silicon	0.863	8	2	960	0.29	0.20
Sodium	0.6217	1	2	540	0.27	0.06
Sulfate	7.22	8	2	960	0.28	0.20
Sulfur	2.565	8	2	960	0.29	0.17
Vanadium	0.0666	8	4	1440	0.29	0.16
Zinc	0.0079	24	6	2880	0.30	0.27
Overall Sampling Plan		12	36	9360		

Table IV. Optimal Sampling Designs, Where Preparation Equals Extraction and Delta = 0.50

Element	Mean Conc. (μg/mL)	Number of Samples	Number of Analyses	Total Sampling Cost ($)	Achieved Precision of Indiv. Plan	Achieved Precision of Overall Plan
Aluminum	0.3945	4	1	480	0.49	0.26
Arsenic	0.0488	2	2	600	0.47	0.21
Barium	0.4021	1	1	300	0.44	0.15
Calcium	5.144	1	1	300	0.49	0.17
Copper	0.0091	1	6	1500	0.47	0.31
Lithium	0.0063	2	14	3480	0.48	0.48
Magnesium	0.9537	1	1	300	0.32	0.11
Manganese	0.0163	1	1	300	0.37	0.10
Molybdenum	0.0856	4	2	720	0.47	0.26
Nickel	0.0039	5	5	1500	0.47	0.30
Potassium	0.4296	1	6	1500	0.49	0.32
Selenium	0.2193	1	1	300	0.38	0.13
Silicon	0.863	3	1	420	0.45	0.26
Sodium	0.6217	1	1	300	0.37	0.10
Sulfate	7.22	3	1	420	0.44	0.26
Sulfur	0.6217	1	1	360	0.50	0.23
Vanadium	0.0666	2	2	600	0.47	0.22
Zinc	0.0079	10	2	1080	0.50	0.37
Overall Sampling Plan		7	14	3780		

precision required in the analysis, the number of samples and analyses required to achieve a given precision in waste composition can be estimated. In the example study, which involved sampling and analysis of coal ash from a power plant, the variability in sample composition was attributable to all three components of variability (time, sampling, and analysis) to different degrees depending on the chemical element. For the majority of elements, time and sampling contributed to a large portion of the overall variability. Therefore, the sampling designs included a greater emphasis on collecting the samples as opposed to analyzing the samples.

Abbreviations and Symbols

δ^2	population variance
AA	atomic absorption (spectrometry)
C	total cost
C*	minimized cost
C_A	cost of sample splitting, preparation, and analysis per split
C_D	cost of daily sampling
d	precision requirement

d_1	relative precision of estimation
EPRI	Electric Power Research Institute
ESP	electrostatic precipitator
ICAP	inductively coupled argon plasma emission (spectrometry)
LPE	linear polyethylene
m	mean concentration
n	appropriate number of samples to be collected
N_A	total number of sampling splits analyzed
N_D	number of sampling days
psig	pounds per square inch gauge
RCRA	Resource Conservation and Recovery Act
S^2	preliminary estimate of variance
$S_A{}^2$	variance between determinations from splits of a single homogenized sample or composite
$S_{AS}{}^2$	variance of analysis and sampling
$S_S{}^2$	variations between samples taken at the same sampling point at the same time
$S_T{}^2$	time variation in the mean concentration of samples taken on different sampling days
T	time interval length
t	Student's t-statistic
V	total sample volume per analysis
V_0	sample volume used in the example study

Acknowledgments

The example study was the result of research on coal ash variability funded by the Electric Power Research Institute. I thank Ishwar Murarka of EPRI for his assistance in the research. Also, coresearchers on the project and coauthors of earlier reports were Barry Eynon of SRI International and Paul Switzer of Stanford University. Janis Carter of Radian reviewed the chapter before submission.

References

1. Holcombe, L. J.; Lorenzen, D.; Johnson, D. In ASTM Special Technical Publication 933; Lorenzen, D.; Conway, R. A.; Jackson, L. P.; Hamza, A.; Perket, C. L.; Lacy, W. J., Eds.; American Society for Testing and Materials: Philadelphia, PA, 1986; Vol. 6.
2. Eynon, B. P.; Switzer, P.; Holcombe, L. J. *Time Variability of Elemental Concentrations in Power Plant Ash*; Electric Power Research Institute: Palo Alto, CA, 1984; EA–3610.
3. *Test Methods for Evaluating Solid Waste*, 2nd ed.; Office of Solid Waste and Emergency Response. U.S. Environmental Protection Agency: Washington, DC, 1982.
4. Lilliefors, H. W. *J. Am. Stat. Assoc.* **1967**, *62*, 399–403.

5. Murarka, I. P. In *Proceedings of Utilization, Treatment, and Disposal of Waste on Land*; Soil Science Society of America: Madison, WI, 1985.
6. McQuaker, N. R.; Kluckner, P. D.; Chang, G. N. *Anal. Chem.* **1979**, *51*, 1082–1084.
7. Switzer, P.; Eynon, B. P.; Holcombe, L. J. *Pilot Study of Time Variability of Elemental Concentrations in Power Plant Ash*; Electric Power Research Institute: Palo Alto, CA, 1983; EA–2959.

RECEIVED for review January 28, 1987. ACCEPTED May 8, 1987.

Chapter 29

Preservation Techniques for Samples of Solids, Sludges, and Nonaqueous Liquids

Larry I. Bone

Although preservation techniques for solid, sludge, or nonaqueous samples have not been established, a few practices such as minimizing holding time, refrigeration, sealing sample bottles, and minimizing headspace are usually helpful. If samples are to be extracted prior to analysis, good preservation practice might involve putting the sample into the extraction medium in the field. Field preservation of samples in methanol prior to a volatile priority pollutant analysis that involves a methanol extract (SW-846 Method 5030) is probably a reasonable practice. The methanol helps preserve the sample and allows the laboratory selection of a more representative subsample. Selection of subsamples for actual analysis is quite critical for heterogeneous materials.

GOOD, RELIABLE ANALYTICAL DATA are very difficult to obtain from solids, sludges, and, to some extent, nonaqueous liquids. Sampling, preservation, and analytical procedures are better understood and are much more reliable for water and air than for solids. Not only are the methods for water and air better, but samples of these media are easier to spike, split, and dilute. The nonhomogeneous nature of the types of samples that are the subject of this section is at the heart of many problems.

Not only do scientists not know how to get good analytical results for solids and sludges, they are not even sure how to define the extent of the sample to be analyzed. For example, do those responsible for such investigations want to find the average concentration of some species over the entire solid, the worst-

1173–6/88/0409$06.00/0 © 1988 American Chemical Society

case concentrations, the best-case concentrations, or something in-between? If investigators want an average over the entire sample, how large would a soil sample be? The entire Earth? Of course not. But where does the sample stop? An additional problem with heterogeneous solids and sludges is that discontinuities or variable contamination can often be seen. Everyone who has ever collected solid samples for chemical analysis has succumbed to the temptation to include "that little green spot" in the portion of the solid chosen for analysis. This inclusion immediately establishes a sample bias and suggests that a prescribed technique must be used regardless of what is seen if the investigator really wants an unbiased, but not necessarily representative, result.

This discussion has quite a bit to do with sample preservation. Before the proper technique can be chosen for any step in the sampling sequence, the investigator must somehow describe the universe to be sampled and decide what question the analytical results are expected to answer. Sometimes for hazard evaluations or plume definition in soils, a worst-case analysis is desired. That is, if a seam is present in the waste or soil, the scientist wants to analyze that seam rather than obtain an average result over the entire sample. In many cases, however, investigators may want the analytical result to be an average over the entire sample. In either case, deciding who is to make that decision is important: the sampler in the field or someone in the laboratory? If someone in the laboratory is chosen, the investigator may want to send an undisturbed sample to the laboratory without preservation. If the sampler in the field is chosen, the investigator will want to subdivide the sample in the field and preserve it. Often, samples are subdivided in the field and again in the lab without any real coordination. The result is that no one knows what kind of bias might be built into the results.

Prior Planning Is Essential

All phases of any sampling exercise need to be thoroughly preplanned. The goal of the sampling exercise must first be defined, and then its extent must be decided upon. The decision must be based on an honest assessment of the real question that the analytical results are supposed to answer. Data gathered simply for the data's sake are seldom of much real value.

The sampling plan must include all phases of the sampling exercise including containers, labels, field logs, sampling devices, blanks, splits, spikes, size of samples, composites, preservation techniques, chain-of-custody procedures, transportation, sample preparation in the field and in the lab, analytical procedures, and reporting. The sampling plan is probably best prepared by involving people from every step in the process, from the field to the laboratory, because decisions to be made on all of the various elements of the plan are interrelated. The plan should, however, have enough flexibility to allow for changes based on information collected in the field. No matter how much thought is given to possible conditions that may be encountered during the field work, unexpected conditions usually arise. The need for field splits, spikes, or composites will have a bearing on preservation methods. Preplanning is particularly important for preservation because preservatives may be put into the sample containers prior to taking them to the field. Other supplies for preservation such as ice, coolers, and sealing tape must also be available.

Standard Preservation Techniques

No prescribed preservation techniques for solids, sludges, or nonaqueous samples exist for either organic or inorganic analysis. A few helpful practices are widely accepted. These practices are to seal containers, minimize headspace, refrigerate samples during storage and transportation, and analyze the samples as soon as possible.

The need to analyze samples of this type as soon as possible cannot be over-emphasized because there is usually no way to preserve them. If the analysis requires an extraction or digestion, carrying out this step as soon as possible is usually acceptable. The sample can then be held for the standard holding time specified by the method. Even if solid samples are analyzed quickly, collecting, transporting, and storing them in an undisturbed condition is usually best if possible. For core samples from waste site investigations, either the entire core or at least a large portion of it is shipped to the lab to be sampled just prior to workup and analysis. These cores can be shipped either in the sampler, sealed in wax, wrapped in foil, or at least sealed in bottles. A 1-pt wide-mouth bottle is particularly useful. A portion of a core can be cut and trimmed so that it nearly fills the bottle. When the sample is to be analyzed, the analyst bisects the core lengthwise and takes the analytical sample from the center of the core. The analyst is to select the sample uniformly over the length of the split core if an average result is desired. Occasionally, however, a worst-case analysis is desired. In this case, the analyst selects a spot or a seam where the sample appears to be the most contaminated.

A few difficulties are involved in handling core samples this way. If a contract laboratory is being used for the analysis, the investigator has very little input into the choice of the actual sample to be tested; the investigator is at the mercy of the analyst. Human nature is such that the analyst may very well bias the sample if the sample is visibly heterogeneous. Of course, so would the person who is asking the analytical question, but at least the investigator would be aware of any bias he or she introduced. Consequently, selection of subsamples should be documented (i.e., a description of the subsample selected should be prepared by a trained observer).

The reason samples are preserved is to prevent any chemical change that might take place between the time the sample is taken and the time it is analyzed. The most frequent causes of these types of changes are volatilization, biodegradation, and oxidation–reduction. Both volatilization and biodegradation can be reduced by storing and transporting the samples at a reduced temperature. The standard ice temperature of about 4 °C is probably the most reasonable to use. Lower temperatures could reduce biological degradation and might reduce volatilization. However, freezing water-containing samples might fracture the sample or cause a slightly immiscible phase to separate and ultimately result in the release of volatile compounds. This opinion is based on the fact that lowering the temperature on some liquid-phase-packed chromatographic columns to dry-ice temperature will decrease retention times. This decrease probably results from solidification of the liquid phase.

Sealing samples and minimizing headspace is always a good idea. Anaerobic samples should not be exposed to air during storage and transport if a possibility of aerobic biodegradation or chemical oxidation exists. Sealing and reducing headspace will minimize the loss of volatile compounds.

Preservation of Samples of Volatile Organic Compounds

The most difficult problem that I have experienced in analyzing solid waste or contaminated soil samples is obtaining reliable reproducible analytical results for volatile organic compounds. In order to characterize wastes and contaminated soils at the Petro-Processors of Louisiana Superfund site, we have been attempting to run standard priority pollutant analysis for the volatile compounds on solid samples. An example of an actual blind split analysis that we carried out probably best illustrates the problem as well as suggesting a preservation technique that might be helpful.

During the field investigation of the Petro-Processors site, we wanted to check the validity of the analytical results received from a contract laboratory by running a blind split in our own Dow Louisiana Division environmental laboratory. I carefully chose a sandy sample from the bottom of a contaminated pond because the sample was easier to homogenize than most of the clay or silt samples we encountered, and the sample was also easy to split. In addition, the sample contained enough contamination to be well within the range of reliable analytical work.

The contract laboratory and the Dow laboratory ran metals and acid and base-neutral extracts by standard priority pollutant protocols and obtained good agreement. However, very poor agreement was obtained on the volatile organic compounds, as can be seen in Table I.

The contract laboratory analyzed for volatile compounds by placing a few tenths of a gram of the sample into 20 mL of water in a 25-mL vial. Volatile compounds were then sparged from the vial onto the 2,4-diphenyl-p-phenylene oxide resin (Tenax) adsorption column of a commercial purge-and-trap apparatus from which the samples were thermally desorbed and analyzed by gas

Table I. Volatile Analysis of a Soil Sample

Compound	Contract Laboratory	In-House Laboratory
Carbon tetrachloride	0.10	1853
1,2-Dichlorobenzene	0.10	7
1,4-Dichlorobenzene	0.10	2
1,3-Dichlorobenzene	0.10	4
1,2-Dichloroethane	0.10	525
1,1-Dichloroethane	0.10	4
1,2-Dichloropropane	0.10	664
Methylene chloride	0.10	8.3
1,1,2,2-Tetrachloroethane	80	1630
Tetrachloroethene	84	664
Toluene	13.5	2.0
1,1,2-Trichloroethane	92	3097
Trichloroethene	1.9	1740

NOTE: All results are in milligrams per kilogram.
SOURCE: Adapted from reference 1.

chromatography–mass spectrometry (GC–MS) Method 624 (2). The Dow laboratory extracted 4 g of the sample into 20 mL of methanol and then put 50–250 μL of the extract into water; this method is similar to SW–846 Method 5030 (3). The Dow samples were screened by purge-and-trap analysis on a gas chromatograph with Hall and photoionization detectors in series. (A Hall detector is an electrolytic conductivity cell.) The water-diluted methanol extract was also analyzed by purge-and-trap Method 624 (2). The contract laboratory reanalyzed their sample by using the Dow extraction method. In this later method, they used purge-and-trap gas chromatography–flame ionization detection (GC–FID) to analyze for the major purgeable compounds. They found higher concentrations than previously, but the results were still not in very good agreement with the Dow results.

We concluded from the results of Table I that the principal reason for the gross differences in the analytical results lies in the laboratory selection of the subsample to be analyzed from this heterogeneous solid. Selection of a larger sample for extraction followed by a split of the more homogeneous extract gives considerably more reliable results. Had the sample been even more heterogeneous, as most of ours were, the results probably would have been more divergent. Of course, even selection of a 4-g sample will give widely scattered results from a very heterogeneous sample. The lesson is that the full sampling process, completely through the laboratory subsampling, must be planned such that the sample selected is large enough to be representative of the universe from which the results are desired.

We also concluded that the base-neutral and acid extract results were a much better measure of the level of soil contamination at the Petro-Processors site than the volatile compounds regardless of the method used for the volatile analysis. In fact, when the U.S. Environmental Protection Agency subsequently analyzed soil samples from the site, they chose to only analyze for base-neutral and acid samples for exactly the same reason.

The experience just described also contains a plausible suggestion for preserving solid samples for volatile analysis. If the sample is to be extracted into methanol, the methanol might as well be added in the field to preserve the sample. A small wide-mouth bottle could be filled about one-half to two-thirds full of methanol in the laboratory prior to going into the field. The bottle and the methanol can then be preweighed so that the weight of sample to be added later can be calculated. In the field, a carefully selected, representative sample of about one-fourth the weight of the methanol is added to the sample bottle. The bottle is sealed, cooled to 4 °C, and transported to the laboratory. The filled bottle is again weighed in the laboratory to calculate the weight of the sample added. The laboratory can also finish the extraction by shaking, tumbling, or sonification. An appropriate ratio of methanol extract to water can be selected to achieve analytical sensitivity in the desired range.

As SW–846 Method 5030 (3) suggests, poly(ethylene glycol) probably could be used in place of distilled-in-glass methanol. In addition, tetraglyme might also be acceptable. Other cases may exist where a solvent to be used in a method might be added in the field to aid in preservation. Another example may be the n-hexadecane method for total volatile content described in SW–846 Method 8240 (3).

References

1. *Remedial Planning Activities Report*; Petro-Processors, Inc., NPC Services, Inc.: Baton Rouge, LA, 1985.
2. *Methods for Organic Chemical Analysis of Municipal and Industrial Wastewater*; U.S. Environmental Protection Agency. U.S. Government Printing Office: Washington, DC, 1982; EPA–600/4–82–057.
3. *Test Methods for Evaluating Solid Waste*; U.S. Environmental Protection Agency: Washington, DC, 1986; SW–846.

RECEIVED for review January 28, 1987. ACCEPTED May 8, 1987.

Chapter 30 ————————————————————————

Sampling and Analysis of Hazardous and Industrial Wastes

Special Quality Assurance and Quality Control Considerations

Larry P. Jackson

The design and conduct of sampling operations for industrial and hazardous wastes must involve the consideration of a series of related problems that are new to the technical community. Among these problems are a changing list of analytes by methods undergoing constant revision, the use of testing protocols and resulting data for nontechnical purposes, and the application of state-of-the-art methods for analytes and matrices in the absence of validation data. This chapter presents a brief discussion of each of these problems and suggests a strategy for dealing with them. The objective is to provide an approach that will allow the results from a sampling and analysis program to meet changing regulatory needs, be technically valid, and provide a means of documenting waste characteristics in a consistent manner so that the value of the data base is retained as regulatory requirements change.

P LANNING AND EXECUTION of a quality assurance (QA) and quality control (QC) program to support an environmental sampling and analysis project of industrial waste streams is complicated by problems arising from an inconsistent maze of regulatory requirements. Foremost among these problems are the rapid rate of introduction of new regulations; varying requirements of the regulatory programs among local, state, and federal agencies; and technical limitations of the methods required to sample and analyze the wastes.

1173–6/88/0415$06.00/0 © 1988 American Chemical Society

These problems are compounded by the lack of guidance and assistance available from the agencies to resolve the problems prior to the submission of environmental permit applications or actual data. This lack of guidance results in the waste of countless hours of labor and many thousands of dollars for industry and regulators alike at a time when permit applications may take 18–24 months for review and when economically and environmentally important projects await a permit before implementation.

This chapter examines the Resource Conservation and Recovery Act of 1976 (RCRA), its subsequent amendments, and the supporting body of environmental regulations for QA and QC implications. This chapter presents examples of QA and QC problems, suggests limited solutions that may save time and money over the life of a project, and proposes a strategy for addressing the majority of the problems.

Institutional Problems

Regulatory Programs

Stated in simple terms, RCRA provides for the identification of hazardous and nonhazardous waste products and establishes guidelines for their disposal. RCRA was the last of the suite of major environmental regulations that grew out of the National Environmental Policy Act of 1972. RCRA was preceded by the Clean Air Act and Clean Water Act. Many of the technologies that grew out of the requirements of these first acts produce wastes that are controlled under RCRA. Examples of these types of technologies are smokestack particulate control, flue-gas desulfurization, and a host of wastewater treatment technologies. All of these processes produce a solid or sludge material that must be disposed. The 1984 Hazardous and Solid Waste Amendments (HSWA) to RCRA now require a minimum level of treatment for many wastes prior to disposal.

Each of the major acts is supported by a body of regulations written and maintained by separate offices within the U.S. Environmental Protection Agency (EPA). The actual implementation of the regulations is conducted by the regional EPA offices or state agencies that have applied for and been granted primacy by the EPA. In some of these states, the requirements are more strict than the federal standards. At each level, the interpretation of the regulations is left to local authorities. The lack of EPA accepted and published methods of analysis, accompanying method validation procedures, and sampling guidelines for many of the unusual matrices encountered in waste streams has resulted in many different local applications of the regulations. This nonuniform application results in procedures acceptable in one state or region not being accepted in another without some degree of modification. For each application of a sampling plan and its attendant QA and QC section, lengthy negotiations with local regulatory authorities must be completed before samples and data are obtained.

In some cases, the initial data submitted as part of a permit application are deemed inadequate upon review and thus require additional sampling and analysis.

The previous points make it appear that the fragmented regulatory structure and lack of published standards have an adverse impact only on industry. This situation is not the case. The public interest in a clean environment is also poorly served. The regulatory community, at all levels, is greatly overextended to issue and monitor permits. The level of technical sophistication required to conduct an intelligent sampling and analysis plan in the myriad of settings encountered in today's industrial and municipal settings is very high. What appear to be unnecessarily strict QA and QC requirements on sampling and analytical plans is an effort to protect the public interest in a clean environment in the absence of guidance. No matter how well-intentioned the applicant and regulator both may be, mistakes have been and will continue to be made. Potentially harmful situations will go undetected. Although good QA and QC standards will not eliminate mistakes, they will decrease the number. The costs, both in terms of time and money, of preparing and implementing an environmental sampling and analysis plan will be reduced.

Solutions

A variety of actions can be taken to address these types of problems. None will work overnight and all will take a commitment of time and resources on the part of private sector parties. The private sector and the regulatory community will have to minimize the adversarial nature of their relationship. Development of technically sound guidelines or standards will require presentation of sufficient supporting material or data to prove the validity of the approach sufficient to gain regulatory acceptance. Basically, the private sector must assume some of the burden of developing QA and QC guidelines for regulatory applications. This task can be achieved in three ways.

Trade associations such as the Chemical Manufacturers Association, Inc.; American Iron and Steel Institute; and American Textile Manufacturers Institute, Inc. should assemble the sampling practices and analytical methods used by their industry along with the QA and QC procedures used to ensure data quality. The methods, practices, and procedures should be written in a form suitable for publication. Descriptions of the intended applications as well as limitations or exclusions (where known) should be included. Data supporting the reliability of the intended application should be provided. The data should be of sufficient quality and quantity to withstand rigorous peer review. Once prepared, the documents should be submitted to EPA for review, potential modification, and eventual approval and inclusion in the list of approved procedures.

An alternative way to accomplish much the same thing is for individual companies or trade associations to join and participate in nationally recognized standard writing organizations such as the American Society for Testing and

Materials (ASTM). The ASTM is actively involved in writing standards for all types of applications and has developed procedures for ensuring the objectivity and technical validity of the standards that they publish. The EPA has a long history of accepting ASTM methods for regulatory application. In most cases where ASTM methods exist when regulations are initially proposed, the ASTM methods are included in the list of accepted methods.

Today, the regulatory process is being driven by timetables established by Congress. Regulations are being written in areas where no standard methods exist. The EPA is very active within the ASTM and seeks objective technical input at every opportunity to aid in developing good methods for regulatory application. This input includes sampling and analytical methods, validation procedures, and QA and QC requirements. Although the time frame of EPA's need for methods frequently is shorter than ASTM's procedural requirements for adoption, the EPA uses the technical information obtained by their participation in the process to develop their proposed methods, which are published for public comment. With more active participation from the public sector, ASTM methods can be developed faster and the necessary methods made available to meet the congressionally driven timetable.

In those cases where the EPA has not formally accepted the trade association or ASTM methods for regulatory use, the permit holder or applicant can still submit the method with its supporting documentation to local regulatory authorities in support of their use of the method. The method will be evaluated and accepted or rejected on the basis of its merits where no accepted method exists. In cases where EPA may have specified a method, the private sector party will have to submit additional data proving equivalency or establishing the lack of validity of the existing method for the particular application being considered. This process is not used as often as it should be to challenge poor practices. The private sector petitioners must provide sufficient data to support their applications, and the regulator must be receptive to a well-developed technical presentation. This approach has not met with much success in the past because many members of the regulated community and public interest groups have lobbied extensively from opposite points of view to influence the regulatory process. Both groups are beginning to accept the value of quality data to support their respective positions, and now is the time to increase the use of this approach.

Technical Problems

The past few years have brought many changes into the regulation of waste disposal practices. The Hazardous and Solid Waste Amendments of 1984 require the banning of the land disposal of certain hazardous wastes. The regulations governing a first group of these wastes went into effect on November 8, 1986, and these regulations include a broad range of new analytical requirements that exemplify the type of technical problems that must be addressed in a sampling and analysis project at an industrial waste site. The new regulations expand the list of analytes considered in making the hazardous–nonhazardous decision and introduce a new waste analysis test. Few of the existing environmental monitoring plans in place at industrial plants or proposed in pending permit applications cover these requirements.

Volatile and semivolatile organic compounds have been added to the list of analytes of concern. The list will continue to grow as the EPA gathers more data on the degree of risk associated with other compounds, and as the EPA can establish levels of these materials that are protective of the environment. The tests required to establish if the wastes are hazardous lack sufficient analytical sensitivity to reliably measure at the level deemed protective of the environment. The regulatory threshold value will decrease as the analytical methods become more sensitive and reliable. Procedures on sampling wastes are recognized within the regulations as a source of an unknown level of error. Sample holding-time limits have been established for the sample after collection and at an intermediate point in the analytical scheme. Reanalysis by the method of standard addition is required for each matrix if spike recoveries do not meet specified standards or if the measured value for the individual analyte falls within 20% of the regulatory threshold.

Before the technical details of a sampling and analysis plan are discussed, two institutional issues must be acknowledged. First, the costs associated with the analytical program establishing the environmentally safe performance of an industrial facility have increased from 10- to 100-fold in the last 15 years. The QA and QC share of the analytical budget is rising and will continue to rise as long as the regulatory requirements are vague. Second, the vagueness of the regulatory requirements and the known deficiencies of many of the required methods may allow for unintentional or intentional bias to influence the outcome of the sampling and analytical program used to establish the degree of hazard for any waste. It is hoped that both public interest groups and industry will realize that the costs for sound, objective decision making are worthwhile and that both groups are vulnerable to the adage, "Pay me now or pay me later."

Traditional QA and QC programs have a single goal: to ensure that a representative sample is available for analysis and that the analytical results are as accurate as the methods allow. Two ancillary goals should be added for QA and QC programs in support of hazardous waste determination. (1) The program should be flexible and anticipate the changes in regulatory direction to minimize permit modifications. (2) The program should be technically rigorous so that the measured values are of adequate precision and accuracy to withstand review on their own merits despite the lack of regulatory QA and QC procedures.

Modification of Quality Assurance and Quality Control Programs

Many QA and QC programs currently approved for use in regulatory settings do not address all the requirements of the new regulations. As they are modified, the QA and QC programs should be expanded to include the new types of samples and analytes covered in the new regulations and those that can be anticipated in the next 3–5 years. New analytes for regulation are likely to be chosen from Appendix VIII of the RCRA regulations. Careful review of existing data on starting materials, process chemistry, and actual waste stream composition should identify all compounds currently listed, and plans can be made at this time to include them as they're added in the future.

Two other changes can be made to existing QA and QC plans. Regulations require the taking of a sufficient number of samples to allow some preliminary testing to be done to decide on the proper protocol for hazardous determination. This determination can be done in the field as the samples are collected, and as a result, the actual number of samples collected and returned to the off-site laboratory can be reduced. The sampling plan and its attendant QA and QC section will have to be modified to document this determination, but the plan will be cost-effective if the total number of samples handled from the field to the laboratory is reduced significantly. Complete directions for analysis by the laboratory may allow a single sample to serve for the analysis of volatile and semivolatile organic compounds as well as inorganic compounds. This plan can also lead to significant reduction in sample acquisition and handling costs. Laboratory documentation practices will have to be instituted to ensure proper sample handling, but these costs should be less than working with multiple samples. In addition, biologically active wastes such as wastewater treatment sludges may contain both biologically resistant and nonresistant compounds. Modern sampling plans should contain directions on how to handle the samples and how long to store them to prevent alteration prior to analysis. The QA and QC plan should contain provisions for on-site analysis or field-spiking procedures to detect any alteration in the samples over time.

At present, the regulations do not require that field spikes be run on waste samples nor that laboratory spikes be run on the waste sample prior to conducting the *toxicity characteristic leaching procedure* (TCLP), which is the new test proposed by the EPA to determine if a waste is hazardous by the characteristic of toxicity. The absence of spiking procedures from the regulatory process is a recognition that nothing is known about how to do this test reliably. This absence of spiking procedures leads to questions of how to defend the value of the TCLP as a regulatory tool if spikes demonstrate that samples are not stable over the prescribed holding period or that analyte recoveries may be very low for certain types of compounds. This situation says nothing about being able to document that the performance of the TCLP is sufficiently precise on real-world samples to warrant its regulatory application. If any of these circumstances could be proven to exist, monumental problems would be caused for the EPA in promoting their whole regulatory scheme at a time when Congress demands regulation of wastes.

The fact that these types of QA and QC procedures are not required by regulation does not in any way relieve the sampling and analytical plan from considering them. These procedures are critical parts of any good sampling and analysis plan and must be included. Multimillion dollar decisions and permits to continue to operate depend on their results. Project sponsors must prove that the procedures proposed in the plan and implemented in the field and laboratory perform as anticipated regardless of whether the regulatory process requires the proof. This proof will require considerable effort on the part of the project sponsor and will constitute a major research project on its own. To make matters worse, proof will be required for each unique sample matrix. Effort of this type and expense may be beyond the resources of all but the largest companies. This type of project will require the commitment of industry through their trade associations. As the methods are developed and validated, they can be used to evaluate proposed regulations and improve the quality of data generated in the regulatory process. If done properly, the methods can be submitted to the EPA for inclusion in the list of accepted procedures as discussed previously.

The major questions to be addressed in this type of research project are as follows:

- How should spikes be added?
- How is the spiking level determined to ensure that appropriate levels of analytes will be found in the TCLP leachate?
- How is uniform distribution of the spike in the bulk sample achieved?
- How long are the spike samples aged prior to analysis?
- How is it determined that the sample-spike mixture is chemically passive so that observed values will be a measure of the sampling and analysis steps and not a measure of sample-spike alteration?

Interferences

As the number of analytes increases and the regulatory thresholds are lowered to near or below reliable quantitation limits, the importance of sample matrix effects will increase. The regulations currently imply that the analytical methods contain adequate discussion of interferences, but this implication is incorrect. The discussion of interferences is limited and often is concerned with those interferences present in relatively simple samples. Nowhere is this situation better illustrated then by EPA's own special analytical services (SAS) contracts issued as part of their contract laboratory program (CLP), which supports the Superfund activity. The SAS program was developed because the analytical complexity of many samples was beyond the capabilities of all but the most sophisticated laboratories. Special procedures are used with the permission of the EPA in these laboratories to analyze difficult samples. These procedures are frequently arrived at by verbal agreement between EPA personnel and the analyst. The procedures must be fully documented, but full validation is not required. This reasonable approach is denied to participants in the private sector when faced with RCRA-related problems.

Interferences are frequently considered to be compounds whose presence obscures the measurement of the analyte of interest by the introduction of an unrelated analytical signal where the analyte is measured. Interferences can also suppress the signal of interest to a level that the analyte cannot be accurately measured. These problems are sample- and method-specific and will have to be solved to the extent possible by existing procedures. The easiest solution may be the substitution of another approved method. The best example is the choice of standard flame atomic absorption spectrometry, graphite furnace atomic absorption spectrometry, or inductively coupled plasma spectrometry for the analysis of inorganic elements. These methods are all accepted methods and can be selected to avoid the most serious interferences.

Analysis of Organic Analytes

The analysis of organic analytes is not as easy. The number of broad-range analytical methods approved for general use is limited. Currently, combined gas chromatography–mass spectrometry (GC–MS) is the only universal method that can measure all volatile and semivolatile organic compounds. GC with specific-element detectors has limited utility in that it is most sensitive for compounds containing specific elements such as chlorine, sulfur, or nitrogen. These methods suffer from the same type of interferences as the inorganic methods, but they are not as interchangeable. In addition, these methods are generally more expensive. These methods suffer from another limitation: They are so sensitive that they are subject to detector overload causing a protective instrument shutdown or prolonged loss of measurement sensitivity. If the sample injected into the instrument has a single component in high concentration, the entire analytical procedure may be invalidated.

Validation of Screening Methods

Sampling and analysis plans will have to take advantage of the screening tests allowed under the regulations to establish that specific analytes are not present at regulatory threshold levels in the waste sample. Then a method can be selected from the approved measurement methods for the remaining material. Each of the screening methods will have to be validated because very few are approved for use at this time. As with the spiking method problem, the validation process requires considerable resources but may offer the only alternative to accurate measurement of many analytes in the presence of high concentrations of other compounds. The EPA is helping in this regard by publishing a general method validation guideline, and many of the new and reapproved methods are being published with matrix evaluation and method validation procedures. As with other discretionary issues, the private sector will have to negotiate with local regulatory authorities for the application of these new procedures. These procedures offer an opportunity to effect considerable cost savings over time once they gain widespread approval.

Abbreviations

ASTM	American Society for Testing and Materials
CLP	contract laboratory program
EPA	U.S. Environmental Protection Agency
GC–MS	gas chromatography–mass spectrometry
HSWA	1984 Hazardous and Solid Waste Amendments
QA	quality assurance
QC	quality control
RCRA	Resource Conservation and Recovery Act
SAS	special analytical services
TCLP	toxicity characteristic leaching procedure

RECEIVED for review January 28, 1987. ACCEPTED June 24, 1987.

GLOSSARY AND INDEXES

Glossary

Accuracy   Difference between measured and referenced values.

Additive interferences   Interferences caused by sample constituents that generate a signal that adds to the analyte signal.

Air quality models   Models that relate emissions and receptors.

Anisotropy   Condition in which the variance structure, as measured by the semivariogram, is often different in the range of correlation in different directions.

Audit sample   Prepared reference sample inserted into the sample processing procedure as close to the beginning as possible.

Bias   Degree of agreement of a measurement (or an average of measurements of the same thing) having an accepted reference or true value.

Binomial distribution   Distribution of a variable that can be classified into one of two categories.

Blank   Sample that contains the analyte of interest but in other respects has, as far as possible, the same composition as the actual sample.

Bootstrap method   Computationally intensive measurement of uncertainty that involves recalculation of statistical parameters, such as those yielded by description, classification, and correlation models, from many randomly selected subsets of the model input data set.

Box plots   Geographical depictions of percentiles.

Cation exchange capacity (CEC)   Test performed on the fine-grained (<2 mm) fraction of the soil.

Classificatory models   Models that define categories based on the achievement of certain measured values for a set of attributes. Each category has a specific set of decisions associated with the members of that category.

Comparability   Confidence with which one data set can be compared to another.

Completeness   Measure of the amount of valid data obtained from a measurement system compared to the amount that was expected to be obtained under correct normal conditions.

Contamination   Something inadvertently added to the sample during the sampling and analytical process.

Continual sampler   Sampler that withdraws a sample constantly from a stream and accumulates the withdrawn volume for collection at a later time.

Control charts   Charts used to provide the most effective mechanism for interpreting blank results. In the control mode, control charts can be used to detect changes in the average background contamination of a stable system.

Collocated samples   Two or more portions collected at the same point in time and space so as to be considered identical.

Composite sampling plan   Combining aliquots from two or more samples and analyzing the resulting pooled sample.

Concentration monitoring approach   Deposition velocity estimation method that focuses on air quality alone, although often with supporting measurement of a selected set of meteorological variables.

Conditional simulation   Statistical simulation conditioned on a relatively sparse data set generated to produce alternate, equally likely scenarios of what the spatial and temporal distribution of values may be. This method allows the consequences of various random field assumptions to be studied in the computer before actual field data are collected.

Confidence interval   Range of values that can be declared with a specified degree of confidence to contain the correct or true value for the population.

Control   Type of sample against which the results of a procedure are judged.

Correlation   Statistical measurement of the intuitive physical fact that samples taken close together are more similar in value than samples taken farther apart.

Covariational models   Models that calculate measures of association between two or more variables and are often used to assign effects to a pollutant species or a pollutant to an emissions source.

Data quality   Magnitude of error associated with a particular data set.

Data quality objectives (DQOs)   Qualitative and quantitative statements developed by data users to specify the quality of data needed from a particular data collection activity.

Decision tree   Way of presenting a classification rule converted from a training set.

Descriptive models   Models that summarize the spatial, temporal, and statistical distributions of individual observations.

Domain   Subject area of an expert system.

Dry deposition   Aerodynamic exchange of trace gases and aerosols from the air to the surface as well as the gravitational settling of particles.

Engineering judgment   Estimate of the characteristics of a waste made on the basis of knowledge of the processes that produce the waste.

Equipment blanks   Special type of field blank used primarily as a qualitative check for contamination rather than as a quantitative measure.

Error of the estimate   Difference between the true population average and the estimate from the particular survey of the average.

Estimation block   In situ volume represented by the estimated value.

Expert   Person who studies or applies information sufficiently well enough to become knowledgeable.

Expert system   Computer program that emulates a human expert in making decisions.

False negative   Error of concluding that an analyte is not present in the media sampled when it is.

False positive   Error of concluding that an analyte is present in the media sampled when it is not.

50-percentile point   Value that has 50% of the total values at or below it.

Field blank   Blank used to provide information about contaminants that may be introduced during sample collection, storage, and transport.

Gaussian distribution   Distribution of values that arises when many small independent fluctuations in components tend to cancel each other out to yield a stable average.

Geostatistical estimation   Process of using space–time data, together with the assumptions and structural characterization of the random field, to estimate parameters or characteristics (properties) related to the random field.

Geostatistical sampling   Sampling procedures that are guided by the assumed properties of the random field and prior, or early, estimates of the covariance or variogram functions.

Geostatistical test of hypothesis methods   Procedures for making decisions concerning conjectures about the parameters of the random field from statistics computed from field data, together with the assumed random field characteristics and estimated covariance or variogram.

Geostatistics   Statistics used to describe the study of regionalized or spatially correlated variables.

Grid unit length   Distance between sample locations.

Gross sample (bulk sample)   Pool of two increments that is reduced or prepared as subsamples for analysis.

Heteroscedastic   When the standard deviation of each component of the measurement error is allowed to depend on concentration.

Heteroscedastic curve (envelope of uncertainty)   Curve that relates inter-laboratory coefficients of variation to concentration.

Hybrid plan   Sampling plan that incorporates elements of intuitive sampling plans and statistically based plans.

Inductive learning   Knowledge representation technique where rules used in an expert system are induced by a computer program from examples provided to it.

Inferential method   Deposition velocity estimation method that relies on measurements made of controlling surface and atmospheric properties.

Inferential models (receptor-oriented models)   Models that can be derived either from fundamental physical and chemical laws or from empirical observations. These models differ from mechanistic models in that their primary set of input data is that which is measured at the receptor, not at the source.

Information   Facts or data obtained from study, investigation, or instruction.

Interferences   Compounds whose presence obscures the measurement of the analyte of interest by the introduction of an unrelated analytical signal where the analyte is measured.

Interlaboratory precision (reproducibility)   Variation associated with two or more laboratories or organizations using the same measurement method.

Intralaboratory precision (repeatability)   Variation associated with a single laboratory or organization.

Intuitive sampling plan   Sampling plan based upon judgment, often by technical experts.

Isokinetic sampling   Method of obtaining a representative sample of an effluent stream containing particulate matter.

Knowledge   Condition of knowing something with familiarity gained through experience or association.

Kriging   Linear-weighted average interpolation technique used in geostatistics to estimate unknown points or blocks from surrounding sample data.

Lag axis   Distance between points in linear units such as meters or kilometers.

Local control site   Control near in time and space to the sample of interest.

Matched-matrix field blank   Most common type of field blank, where the blank simulates the sample matrix.

Matched pair   Two observations, one with treatment and the other serving as control.

Matched-pair Wilcoxon test   Test used to determine whether a treatment effect is significant as compared with another observation that is similar in all ways with the first except that the treatment was not received.

Matrix control (field spike)   Control used to estimate the magnitude of interferences caused by a complex sample matrix.

Measurement process   Process that includes the sampling design phase, sampling implementation phase, and analytical phase.

Measurement uncertainty   Uncertainty that results from the variability of the environmental measurements used as model input data.

Mechanistic models (source-oriented models)   Models that contain mathematical descriptions of the interactions among variables derived from fundamental physical and chemical laws.

Median   Value that has 50% of the values below it and 50% of the values above it.

Method 18   Set of "loose" procedures for measurement of gaseous organic compound emissions by gas chromatography.

Model uncertainty   Uncertainty that is caused by deviations from the model assumptions.

Monte Carlo method   Computationally intensive measurement of uncertainty that involves perturbation of all input data with random numbers drawn from an assumed error distribution.

Multiplicative interferences   Interferences that either increase or decrease the analyte signal by some factor without generating a signal of their own.

Nonparametric procedures   Statistical techniques that can be applied without concern for the actual distribution of the underlying population from which the data were collected.

Nonvolatile organic compounds   Organic compounds that have a vapor pressure range $<10^{-8}$ kPa.

P-percentile point   Value at or below which P percent of all population values lie.

Parameters   Constants that are not supplied to a model on a case-by-case basis, as are input measurements. These constants are obtained by a theoretical calculation, by measurements made elsewhere and assumed to be appropriate for the place and time being studied, or by tacit assumption that the value of a variable is constant or negligible.

Precision   Measure of the variability of measurements of the same quantity by the same method.

Primary chemical constituents   Category of chemical constituents that may represent known species that have been identified in the sampling matrix or are required to be determined by regulation.

Primary sample   First collection of any material for the analysis of one or more analytes.

Primary sampling unit   Time period or volume of waste for which a composite estimate is desired. The unit may be a time period if the waste is a flowing stream, or a volume if the waste is an impoundment.

Protocol   Thorough, written description of the detailed steps and procedures involved in the collection of samples.

Random field sampling (geostatistical sampling)   Method in which the goal is to arrange sampling so as to make every individual of the total population equally likely to be selected in the sample.

Random error   Error that can be estimated by the use of standard statistical techniques.

Range   Spread of values calculated by subtracting the lowest value from the highest value.

Range of correlation   Distance on the lag axis at which the semivariogram's curve reaches the sill.

Reagent blank (method blank)   Blank that contains any reagents used in the sample preparation and analysis procedure.

Relative standard deviation   Estimate of the average error in the measurement due to unassignable causes and usually expressed as a percentage of the average sample concentration.

Representativeness   Correspondence between the analytical result and the actual environmental quality or the condition experienced by a contaminant receptor.

Routine-duplicate pair   Sample pair consisting of a duplicate sample collected coincidentally with the routine sample taken at some location.

Sample collection   Contact of the sample with the sampling device and its materials of construction.

Sample support   In situ description of the physical sample.

Sampling   Attempt to choose and extract a representative portion of a physical system from its surroundings.

Sampling media blank (trip blank)   Blank consisting of the sampling media used for collection of field samples.

Scientifically true value   The true value (as compared with the statistically true value).

Secondary chemical constituents   Category of chemical constituents that may include transformation products of primary chemical species, environmental variables needed to characterize conditions or meet criteria for representativeness, or other chemical species that may be indicators of sample integrity.

Semivariogram   Graph in which spatial correlation can be measured and displayed.

Semivolatile organic compounds   Organic compounds that have a vapor pressure range from 10^{-2}–10^{-8} kPa.

Sign test   Test designed to determine whether the true theoretical median equals a specified value.

Signed Wilcoxon test   Test for the median in which the sample is taken to contain independent, identically distributed observations from a reference population having a symmetric probability density.

Sill   Upper bound of the rise in variance on a semivariogram.

Solvent blank   Blank that consists only of the solvent used to dilute the sample.

Spatial gradient technique   Technique especially appropriate for sampling rivers for chemical constituents. By applying this technique, the distance between points on a transect or grid in a grid pattern can be determined.

Spatial outlier   Sample value that does not agree in magnitude with the values of its neighboring samples, especially the samples within a range of correlation.

Spatial variables   Field samples closer together than the range of correlation.

Standard deviation   Square root of the variance.

Standard error   Standard deviation of the average.

Stationarity   Decision to proceed with averaging and inference over a predetermined population or area in space.

Statistically based plan   Sampling plan that provides the basis for making probabilistic conclusions that are independent of personal judgment.

Statistically true value   Value that the mean of a population approaches as the number of samples increases.

Stratified random sampling   Sampling technique in which estimates of strata means are combined to yield estimates of the population mean.

System blank (instrument blank)   Measure of the instrument background, or baseline, response in the absence of a sample.

Systematic error   Error that usually results in a consistent deviation (bias) in a final result and cannot be estimated statistically.

Systematic sampling   Use of random numbers to lay a Cartesian grid over a region so that the coordinate origin and axis orientation are random. Samples are then taken at each grid intersection.

Time resolution (of an in situ measurement)   Time required for the real-time sensor to reach 90% of the final response to a step change in concentration of the measured species.

Toxicity characteristic leaching procedure (TCLP)   Test proposed by the EPA to determine if a waste is hazardous by the characteristic of toxicity.

Training set   Set of exemplary situations used by an expert to "teach" a computer.

Two-sample Wilcoxon test   Test designed to determine whether the mean of one population differs from the mean of another population.

Variance   Measure of the variability in a population.

Volatile organic compounds   Organic compounds that have a vapor pressure range $>10^{-2}$ kPa.

Volatilization   Physical process in which volatile species can be lost to the atmosphere.

Wind rose   Diagram that summarizes statistical information about the wind.

Windows   Directional classes used for grouping sample pairs.

Worst case   Situation that results in the most conservative results.

Affiliation Index

Subject Index

A

H

I

Copy editing and indexing by Keith B. Belton
Production by Cara Aldridge Young
Book design by Janet S. Dodd
Jacket design by Carla L. Clemens

Typeset by Hot Type Ltd., Washington, DC
Printed and bound by Maple Press Company, York, PA

RECENT ACS BOOKS

Personal Computers for Scientists: A Byte at a Time
By Glenn I. Ouchi
276 pp; clothbound; ISBN 0–8412–1000–4

The ACS Style Guide: A Manual for Authors and Editors
Edited by Janet S. Dodd
264 pp; clothbound; ISBN 0–8412–0917–0

Silent Spring Revisited
Edited by Gino J. Marco, Robert M. Hollingworth, and William Durham
214 pp; clothbound; ISBN 0–8412–0980–4

Phosphorus Chemistry in Everyday Living, Second Edition
By Arthur D. F. Toy and Edward N. Walsh
362 pp; clothbound; ISBN 0–8412–1002–0

Pharmacokinetics: Processes and Mathematics
By Peter G. Welling
ACS Monograph 185; 290 pp; ISBN 0–8412–0967–7

Synthesis and Chemistry of Agrochemicals
Edited by Don R. Baker, Joseph G. Fenyes, William K. Moberg,
and Barrington Cross
ACS Symposium Series 355; 474 pp; 0–8412–1434–4

Nutritional Bioavailability of Manganese
Edited by Constance Kies
ACS Symposium Series 354; 155 pp; 0–8412–1433–6

Supercomputer Research in Chemistry and Chemical Engineering
Edited by Klavs F. Jensen and Donald G. Truhlar
ACS Symposium Series 353; 436 pp; 0–8412–1430–1

Sources and Fates of Aquatic Pollutants
Edited by Ronald A. Hites and S. J. Eisenreich
Advances in Chemistry Series 216; 558 pp; ISBN 0–8412–0983–9

Nucleophilicity
Edited by J. Milton Harris and Samuel P. McManus
Advances in Chemistry Series 215; 494 pp; ISBN 0–8412–0952–9

For further information and a free catalog of ACS books, contact:
American Chemical Society
Distribution Office, Department 225
1155 16th Street, NW, Washington, DC 20036
Telephone 800–227–5558